KLOE-2 Workshop on e+ e- Collision Physics at 1 GeV

EPJ Web of Conferences Volume 166 (2018)

Frascati, Italy
26 - 28 October 2016

Editor:

Antonio Di Domenico

ISBN: 978-1-5108-5572-4

Printed from e-media with permission by:

Curran Associates, Inc.
57 Morehouse Lane
Red Hook, NY 12571

proceedings.com

Some format issues inherent in the e-media version may also appear in this print version.

Printed by Curran Associates, Inc. (2018)

For additional information, please contact EDP Sciences – Web of Conferences
at the address below.

EDP Sciences – Web of Conferences
17, Avenue du Hoggar
Parc d'Activité de Courtabœuf
BP 112
F-91944 Les Ulis Cedex A
France

Phone: +33 (0) 1 69 18 75 75
Fax: +33 (0) 1 69 28 84 91

contact-edps@webofconferences.org

Additional copies of this publication are available from:

Curran Associates, Inc.
57 Morehouse Lane
Red Hook, NY 12571 USA
Phone: 845-758-0400
Fax: 845-758-2633
Email: curran@proceedings.com
Web: www.proceedings.com

TABLE OF CONTENTS

Preface

The KLOE-2 workshop on e^+e^- collision physics at 1 GeV was held at the Frascati laboratories of INFN from 26 to 28 October 2016. It has been devoted to a thoroughly discussion of the KLOE-2 Physics program at DAΦNE, and more in general about topics related to e^+e^- Physics at 1 GeV c.m energy:

- CP and T violation, CPT and QM tests
- K_S decays, η-meson decays and chiral lagrangians
- ϕ-meson decays, light hadron spectroscopy and transition form factors
- γγ Physics
- Dark force searches
- Hadronic contribution to the anomalous magnetic moment of the muon
- Low energy kaon interactions
- Future machines and new detectors

A special session has been devoted to new proposals for the measurement of the hadronic contribution to the muon g-2.

The Workshop consisted of plenary sessions held at the Bruno Touschek Auditorium Hall of the INFN Frascati Laboratories, with invited talks. The enthusiastic and interested participation of more than 60 people, including 28 invited speakers coming from all Europe, Russia, China, Canada and USA, contributed to the success of the workshop. All presentations can be found at the web site http://agenda.infn.it/event/kloe2ws. Written contributions from most of the invited speakers have been collected in this volume.

Frascati, Fall 2017

The editor
Antonio Di Domenico (Sapienza University and INFN - Rome)

The Organizing Committee

Caterina Bloise (LNF)
Eryk Czerwinski (Krakow)
Erika De Lucia (LNF)
Antonio Di Domenico (Roma) - Chair
Paolo Gauzzi (Roma)
Simona Giovannella (LNF)
Enrico Graziani (Roma Tre)
Andrzej Kupsc (Uppsala)
Luca Trentadue (Parma)
Graziano Venanzoni (LNF)

Invited speakers

Patrik Adlarson (Mainz)
Giovanni Amelino-Camelia (Rome)
Michele Arzano (Rome)
Jose Bernabeu (Valencia)
Catalina Curceanu (Frascati)
Giancarlo D'ambrosio (Naples)
Achim Denig (Mainz)
Simon Eidelmann (Novosibirsk)
Shuangshi Fang (Beijing)
Paola Gianotti (Frascati)
Beatrix Hiesmayr (Vienna)
Fred Jegerlener (Berlin)
Karol Kampf (Prague)
Marc Knecht (Marseille)
Bastian Kubis (Bonn)
Leonard Lesniak (Krakow)
Stefan Leupold (Uppsala)
Guido Martinelli (Rome)
Nick Mavromatos (London)
Catia Milardi (Frascati)
Emilie Passemar (Indiana)
Paride Paradisi (Padua)
Massimiliano Procura (CERN)
Christoph Redmer (Mainz)
Olga Shekhovtsova (Krakow)
Eugeny Solodov (Novosibirsk)
Natalia Toro (SLAC/Perimeter)
Sean Tulin (York)

USCITA

edp sciences

EPJ Web of Conferences

Statement of Peer review

In submitting conference proceedings to EPJ Web of Conferences, I certify to the Publisher that I adhere to the **Policy on Publishing Integrity** of the journal in order to safeguard good scientific practice in publishing.

1. All articles have been subjected to peer review administered by the proceedings editors.
2. Reviews have been conducted by expert referees, who have been requested to provide unbiased and constructive comments aimed, whenever possible, at improving the work.
3. Proceedings editors have taken all reasonable steps to ensure the quality of the materials they publish and their decision to accept or reject a paper for publication has been based only on the merits of the work and the relevance to the journal.

Title, date and place of the conference

KLOE-2 workshop on e+e- collision physics at 1 GeV
26-28 October 2016
INFN - Laboratori Nazionali di Frascati, Frascati, Italy

Proceedings editor(s):

Antonio Di Domenico

Date and editor's signature

4 Dec 2017

17, avenue du Hoggar - PA de Courtabœuf – BP 112 - 91944 Les Ulis Cedex A (France)
Tél. : 33 (0)1 69 18 75 75 - Fax : 33(0)1 69 07 45 17 – www.edpsciences.org

 https://doi.org/10.1051/epjconf/201816600002

Status of the KLOE-2 experiment

Danilo Domenici[1,⋆] on behalf of the KLOE-2 Collaboration

[1] *INFN - Laboratori Nazionali di Frascati, via E. Fermi 40, 00044 Frascati, Italy*

Abstract. The KLOE-2 experiment is steadily taking data since November 2014 at DAFNE, the Φ-factory in the Frascati National Laboratories. 3.5 fb^{-1} of integrated luminosity have been collected up to now, with an average rate of 10 pb^{-1}/day. Performance of the detector, data taking conditions and data quality monitoring are briefly summarized.

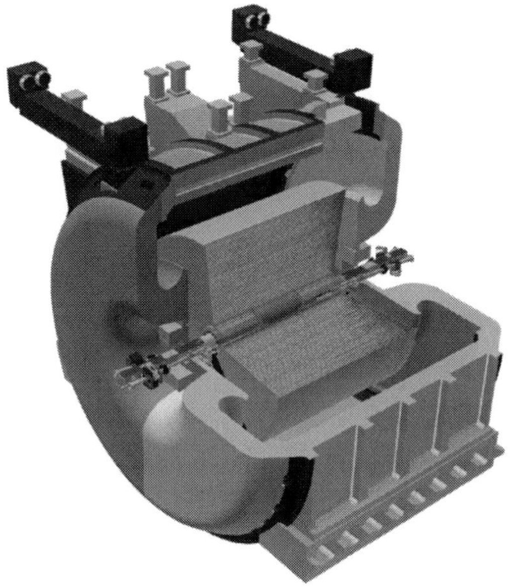

Figure 1. A sketch of the KLOE-2 detector.

1 Introduction

DAΦNE (Double Annular Φ-factory Nice Experiment) [1] is the last of a long tradition of electron-positron colliders realized at the Frascati National Laboratory of the Italian Institute for Nuclear Physics (INFN). It started its operations in 1999 and provided collisions for KLOE [2], DEAR [3]

⋆e-mail: Danilo.Domenici@lnf.infn.it

and FINUDA [4] experiments until 2007. In this period KLOE collected 2.5 fb^{-1} of integrated luminosity at 1019.4 MeV of energy, corresponding to the Φ peak, and additional 250 pb^{-1} at 1000 MeV of energy.

In 2008 DAΦNE has been upgraded with a deep change of the interaction region, implementing a large-Piwinski angle optics with crab waist, an innovative scheme developed in Frascati and tested here for the for the first time [5, 6]. Profiting from the improved performance in terms of luminosity, the KLOE detector has been upgraded with new detectors and started a second run as KLOE-2 at the end of 2014, with a refreshed physics program based on precise measurements achievable with DAΦNE increased intensity [7]. The aim is indeed to collect 5 fb^{-1} of integrated luminosity, enough to significantly increase the accuracy in the study, for instance, of K_S, η and η' rare decays, as well as to set new limits on CPT symmetry violation and searches for hidden particles.

2 The KLOE-2 detector

KLOE-2 is general purpose detector, covering the full solid angle, composed by three main sub-detector systems:

- The tracking system. The tracking is performed by a Drift Chamber (DC) and an Inner Tracker (IT) both immersed in an axial magnetic field of 0.52 T created by a superconductive coil.
- The calorimeter system. The central electromagnetic calorimeter (EMC) is composed by a barrel and two end-caps covering 98% of the solid angle. The Quadrupole calorimeter (QCALT) instruments the two permanent focusing quadrupoles aside of the interaction point. The Crystal calorimeter (CCALT) has an acceptance for very low polar angles.
- The tagging system. There are a Low-energy tagger (LET) and a High-energy tagger (HET) devoted to identify events where photons are radiated by the beams.

2.1 The tracking system

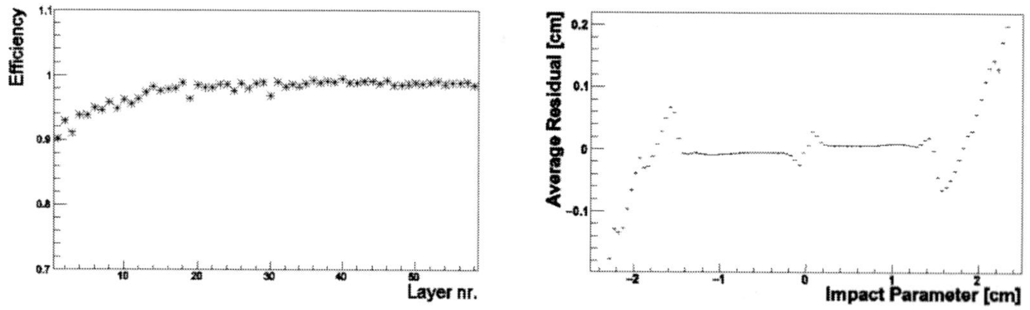

Figure 2. Performance of the Drift Chamber. Left: efficiency vs layer. Right: residuals vs impact parameter.

The KLOE Drift Chamber [8] is the largest detector of this kind ever built. It has a diameter of 4 m and a length of 3.7 m, resulting in a huge volume of almost 50 m^3 filled with a very light mixture of Helium:Isobuthane (90:10). The mechanical rigidity is provided by two plates of carbon fiber connected with 12000 sense wires and 40000 field wires creating a structure of 58 layers of drift cells with a stereo geometry allowing to extract information on the longitudinal coordinate as well. The spatial resolutions on the single hit are $\sigma_{r\phi} = 150\ \mu$m and $\sigma_Z = 2$ mm respectively for the bending and non-bending planes, while a decay vertex inside the detector volume is reconstructed with a resolution of $\sigma_V = 1$ mm.

The performance of the DC are continuously monitored using both cosmic rays muons and Bhabha scattering particles. Figure 2 shows the efficiency of the single layers and the residuals as a function of the track impact parameter.

The DAΦNE beam-pipe around the interaction point is shaped as a sphere with a radius of 12 cm, in order to keep most of the K_S decays inside the vacuum and to avoid the K_L regeneration rate depending by the emission angle. The innermost detecting layer of the DC is placed at 26 cm from the beam line. In order to reduce the lever arm and improve the resolution for reconstruction of vertexes near the interaction point, an Inner Tracker has been inserted inside the Drift Chamber. It is composed by four independent cylindrical layers of GEM detectors providing two-dimensional space point information with resolutions of $\sigma_{r\phi} = 250\ \mu$m and $\sigma_Z = 400\ \mu$m respectively for the bending and non-bending planes. The Inner Tracker and the Drift Chamber are used as an integrated tracking device, with a reconstruction strategy which starts from the long track in the DC and adds the four point in the IT implementing a Kalman Filter algorithm to update the track parameters. The innermost point placed at 13 cm from the beam line and the very good resolution in Z contribute to improve the resolution for $K_S \rightarrow \pi^+\pi^-$ decay vertexes down to $\sim 0.5\ \tau_S$.

Figure 3. The KLOE-2 Inner Tracker before the insertion in the Drift Chamber.

The cylindrical GEM detectors have been realized by splicing and rolling single GEM foils with a technique specifically developed in Frascati [9]. The dead zones are thus reduced at minimum and the detector is very light, with an overall material budget corresponding to 2% of a radiation length, mandatory to minimize the multiple scattering, the photon conversion and the kaon regeneration. In order to achieve this result all the front-end electronics boards are placed at the two sides of the detector. The readout electrode has strips along the cylinder axis spaced with a pitch of 650 μm, straightforwardly providing the rϕ coordinate. The coordinate along the beam axis is obtained combining the rϕ coordinate with a V coordinate given by helicoidal strips crossing at an angle of 20°.

Figure 3 shows a preliminary distribution of the residuals for Bhabha scattering events reconstructed in DC and extrapolated to IT Layer 3 in rϕ coordinate.

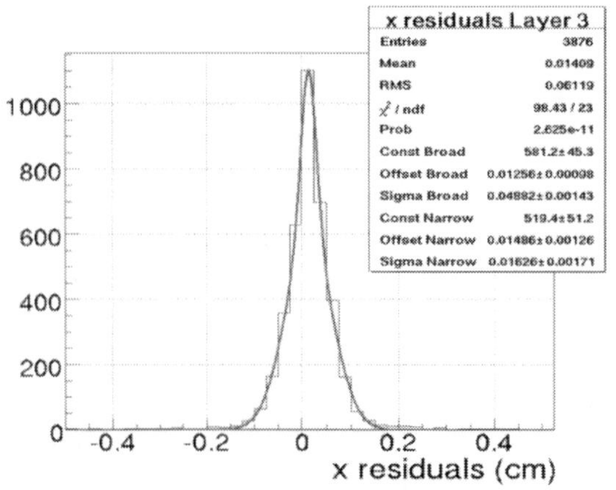

Figure 4. Residuals in x (corresponding to $r\phi$ coordinate) of the Layer 3 of Inner Tracker with respect to DC tracks of Bhabha scattering events.

2.2 The calorimeter system

The KLOE electromagnetic calorimeter [10] is composed by a barrel and two end-caps with an hermeticity of 98% of the whole solid angle. The sampling structure is made by modules of five planes realized with a sandwich of scintillating fibers and 0.5 mm thin Lead foils. The resulting thickness of 15 radiation lengths allows a good efficiency also for neutral kaons and neutrons. The fibers are read-out by both sides of the modules by almost 5000 photo-multipliers coupled via light-guides.

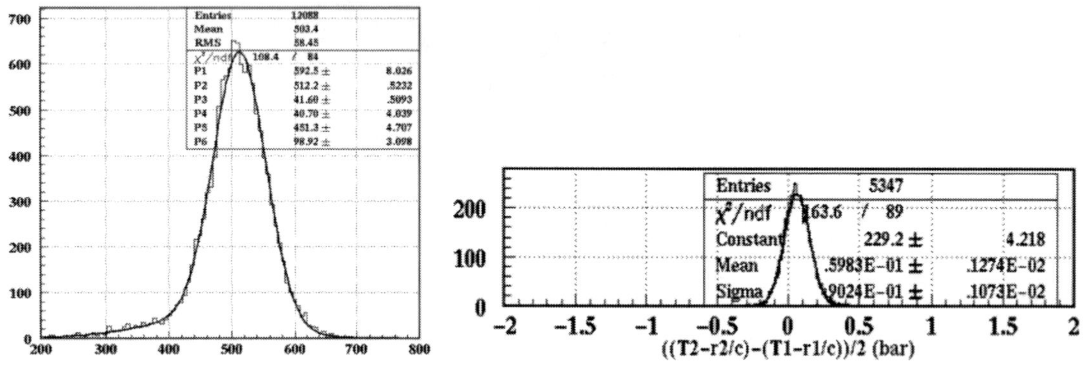

Figure 5. Performance of the Electromagnetic Calorimeter. Left: energy resolution. Right: time resolution.

Both energy and time resolutions (see Figure 5) are evaluated using 510 MeV photons from $\gamma\gamma$ scattering events. The outstanding timing capabilities allow to perform particle identification with time of flight techniques.

At DAΦNE the last focusing quadrupole are at only 1 m from the collision point, thus being placed inside the KLOE detector volume. In order to recover this space as an active area, the quadrupoles are surrounded by 2 structures made by 12 towers of Tungsten and scintillating tiles, the QCALT [11]. The light produced by the plastic scintillator is collected by wavelength-shifting fibers and read-out by Silicon Photo-multipliers. The QCALT is essential to increase the hermeticity of the calorimeter system, mostly for K_L neutral decays.

To increase the acceptance for neutral decays of short-lived particles (e.g. $K_S \rightarrow \gamma\gamma$), two small calorimeters are placed just at 20 cm from the interaction point. Each CCALT is composed of 48 LYSO crystals directly coupled to a Silicon Photo-multiplier (SiPM) for the read-out. This detector can be also used as a luminosity monitor exploiting the abundant Bhabha scattering events at low angle and the good timing performance of the LYSO giving a good background rejection.

2.3 The tagging system

One of the main part of the KLOE-2 physics program is the study of mesons generated by a $\gamma\gamma$ fusion. The tagging of the events where both an electron and a positron undergo a bremsstrahlung emission is accomplished by two different systems [12]. Particles with an energy between 150 and 400 MeV are detected by the Low Energy Taggers (LET), two stations placed at 1 m from the interaction point composed of 20 LYSO crystals read-out by SiPM. Particles with an energy above 400 MeV are detected by the High Energy Taggers (HET), two plastic scintillator hodoscopes placed at 11 m from the interaction point, after the first bending dipoles of DAFNE.

3 KLOE-2 data-taking

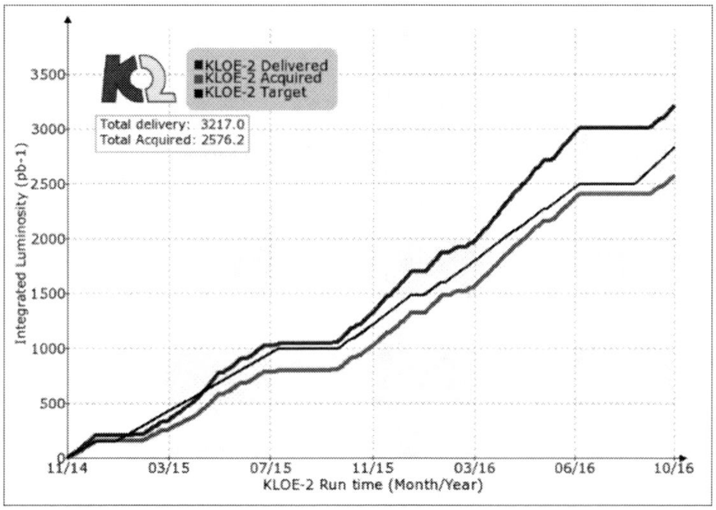

Figure 6. Summary of integrated luminosity of KLOE-2. Blue: delivered luminosity. Red: acquired luminosity. Black: goal for 5 fb^{-1} delivery in March 2018.

Figure 6 shows the integrated luminosity of KLOE-2 starting from the beginning of the run, in November 2014. The ~1 fb^{-1} delivered in the first year has been doubled in the second year thanks to a deep optimization of the collider, and the progression is now steadily pointing towards the goal of 5 fb^{-1} acquired by the end of the data-taking period, what required to accomplish our physics program.

The typical operational characteristics of DAFNE are:

- beam currents of 1.5 A for electrons and 1 A for positrons
- 105 bunches stored with 2.7 ns spacing
- top-up injection of both beams every 10 minutes

Figure 7 shows beam currents and instantaneous luminosity during a typical hour of operation.

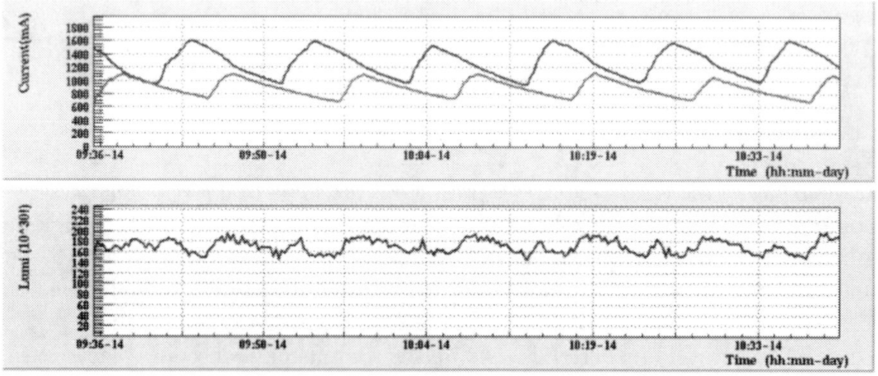

Figure 7. Currents and luminosity in a typical hour of collisions.

The best luminosity achievements of DAFNE in KLOE-2 data-taking are:

- 2.21×10^{32} cm^{-2} s^{-1} as maximum instantaneous luminosity
- 651 nb^{-1} h^{-1} as maximum hourly integrated luminosity
- 13.4 pb^{-1} d^{-1} as maximum daily integrated luminosity
- 76.3 pb^{-1} w^{-1} as maximum weekly integrated luminosity

The KLOE-2 trigger is based on energy deposits in the Electromagnetic calorimeter. It requires two separated sectors with E>50 MeV in the barrel and E>150 MeV in the end-caps. At an average luminosity of 1.6×10^{32} cm^{-2} s^{-1} the trigger rate is 7.5 kHz, with the following distribution:

- 500 Hz are due to Φ decays;
- 2.5 kHz are due to Bhabha scattering events;
- 3 kHz are due to cosmic rays;
- 1.5 kHz are due to machine induced background, mostly Touschek scattering events, i.e. intra-bunch scattered particles.

3.1 Data quality monitor

Beam parameters like space position of the luminous point and transverse beam momentum are precisely measured with the KLOE detector and used as a monitor for the collider operation. The center-of-mass energy of the beams is also monitored using benchmark counters like e.g. $K_S \to \pi^+\pi^-$ (see Figure 8)

In order to continuously monitor the quality of the collected data, four benchmark analysis are used:

- $\Phi \to \eta\gamma$ with $\eta \to 3\pi^0$

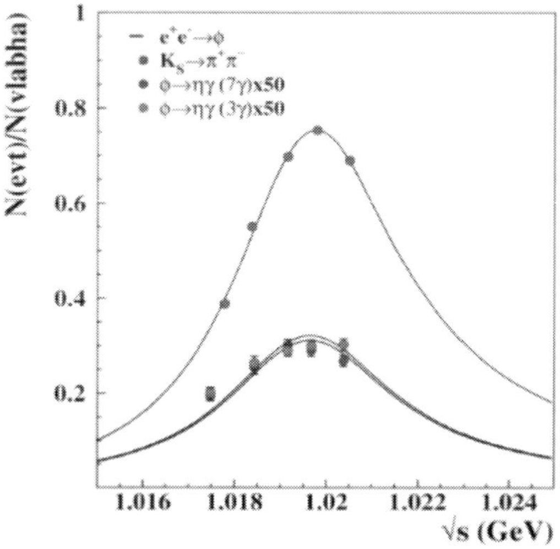

Figure 8. Scan of center-of-mass energy of the beams in steps of 550 keV used to precisely set the operation point to the Φ mass peak at 1019.4 MeV.

- $\Phi \to \eta\gamma$ with $\eta \to \gamma\gamma$
- $K_S \to \pi^+\pi^-$
- $K_L \to \pi^+\pi^-$

The former are fully neutral channels useful to monitor mostly the performance of the calorimeter, while the latter are fully charged channels useful to monitor the performance of the tracking system.

References

[1] A. Gallo *et al.*, Conf. Proc. C060626 (2006) 604, SLAC-PUB-12093
[2] KLOE Collaboration, Nuovo Cimento 30 (2008) 10
[3] DEAR Collaboration, Nucl. Phys. A **691** (2001) 278
[4] FINUDA Collaboration, Int. J. Mod. Phys. A **26** (2011) 420
[5] M. Zobov et al., Phys. Rev. Lett. **104**, 174801 (2010).
[6] C. Milardi et al., JINST, 7, 2012,T03002.
[7] G. Amelino Camelia *et al.*, Eur. Phys. J. C 68 619 (2010)
[8] M. Adinolfi *et al.*, Nucl. Inst. & Meth. **A 488** (2002) 51
[9] G. Bencivenni and D. Domenici, Nucl. Inst. & Meth. **A 581** (2007) 221
[10] M. Adinolfi *et al.*, Nucl. Inst. & Meth. **A 482** (2002) 51
[11] F. Happacher and M. Martini, Acta Phys. Pol. B, 46 (2015) 87
[12] Babusci D. *et al.*, Acta Phys. Pol. B, 46 (2015) 81

Status of the KLOE-2 Inner Tracker

Erika De Lucia[1,★] on behalf of the KLOE-2 Collaboration

[1] *Laboratori Nazionali di Frascati dell'INFN, via E. Fermi 40, 00044 Frascati (RM), Italy*

Abstract. KLOE-2 at the DAΦNE Φ-factory is the main experiment of the INFN Laboratori Nazionali di Frascati (LNF) and is the first high-energy experiment using the GEM technology with a cylindrical geometry, a novel idea developed at LNF. Four concentric cylindrical triple-GEM detectors compose the Inner Tracker, inserted around the interaction region and before the inner wall of the pre-existing KLOE Drift Chamber to improve the resolution on decay vertices close to the interaction point. State-of-the-art solutions have been expressly developed or tuned for this project: single-mask GEM etching, multi-layer XV patterned readout, PEEK spacer grid, GASTONE front-end board, a custom 64-channel ASIC with digital output, and the Global Interface Board for data collection, with a configurable FPGA architecture and Gigabit Ethernet. Alignment and calibration of a cylindrical GEM detector was never done before and represents one of the challenging activities of the experiment. The Inner Tracker detector construction, operation, calibration and performance obtained with cosmic-ray muons and Bhabha scattering events will be reported.

1 Introduction

The KLOE-2 experiment at the INFN Laboratori Nazionali di Frascati (LNF) is the continuation of the KLOE experiment[1], upgraded with state of the art technology to improve its discovery potential. Data taking campaign started in November 2014 and will allow to pursue a physics program [2] including: i) CPT symmetry and quantum coherence tests with neutral kaons with an unprecedented precision [3], ii) high precision studies of $\gamma\gamma$-physics processes [4] and iii) search for new exotic particles that could constitute the dark matter [5, 6]. The previous KLOE setup consisting of a huge Drift Chamber (DC) [7] and an Electromagnetic Calorimeter (EMC) [8], both immersed in a 0.5 T axial magnetic field, underwent several upgrades: i) state-of-the-art cylindrical GEM detector, the Inner Tracker (IT), to improve vertex reconstruction capabilities near the interaction region, ii) LET [9] and HET [10, 11] e^+e^- taggers for $\gamma - \gamma$ physics, iii) CCALT and QCALT detectors [12], crystal and tile calorimeters positioned near the interaction point and along the beam-pipe with the goal of improving multi-photon detection in rare decays and the acceptance for rejecting $K_L \to 3\pi^0$ background events in CP-violating $K_L \to 2\pi^0$ decays respectively.

Starting from November 2014, KLOE-2 acquired 2.4 fb^{-1} by July 2016 and restarted data taking in September 2016 with the aim of collecting more than 5 fb^{-1} in the next 1-1.5 years [13].

KLOE-2 is the first high-energy experiment using the GEM technology with a cylindrical geometry: the IT was mounted on DAΦNE beam pipe and inserted inside the KLOE apparatus in July 2013

★e-mail: erika.delucia@lnf.infn.it

and started its commissioning phase January 2014, defining its optimal working point and measuring its performance. Alignment and calibration of a cylindrical GEM detector was never done before and represents one of the challenging activities of the experiment.

2 Detector description

This innovative detector shown in Fig.1 is composed by four concentric cylindrical triple-GEM (CGEM) detectors with 70 cm total active length and at radii from 13 cm, to preserve the K_S-K_L quantum interference region, to 20.5 cm due to the constraints from DC inner wall. Each CGEM [14] is a triple-GEM detector composed by five concentric cylindrical electrodes (Fig.2): the cathode, to set the drift field, 3 GEM foils for the electron multiplication and the anode, acting also as readout circuit.

Figure 1. The Inner Tracker detector.

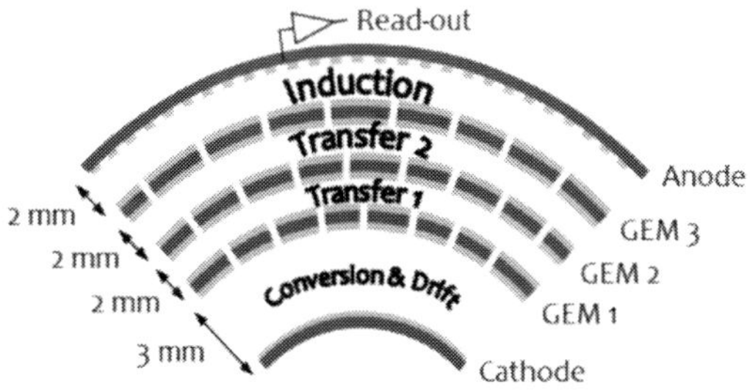

Figure 2. Cross-section of the triple-GEM detector.

The feasibility of a cylindrical triple-GEM detector with an XV readout was demonstrated during three years of R&D [15–17]. These activities included the tuning and development of a new GEM manufacturing procedure of GEM foils of unprecedented size (up to 50x100 cm^2) with a single-mask electro-chemical etching of the micro-holes, produced with the TE-MPE-EM CERN group and followed and supported within the RD51 Collaboration [18]. The anode readout is patterned with longitudinal X strips with 650 μm pitch, interleaved, on the same substrate and at the same level, with pads connected through internal vias to form V strips at an angle within 25°÷27° and with 600 μm pitch, for a total of about 30,000 FEE channels. Space coordinates are reconstructed using this dedicated XV strips patterned readout coupled to the GASTONE front-end, the 64 channel ASIC with digital output developed for the KLOE-2 experiment [19]. Data collection is then performed with a General Interface Board (GIB) with a configurable FPGA architecture, Gigabit Ethernet and the readout driver (ROD), also developed for KLOE-2 [20]. An Ar/iC_4H_{10} gas mixture was chosen, accounting for the most important operating parameters, including the discharge probability measured with α-particles.

The IT has been inserted in the free space between the beam pipe and the inner wall of KLOE DC [7] providing $\simeq$ 150 μm spatial resolution in the bending plane, $\sigma_z \simeq 2$ mm along the beam line and $\simeq$ 6 mm on decay vertices close to the IP. The resolution on vertices close to the IP reconstructed from low-momentum charged secondaries is expected to improve of about a factor 2–3 [21].

3 Detector Operation

The construction of the KLOE-2 Inner Tracker was completed in March 2013 and in June 2013 this new sub-detector was integrated on DAΦNE beam pipe and equipped with two cooling systems: copper coils as water radiators on the front-end electronics and air blowing on the IT end-caps and between the beam-pipe and the IT innermost layer. The Inner Tracker was finally inserted inside the KLOE-2 apparatus in July 2013.

The optimization of the IT operational parameters with colliding beams started in late 2014 during the first KLOE-2 data taking campaign, as a function of the e^+e^- beam currents and machine background conditions. Dedicated online monitoring procedures were developed and are currently used to check IT temperature, currents and voltages, together with occupancy and clustering performance [22] with DAΦNE delivering collision data. To safely operate all KLOE-2 sub-detectors and for e^+e^- injection optimization, the information from three indicators of the machine background level is sent online to DAΦNE operators: the current measured on the first IT layer, the total current measured in the DC and the background level measured in the EMC end-caps. This allowed machine background conditions to be kept at an acceptable level having at the same time good collider uptime and luminosity performance. Then in order to measure efficiency and resolutions, dedicated procedures were developed exploiting the Drift Chamber track reconstruction and using cosmic-ray muon data and Bhabha scattering events.

Cosmic-ray muon tracks reconstructed in the Drift Chamber with 0.52 T magnetic field are used to evaluate IT efficiency. Tracks reconstructed in the DC are extrapolated to the Inner Tracker with straight-line approximation. The normalization sample for the efficiency evaluation is given by the events with two crossing points with each IT layer. We then look for reconstructed 3D clusters in the IT, close to expected position from the DC extrapolated track. High voltage scans have been performed setting different induction field values, namely 5 kV/cm and 6 kV/cm. A single-view efficiency up to 98% has been measured by setting the induction field to 6 kV/cm, as reported in Fig. 3 for Layer 1, together with a satisfactory 95% two-view efficiency. All layers provide similar results. Presently

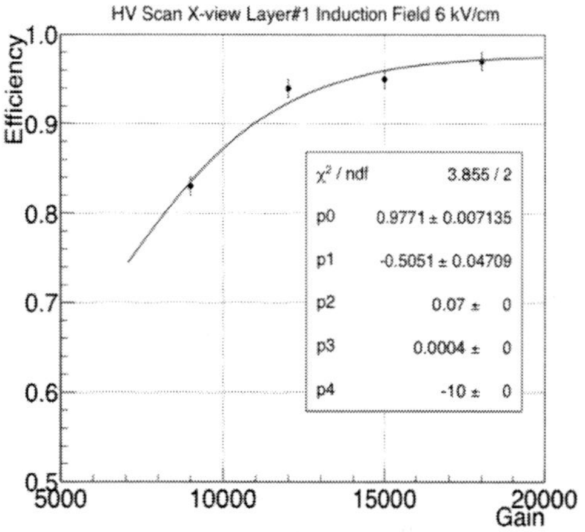

Figure 3. Layer 1 single-view efficiency as a function of the effective gain. Black points are experimental data, while the solid line is the Fermi-Dirac best fit curve.

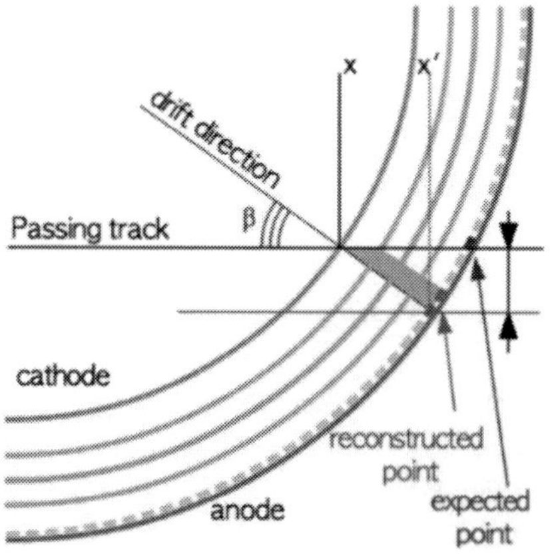

Figure 4. Non-radial tracks effect on the position measured with a cylindrical GEM prototype.

the IT is operated with a nominal gain of 1.2×10^4, with electric fields set at 1.5/3/3/6 kV/cm and

GEM1/GEM2/GEM3 voltages set at 285/280/265 V. These values were chosen as a good compromise between detector efficiency and safe operation with collisions.

3.1 Alignment and Calibration

To properly perform track reconstruction using the IT information, alignment and calibration are first performed. During the R&D phase of the project, two effects with $\mathcal{O}$(1–2 mm) average corrections were observed and measured with planar XV prototypes at the test beams [21]. The first effect is related to non-radial tracks crossing the detector (Fig. 4): due to the angle between the electric field drift line direction and the incident track, the *expected position* will be shifted to the *reconstructed point* and the signal electron cloud will be spreaded. The second effect is due to magnetic field: the signal electron cloud experiencing the Lorentz force will be shifted and spreaded. According to the impact parameter of the track on the cylindrical detectors, these two effects will produces a focusing or de-focusing of the electron cloud. Cosmic-ray muon data have been used to evaluate the non-radial correction (without B-field) and the magnetic field influence (with B-field), and Bhabha scattering events to check both.

Alignment and calibration of the IT is performed with respect to DC reconstructed tracks, providing the *expected position* on each IT layer, and exploiting the residual distribution along the three coordinates. A first set of alignment and calibration parameters was obtained with cosmic-ray muons without B-field[23]. Before any alignment or calibration, the residual widths and mean values for Layer #4 are: Res(x)$\sim$ 1.5 mm, Res(y) $\sim$ 280 μm, Res(z) $\sim$ 3.6 mm and $\Delta x \sim$ 860 μm, $\Delta y \sim$ -50 μm $\Delta z \sim$ 2 mm. After the first alignment and calibration, residual distributions exhibit average widths along the x-axis Res(x) $\sim$ 400 μm for all layers and are all centered around zero within 50 μm. Results which are within expectations for the Inner Tracker resolution. Then starting from the B-field Off sample parameters, also a first set of calibration parameters for the cosmic-ray muon sample with B-field was obtained with average value Res(x) $\sim$ 550 μm for all layers. Finally the average residuals have been cross-checked with Bhabha scattering events measuring Res(x) $\sim$ 400 μm for all layers. Fig. 5 shows the path towards the first alignment and calibration results of the IT Layer #4. This

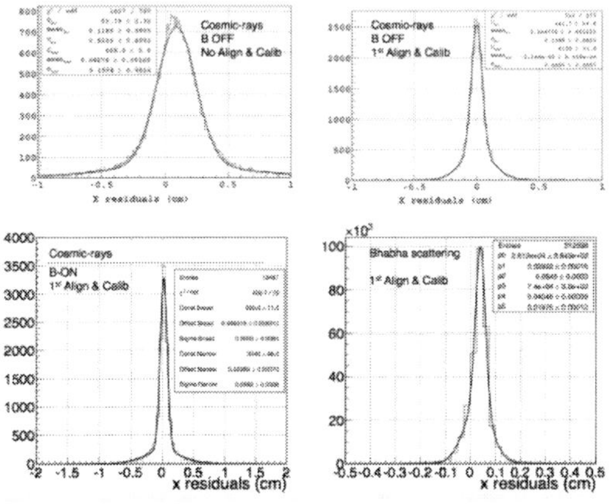

Figure 5. Layer #4 residuals for the x coordinate: cosmic-ray muon B-field Off without (top left) and with (top right) alignment and calibration, cosmic-ray muon B-field On sample (bottom left) and Bhabha scattering events (bottom right).

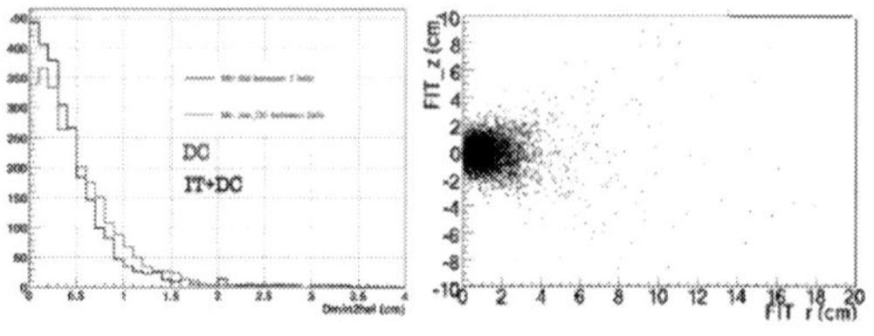

Figure 6. Very first comparison between simple vertex finder position obtained with DC (red) and IT+DC tracking (black) with two-pion Kaon decays. Left: two-helix distance. Right: simple vertex fit z position vs transverse radius

validates our method and also suggests how to improve the alignment and calibration procedure in order to reach the best detector performance.

IT clusters are included in track reconstruction, using the Kalman filter technique and starting from DC hits and track parameters. Bhabha scattering events, selected as in the alignment and calibration procedure, are used as benchmark: i) alignment and calibration parameters obtained with the cosmic-ray muon sample are inserted, ii) then calibrated IT clusters are used for IT+DC integrated tracking.

First studies have been performed using both Bhabha scattering events and two-pion Kaon decays, and with a simple vertex finder based on minimum distance between extrapolated tracks, with DC only and using the integrated IT+DC tracking. A very preliminary comparison is shown in Fig. 6: left panel shows the two-helix minimum distance as obtained with a simple vertex finder using DC tracks only (red) and using the very first version of the integrated tracking IT+DC (black); in the right panel, with the same colour code, the z position vs transverse radius of the reconstructed vertex is shown. Presently we are setting up the official reconstruction with the IT+DC integrated tracking and its interface with the official vertex finder routine, tests both Bhabha scattering event and $K_S \to \pi^+\pi^-$ decays are ongoing.

4 Conclusions

KLOE-2 is the first high-energy physics experiment to build and operate a cylindrical GEM detector. Results reported are very positive and encouraging for the integrated tracking performance optimization phase and are also of extreme interest for the other experiments which will use cylindrical GEM detectors (BESIII) or are planning to (Novosibirsk).

The challenging task of aligning and calibrating this innovative detector was performed, with cosmic-ray muon data and Bhabha scattering events. Starting from Res(x) ~1.5 mm width of the residual distribution with cosmic-ray muon B-field Off sample, the first alignment and calibration of the IT allowed Res(x) ~400 μm width to be reached, also with Bhabha scattering events. Further improvements are under study: we are presently working on refining the parameters from both the cosmic-ray muon B-field Off and On samples.

Alignment and calibration parameters obtained so far have been inserted in the IT+DC integrated tracking. First studies, performed using both Bhabha scattering events and two-pion kaon decays, show improvements already with the first and not yet refined alignment and calibration parameters. Detector reconstruction performance optimization phase started using both Bhabha scattering event and $K_S \to \pi^+\pi^-$ decays for validation.

5 Acknowledgements

We warmly thank our former KLOE colleagues for the access to the data collected during the KLOE data taking campaign. We thank the DAΦNE team for their efforts in maintaining low background running conditions and their collaboration during all data taking. We want to thank our technical staff: G.F. Fortugno and F. Sborzacchi for their dedication in ensuring efficient operation of the KLOE computing facilities; M. Anelli for his continuous attention to the gas system and detector safety; A. Balla, M. Gatta, G. Corradi and G. Papalino for electronics maintenance; C. Piscitelli for his help during major maintenance periods. This work was supported in part by the Polish National Science Centre through the Grants No. 2013/08/M/ST2/00323, 2013/11/B/ST2/04245, 2014/14/E/ST2/00262, 2014/12/S/ST2/00459.

References

[1] Bossi F., De Lucia E., Franzini J. L., Miscetti S., Palutan M. and KLOE Collaboration, Nuovo Cimento, 30 (2008) 10.

[2] G. Amelino Camelia *et al.*, Physics with the KLOE-2 experiment at the upgraded DAΦNE, *Eur. Phys. J.* C 68 619 (2010).

[3] E. Czerwinski for the KLOE-2 Collaboration, *Overview of KLOE results on kaon physics and KLOE-2 perspectives*, these proceedings

[4] S. Giovannella for the KLOE-2 Collaboration, *KLOE results on hadron physics and perspectives for KLOE-2*, these proceedings

[5] E. Graziani for the KLOE-2 Collaboration, *Search for the dark boson at KLOE and perspectives for KLOE-2*, these proceedings

[6] G. Venanzoni E. Graziani for the KLOE-2 Collaboration, *KLOE results and perspectives for KLOE-2: running of alpha e.m., hadronic cross section*, these proceedings

[7] M. Adinolfi *et al.*, Nucl. Inst. & Meth. **A 488** (2002) 51.

[8] M. Adinolfi *et al.*, Nucl. Inst. & Meth. **A 482** (2002) 51.

[9] Babusci D. et al., Acta Phys. Pol. B, 46 (2015) 87.

[10] Babusci D. et al., Acta Phys. Pol. B, 46 (2015) 81.

[11] D. Moricciani for the KLOE-2 Collaboration, *Status of the KLOE-2 high energy taggers*, these proceedings

[12] Happacher F. and Martini M., Acta Phys. Pol. B, 46 (2015) 87.

[13] D. Domenici for the KLOE-2 Collaboration, *Status of the KLOE-2 experiment*, these proceedings

[14] G. Bencivenni, D. Domenici, *The Ultra-Light Cylindrical-GEM detector as Inner Tracker at KLOE-2*, Nucl. Inst. & Meth. **A 581** (2007) 221.

[15] G. Bencivenni *et al.*, *A Time Domain Reflectometer with 100 ps precision implemented in a cost-effective FPGA for the test of the KLOE-2 Inner Tracker readout anodes*, Nucl. Inst. & Meth. **A 698** (2013) 185.

[16] A. Balla *et al.*,*The cylindrical GEM detector for the KLOE-2 Inner Tracker*, JINST **9** C01014 (2014).

[17] A. Balla *et al.*, *Status of the cylindrical-GEM project for the KLOE-2 inner tracker*, Nucl. Inst. & Meth. **A 628** (2011) 18.

[18] M. Alfonsi *et al.*, *Activity of CERN and LNF groups on large area GEM detectors*, Nucl. Inst. & Meth. **A 617** (2010) 151-154; presented at the 11th Pisa Meeting on Advanced Detector, May 24-30, 2009, La biodola - Isola d'Elba (Italy).

[19] A. Balla *et al.*, *GASTONE64: A new front-end ASIC for the cylindrical GEM Inner Tracker of KLOE-2 experiment at DAΦNE*, Nucl. Inst. & Meth. **A 721** (2013) 523.

[20] P. Branchini *et al.*, *Front-end DAQ strategy and implementation for the KLOE-2 experiment*, JINST **8** T04004 (2013).

[21] F. Archilli *et al.*, KLOE-2 Collaboration, Technical design report of the inner tracker for the KLOE-2 experiment, *arXiv:1002.2572v1 and LNF-10/3(P)* INFN-LNF, Frascati,2010.

[22] A. Di Cicco, G. Morello, *Commissioning of the Inner Tracker of the KLOE-2 experiment*, Acta Phys. Polon. B 46 (2015) 73.

[23] A. Balla *et al.*, *The KLOE-2 Inner Tracker: Detector commissioning and operation*, Nucl. Inst. & Meth. **A 845** (2017) 266.

Discrete symmetries with neutral mesons

José Bernabéu

Departamento de Física Teórica, Universidad de Valencia,
e IFIC, Joint Centre CSIC-UV, E-46100 Burjassot, Spain

Abstract. Symmetries, and Symmetry Breakings, in the Laws of Physics play a crucial role in Fundamental Science. Parity and Charge Conjugation Violations prompted the consideration of Chiral Fields in the construction of the Standard Model, whereas CP-Violation needed at least three families of Quarks leading to Flavour Physics. In this Lecture I discuss the Conceptual Basis and the present experimental results for a Direct Evidence of Separate Reversal-in-Time T, CP and CPT Genuine Asymmetries in Decaying Particles like Neutral Meson Transitions, using Quantum Entanglement and the Decay as a Filtering Measurement. The eight transitions associated to the Flavour-CP eigenstate decay products of entangled neutral mesons have demonstrated with impressive significance a separate evidence of TRV and CPV in B_d-physics, whereas a CPTV asymmetry shows a 2σ effect interpreted as an upper limit. Novel CPTV observables are discussed for K physics at KLOE-2, including the difference between the semileptonic asymmetries from K_L and K_S, the ratios of double decay rate Intensities to Flavour-CP eigenstate decay products and the ω-effect. Their observation would lead to a change of paradigm beyond Quantum Field Theory, however there is nothing in Quantum Mechanics forbidding CPTV.

1 The Fundamental Role of Symmetry Breaking

Symmetries of the Physical Laws, and their Breakings, have played a crucial role in the advance of knowledge. Conservation Laws lead to regularity and order in the description of physical phenomena even when there is no input from dynamics. In the second half of the XXth century, a conceptual revolution appeared when the requirement of local gauge symmetries was understood as originating the interactions themselves with their charges. The emergence of the Standard Model(SM) $SU(3)_C$x $SU(2)_L$x $U(1)_Y$ prompted a spectacular development of particle physics with precise experimental results in agreement with theoretical predictions. The gauge symmetry is exact for massless particles only, which is evidently wrong in Nature. This apparent contradistinction was solved by the Brout-Englert-Higgs mechanism [1] of Spontaneous Breaking of the ElectroWeak Gauge Symmetry $SU(2)_L$x $U(1)_Y$ into $U(1)_{e.m.}$ This methodology for explaining the Origin of Mass leaves as a remnant the Higgs Boson, discovered [2] at CERN by the ATLAS and CMS Collaborations.

With respect to the Discrete Symmetries, P, C, CP and T have been observed to be separately broken by weak interactions. Their breakings have had a fundamental imprint in the construction of the SM:

- Parity Violation, first observed in nuclear beta decay [3] and later, together with the Violation of Charge Conjugation (C), in charged pion decays [4], led to the V-A theory for charged current weak interactions [5] and the role of chirality as the essential ingredient of requiring different transformation properties under the gauge group for left- and right-handed fermionic fields. The chiral fields are the starting components of the theory in the SM. The electroweak unification [6, 7, 8] culminated with the discovery [9] of neutral current weak interactions by the Gargamelle experiment at CERN.
- CP Violation was unexpectedly discovered in 1964 [10] first in the decay of the long lived neutral Kaon K_L into two pions, a decay forbidden by the CP Conservation Law. The precision measurements in the Kaon system with secondary Kaon beams and at CPLEAR [11], and specially the observation of CPV in the B_d system in the B-Factories [12, 13] in 2001 prompted the conviction that the Cabibbo-Kobayashi-Maskawa mechanism of Quark Mixing [14, 15], with at least three families of fermions, is a good description of CPV in the SM. As seen, the origin of the Breaking is here advocated to be in the particle content of the theory. Physical phenomena needing more than one family are included in what we call Flavour Physics.
- A direct evidence of Time Reversal Violation (TRV) is, however, more difficult to be observed. The operator implementing Reversal-in-Time T in the space of quantum states is Antiunitary [16], in such a way that a true test needs the comparison between the probability for one transition $i \rightarrow f$ and that for the $f \rightarrow i$ transition. In agreement with the "CPT Theorem" [17] establishing that, under very general assumptions, the description of interactions by means of a Quantum Field Theory (QFT) leads to exact CPT invariance, we should expect TRV for the $K^0 - \bar{K}^0$ and $B_d - \bar{B}_d$ systems as well. However, these neutral mesons are unstable particles and the decay is irreversible. In words of Wolfenstein [18], "the T-reverse of a decaying particle state is not a physical state", so a NO-GO argument is operating and a direct evidence for TRV appears to be "impossible" for decaying particles. The bypass to this argument came [19, 20] in 1999 from the strategy of not including explicitly the decay products in the TRV asymmetry and using: i) the decay as a filtering measurement of the projected meson state only; ii) Quantum Entanglement of the neutral meson system in order to transfer the information from the decaying meson to its (still alive) orthogonal partner; iii) the asymmetry is searched in the time evolution of this last partner. Under a deep scrutiny of this conceptual basis by many authors, the first observation [21] of TRV in the time evolution of B_d's was announced in 2012 by the BABAR Collaboration, using the methodology that was previously defined in [22]. Its observation triggered [23, 24] a revisit to the conditions under which the asymmetry becomes a genuine observable of TRV. For the neutral Kaon system there is a proposal in the DAPHNE-Factory using Entanglement.
- Although P, C, CP and T symmetries have been observed to be separately broken, there is no experimental evidence of CPT Violation. Its observation would imply a change of paradigm to beyond QFT. There is, however, nothing at the level of Quantum Mechanics which forbids CPTV.

In Section 2 we discuss results and proposals for the search of CPTV in Semileptonic Decays of Neutral Kaons. The state of the art for Genuine T, CP, CPT Separate Asymmetries in B_d transitions is presented in Section 3. Direct Measurements of TRV in neutral Kaon transitions at KLOE-2 are discussed in Section 4 and the Direct tests of CPT in neutral Kaon transitions at KLOE-2 in Section 5. For entangled Kaons, the possibility of a ω-Effect weakening Entanglement, due to an ill-defined CPT operator, is analysed in Section 6. Conclusions and Outlook are given in Section 7.

2 Breaking CPT in SemiLeptonic Decays of Neutral Kaons

I discuss Tests of the CPT symmetry using the semileptonic decays of neutral Kaons. Assuming the ΔS = Δ Q rule, as implied by the description of flavoured mesons in terms of quarks, the allowed transitions are pictured by the diagrams

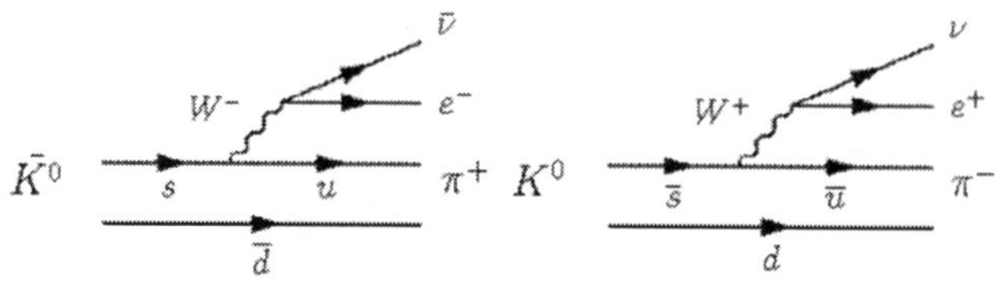

Figure 1- Two decays are allowed according to elementary Quarks (ΔS=ΔQ rule)

From the flavour states $K^0, \bar{K}^0$ the effective Hamiltonian in a Weisskopf-Wigner approach [25] includes a non-diagonal CPV parameter and a diagonal CPTV parameter, so that the states with definite time evolution are

$$|K_s\rangle = \frac{1}{\sqrt{2(1+|\epsilon_s|^2)}}\left((1+\epsilon_s)|K^0\rangle + (1-\epsilon_s)|K^0\rangle\right)$$

$$|K_L\rangle = \frac{1}{\sqrt{2(1+|\epsilon_L|^2)}}\left((1+\epsilon_L)|K^0\rangle - (1-\epsilon_L)|K^0\rangle\right) \tag{1}$$

where

$$\epsilon_{S/L} = \epsilon_K \pm \delta_K \tag{2}$$

where "ϵ_K" is the parameter describing CPV whereas "δ_K" is the parameter describing CPTV.

At CPLEAR, the preparation of an initial quantum state of flavour was used. With this tag, the following time-dependent combination of semileptonic decay rates was constructed as an asymmetry

$$\begin{cases} A_\delta(\tau) = \frac{\bar{R}_+(\tau)-\alpha R_-(\tau)}{\bar{R}_+(\tau)+\alpha R_-(\tau)} + \frac{\bar{R}_-(\tau)-\alpha R_+(\tau)}{\bar{R}_-(\tau)+\alpha R_+(\tau)} \\ R_{+(-)}(\tau) = R(K^0_{t=0} \to (e^{+(-)}\pi^{-(+)}\nu)_{t=\tau} \\ \bar{R}_{-(+)}(\tau) = R(\bar{K}^0_{t=0} \to (e^{-(+)}\pi^{+(-)}\nu)_{t=\tau} \\ \alpha = 1 + 4\Re\, \epsilon_L \end{cases} \tag{3}$$

At times outside the interference region, i. e., longer than the short lifetime of K_S, this asymmetry becomes a pure CPT test and determined by the $\Re\,(\delta)$ parameter as

$$A_\delta(\tau \gg \tau_S) = 8\Re\delta \tag{4}$$

It is worth noting that $\Re$ (δ) does not need any absorptive part and it is determined by the difference of diagonal mass terms of $K^0 and\ \bar{K}^0$. The result of the measurement of the asymmetry (3) by the CPLEAR Collaboration [26] is given in Fig. 2

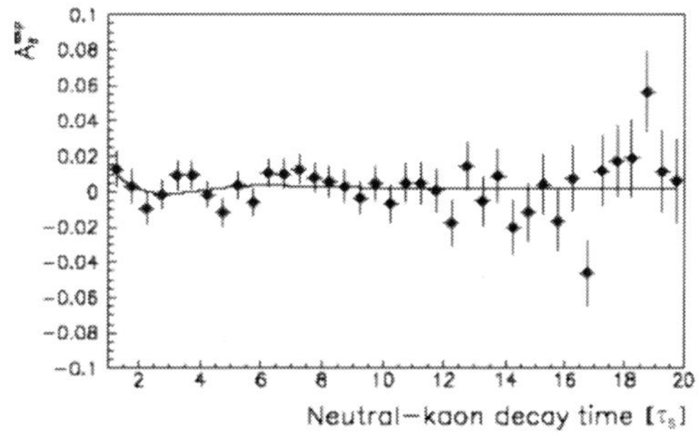

Figure 2-The asymmetry $A_\delta(\tau)$ as defined in Eq.3, measured by the CPLEAR Collaboration.

The fit with a constant term produces the result

$$\Re\delta = (0.30 \pm 0.33 \pm 0.06) \times 10^{-3} \quad (5)$$

where the first error is statistical and the second systematic.

Its interpretation in terms of the mass matrix elements of the $K^0 - \bar{K}^0$ system gives the upper bound

$$|m_{K^0} - m_{\bar{K}^0}|/m_K < 10^{-18} \quad (6)$$

If we take as figure of merit this fractional difference between the masses of a particle and its antiparticle, the result (5) appears to be the best CPT Violation Limit.

At the ϕ-Factory DAPHNE, the Entanglement between the two neutral Kaon states induced by their antisymmetric exchange allows a separate selection of observable asymmetries from either K_L or K_S states. The Charge Asymmetry from these states is given by

$$A_{S,L} = \frac{\Gamma(K_{S,L}\to\pi^- e^+\nu) - \Gamma(K_{S,L}\to\pi^+ e^-\bar{\nu})}{\Gamma(K_{S,L}\to\pi^- e^+\nu) + \Gamma(K_{S,L}\to\pi^+ e^-\bar{\nu})} = 2\ [Re\ (\epsilon_K) \pm Re\ (\delta_K)\ \text{-}\ Re\ (y)] \quad (7)$$

where, in addition to the CPV and CPTV parameters in the Hamiltonian matrix, a possible CPTV in the flavour decay amplitudes of Figure 1, parameterised by "y", is included. As seen, the difference between these two Charge Asymmetries A_S - A_L is able to isolate CPTV effects in the diagonal matrix elements of the Mass Matrix.

The value of A_L is well known from hadronic machines, with the best precise result from the KTeV Collaboration [27]

$$\mathbf{A_L = (3.332\ \pm 0.058_{stat}\ \pm 0.047_{syst}) \cdot 10^{-3}} \quad (8)$$

The measurement of A_S was undertaken in the first period of the KLOE experiment with the result [28]

$$A_S = (1.5 \pm 9.6_{stat} \pm 2.9_{syst}) \cdot 10^{-3} \quad (9)$$

An important improvement of this result is expected at KLOE-2.

3 Genuine Separate T, CP, CPT Asymmetries in B_d transitions

In Quantum Mechanics, there is an operator U_T implementing the T-symmetry acting on the states of the physical system, such that

$$U_T\, \vec{r}\, U_T^+ = \vec{r},\ U_T\, \vec{p}\, U_T^+ = -\vec{p},\ U_T \vec{s}\, U_T^+ = -\vec{s} \quad (10)$$

By considering the commutator of the Correspondence Principle,

$$[r_j, p_K] = i\hbar\delta_{jK} I \quad (11)$$

the operator U_T, contrary to the case of CP, must be ANTIUNITARY; UNITARY for conserving probabilities, ANTI- for complex conjugation. ANTIUNITARITY introduces many important subtleties, in particular the transition amplitude for i → f is transformed under T to

$$S_{i \to f} \xrightarrow{T} S_{U_T f \to U_T i} \quad (12)$$

so that the construction of a Genuine TRV Asymmetry needs the interchange between in and out states.
Something similar can be discussed for the CPT operator, which is ANTIUNITARY too, which needs not only the exchange of in and out states in transitions, but also going from particles to antiparticles.

For unstable particles, the decay products should not be included in the states building the asymmetries for T or CPT. However, these decay products are what we observe. Therefore, the problem is in the preparation and filtering of the appropriate initial and final meson states from the observed two decay channels. For (Flavour, CP) eigenstate decay products, the Editors of Rev. Mod. Phys. decided to put "what is T-transformation experimentally" in the cover page of the Volume in which the Colloquium article [29] was published. This Figure provides the explanation of what has to be measured in order to build a true genuine TRV asymmetry. With appropriate changes, we reproduce this Figure here, not for TRV but for CPTV.

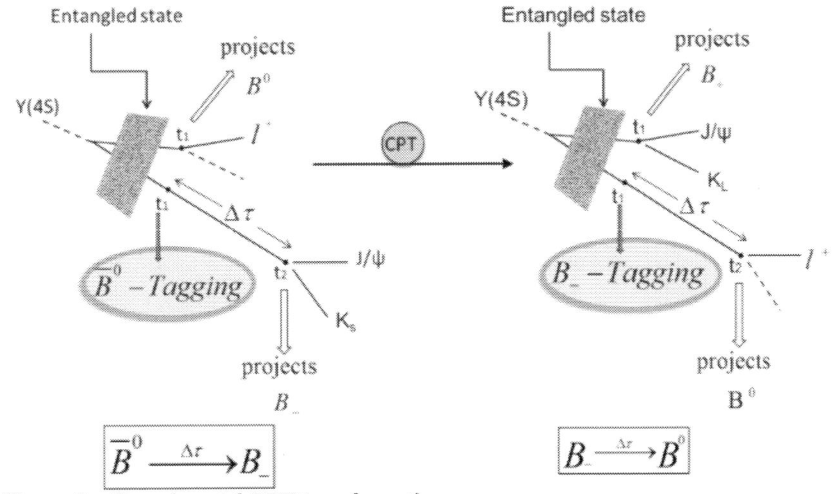

Figure 3 – Experimental CPT transformation

The Figure has illustrated the Reference $\bar{B}^0 \to B_-$ transition, transformed by CPT into the transition $B_- \to B^0$. The named neutral meson states B_+, B_- correspond to the filtered states by the CP = +, - eigenstate decay products J/ψ K_L, J/ψ K_S.

How many (Flavour, CP) decay channels we have at our disposal experimentally? 2 Flavours x 2 CP x 2 time-ordering = 8.

These 8 transitions may be grouped in two blocks with extremely important properties. Within each block, the 4 transitions correspond to one Reference and the three different transitions transformed from the Reference by the T, CP and CPT transformations separately. These amazing properties for (Flavour, CP) eigenstate decay channels of the entangled system of neutral mesons are represented in Fig.4, where the time-dependent Intensities of the 8 transitions are organised appropriately in order to make the direct connection of the transitions by the three Discrete Symmetries apparent. The red (left) and blue (right) blocks connected by time exchange are NOT connected by the symmetries.

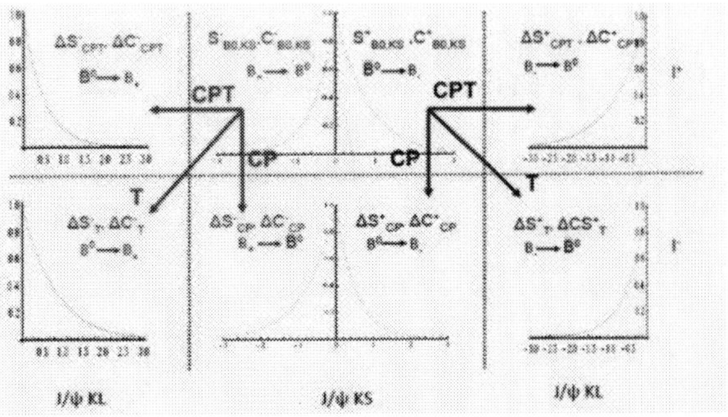

Figure 4 – Reference transition and Separate T, CP, CPT transformed T transitions

Using Quantum Mechanics for the time evolution of the two-state $B^0 - \bar{B}^0$ system and no particular Hamiltonian dynamics, each time dependent Intensity has the following general form

$$I\ (\Delta t) \sim e^{-\Gamma\Delta t}\{C_C \cos(\Delta m \Delta t) + S_c \sin(\Delta m \Delta t) + C_h \cosh(\Delta\Gamma\Delta t) + S_h \sinh(\Delta\Gamma\Delta t)\ \} \qquad (13)$$

where the coefficients C_c, S_c, C_h and S_h depend of the selected transition. Without more dynamical theoretical framework, these coefficients are unknown. However, we can construct Asymmetry Parameters ΔC_c, ΔS_c and ΔC_h (in the B_d-system $\Delta\Gamma = 0$) for each Discrete Symmetry T, CP, CPT. In total, we have 9 Asymmetry Parameters with different information content within each block. Any non-vanishing value of the 3 Asymmetry Parameters for each Symmetry is a bona-fide genuine proof of the Violation of that Symmetry.

In Ref. [30] the authors have worked out the explicit effective Hamiltonian dynamics associated to a Weisskopf-Wigner (WW) quantum description [25] of the meson-antimeson system. In this way the 9 asymmetry parameters can be written in terms of the parameters of the Hamiltonian matrix and the decay amplitudes. Two results are worth mentioning:
i) the Asymmetries are genuine signals of Symmetry Breaking, as they should for any dynamics; ii) the 9 Asymmetry parameters appear with different information content.
Using data from the BABAR analysis [21], in which the Intensities (13) were normalised by dividing by C_h with $\Delta\Gamma = 0$, the authors of [30] have re-derived the 9 Asymmetry Parameters with some mild additional information. The resulting TRV and CPV asymmetries give the result

$$\Delta S_c^T = -0.687 \pm 0.020\,; \;\; \Delta S_c^{CP} = -0.680 \pm 0.021 \qquad (14)$$

These results constitute an impressive separate evidence of TRV and CPV. We emphasize that they are independent results and TRV is established without any assumption on either CPV or CPT invariance.
The analysis performed assuming perfect Entanglement of the neutral meson system in which the two-body state is antisymmetric under the permutation exchange of the two mesons, has also been used to extract the CPTV Asymmetry parameters. The unique antisymmetric state remains unaltered under the time evolution of the entangled system before the first decay, even with Mixing. Under these conditions, given the time ordered (f, g) decays, the reversed time ordering (g, f) decays satisfy the exchange symmetry conditions

$$\mathbf{C_h(f,g) = C_h(g,f)\,;\; C_c(f,g) = C_c(g,f)\,;\; S_c(f,g) = -S_c(g,f)} \qquad (15)$$

Notice that the exchange symmetry has nothing to do with the discrete symmetries T, CP, CPT associated to the dynamics of the system. As a consequence, one has to consider the measurements associated with (f, g) and (g, f) decays as two independent experimental processes measuring the **same** properties, once Eqs. (15) are imposed. The CPTV Asymmetry is generated in the WW approach by the θ parameter in the diagonal Hamiltonian matrix elements distinguishing particle and antiparticle properties. The correlated fit to $\mathrm{Re}(\theta)$ and $\mathrm{Im}(\theta)$ is presented [30] in Fig. 5

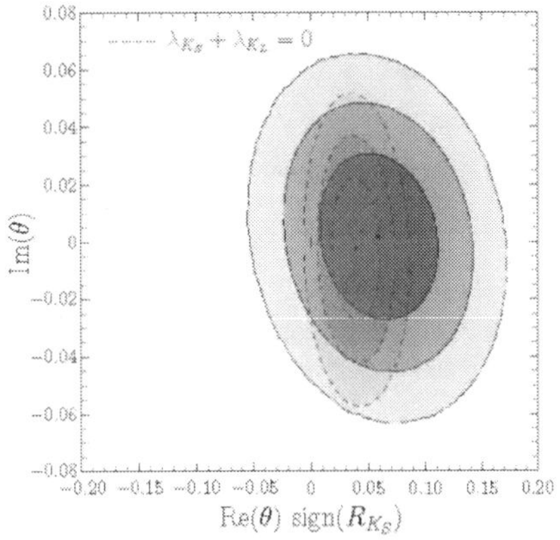

Figure 5 – Correlated fit to the CPTV Re(θ) and Im(θ)

and gives an intriguing 2σ effect for CPTV in Re(θ).

Under an ill-defined CPT operator expected in models of quantum gravity, the perfect Entanglement can be weakened [31] by the presence of an ω-effect associated to the "wrong" exchange symmetry of the neutral meson and antimeson. Eqs. (15) are then no longer valid and their violation would then be a signal of this new form of CPT breaking.

4 Time reversal violation in neutral Kaon transitions

Similarly to the $B^0 - \bar{B}^0$ system, EPR correlations at a ϕ Factory can be exploited to study T-conjugated TRANSITIONS between $K^0, \bar{K}^0$ states and the orthogonal K_+, K_- states filtered by CP eigenstate decay products

$$\begin{aligned} |i\rangle &= \frac{1}{\sqrt{2}}\left[|K^0(\vec{p})\rangle|\bar{K}^0(-\vec{p})\rangle - |\bar{K}^0(\vec{p})\rangle|K^0(-\vec{p})\rangle\right] \\ &= \frac{1}{\sqrt{2}}\left[|K_+(\vec{p})\rangle|K_-(-\vec{p})\rangle - |K_-(\vec{p})\rangle|K_+(-\vec{p})\rangle\right] \end{aligned} \tag{16}$$

Using the decays as filtering measurements of the meson states and Entanglement for the preparation of the living partner initial state, we identify the Reference process, for example $K^0 \to K_-$

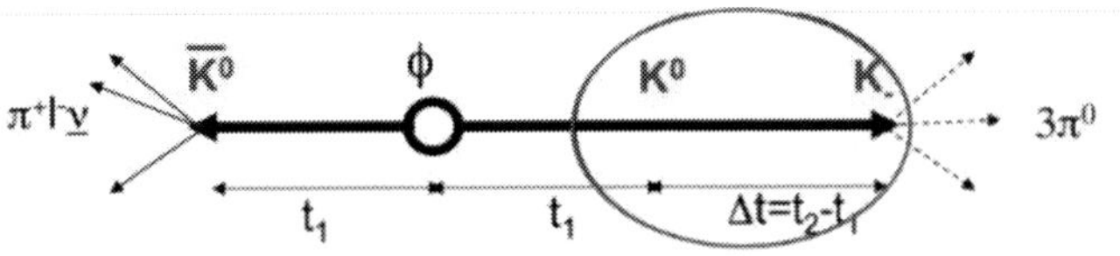

Figure 6 - Neutral Kaon transition $K^0 \to K_-$ filtered by the l^- decay first, $3\pi^0$ decay later.

For this Reference, the T-conjugated process is $K_- \to K^0$, separated from the CP conjugated $\bar{K}^0 \to K_-$ and the CPT-conjugated $K_- \to \bar{K}^0$. The T-conjugated process is associated to the time-ordered 2π decay first, l^+ decay later

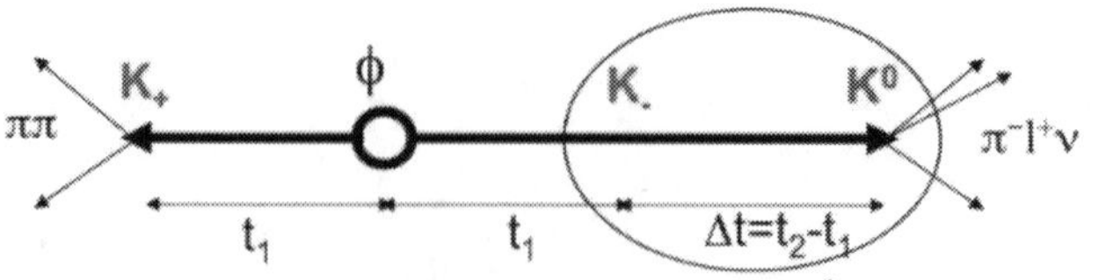

Figure 7 - The Reversal-in-Time Neutral Kaon transition $K_- \rightarrow K^0$

For a T-symmetry test one has a priori 4 possible comparisons of transitions as given in Table 1

T symmetry test

Reference		*T*-conjugate	
Transition	Decay products	Transition	Decay products
$\bar{K}^0 \rightarrow K_-$	$(\ell^+, \pi^0\pi^0\pi^0)$	$K_- \rightarrow \bar{K}^0$	$(\pi^0\pi^0\pi^0, \ell^-)$
$K_+ \rightarrow K^0$	$(\pi^0\pi^0\pi^0, \ell^+)$	$K^0 \rightarrow K_+$	$(\ell^-, \pi\pi)$
$\bar{K}^0 \rightarrow K_+$	$(\ell^+, \pi\pi)$	$K_+ \rightarrow \bar{K}^0$	$(\pi^0\pi^0\pi^0, \ell^-)$
$K_- \rightarrow K^0$	$(\pi\pi, \ell^+)$	$K^0 \rightarrow K_-$	$(\ell^-, \pi\pi)$

Table 1- Neutral Kaon Transitions to be compared for a T symmetry test

One can then define [33] the following ratios of time dependent probabilities

$$\begin{aligned}
R_1(\Delta t) &= P\left[K^0(0) \rightarrow K_+(\Delta t)\right] / P\left[K_+(0) \rightarrow K^0(\Delta t)\right] \\
R_2(\Delta t) &= P\left[K^0(0) \rightarrow K_-(\Delta t)\right] / P\left[K_-(0) \rightarrow K^0(\Delta t)\right] \\
R_3(\Delta t) &= P\left[\bar{K}^0(0) \rightarrow K_+(\Delta t)\right] / P\left[K_+(0) \rightarrow \bar{K}^0(\Delta t)\right] \\
R_4(\Delta t) &= P\left[\bar{K}^0(0) \rightarrow K_-(\Delta t)\right] / P\left[K_-(0) \rightarrow \bar{K}^0(\Delta t)\right]
\end{aligned} \tag{17}$$

Any deviation from R = 1 constitutes a direct evidence of a Violation of T-symmetry.

These theoretical ratios for meson state transitions can be connected to experimental ratios of Intensities for pairs of decay channels as

$$\begin{aligned}
R_1^{\text{exp}}(\Delta t) &\equiv \frac{I(\ell^-, \pi\pi; \Delta t)}{I(3\pi^0, \ell^+; \Delta t)} = R_1(\Delta t) \times \frac{C(\ell^-, \pi\pi)}{C(3\pi^0, \ell^+)} \\
\boxed{R_2^{\text{exp}}(\Delta t) \equiv \frac{I(\ell^-, 3\pi^0; \Delta t)}{I(\pi\pi, \ell^+; \Delta t)} = R_2(\Delta t) \times \frac{C(\ell^-, 3\pi^0)}{C(\pi\pi, \ell^+)}} \\
R_3^{\text{exp}}(\Delta t) &\equiv \frac{I(\ell^+, \pi\pi; \Delta t)}{I(3\pi^0, \ell^-; \Delta t)} = R_3(\Delta t) \times \frac{C(\ell^+, \pi\pi)}{C(3\pi^0, \ell^-)} \\
\boxed{R_4^{\text{exp}}(\Delta t) \equiv \frac{I(\ell^+, 3\pi^0; \Delta t)}{I(\pi\pi, \ell^-; \Delta t)} = R_4(\Delta t) \times \frac{C(\ell^+, 3\pi^0)}{C(\pi\pi, \ell^-)}}
\end{aligned} \tag{18}$$

where the co-factors of Eq. (18) are related to the corresponding Branching Ratios. In practice, one has two measurable ratios with the two time-orderings either $\Delta t > 0$ or $\Delta t < 0$, due to the inverse connections

$$\begin{aligned}
R_2^{\text{exp}}(-\Delta t) &= \frac{1}{R_3^{\text{exp}}(\Delta t)} = \frac{1}{R_3(\Delta t)} \times \frac{C(3\pi^0, \ell^-)}{C(\ell^+, \pi\pi)}, \\
R_4^{\text{exp}}(-\Delta t) &= \frac{1}{R_1^{\text{exp}}(\Delta t)} = \frac{1}{R_1(\Delta t)} \times \frac{C(3\pi^0, \ell^+)}{C(\ell^-, \pi\pi)}.
\end{aligned} \tag{19}$$

In Figs. 8 we give [32] the Standard Model expectations for the two experimental ratios R_2^{exp} and R_4^{exp}, in arbitrary units, as function of Δt for the two time orderings

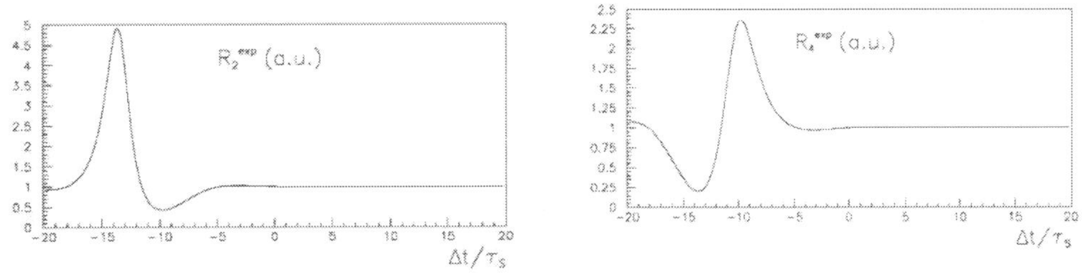

Figure 8 – The experimental ratios R_2^{exp} (left) and R_4^{exp}(right) for TRV expected in the Standard Model

One reaches the conclusion that this T-test could be feasible in KLOE-2 at DAPHNE with a Luminosity of about 10 fb^{-1}.

5. Direct test of CPT in neutral Kaon transitions

EPR correlations at a ϕ-Factory can also be exploited to study CPT-conjugated TRANSITIONS involving Flavour $K^0, \bar{K}^0$ states and the orthogonal K_+, K_- states filtered by CP eigenstate decay products.

By means of Eq. (16), using the decays as filtering measurements of the meson states and Entanglement for the preparation of the living partner initial state, we identify the Reference process, for example $K^0 \to K_-$

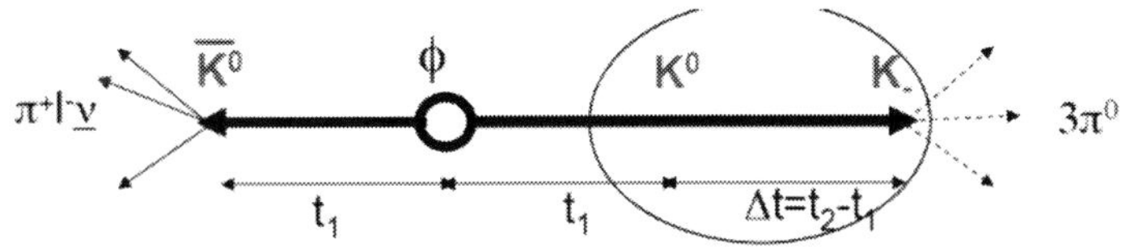

Figure 9- Reference Neutral Kaon Transition $K^0 \to K_-$ associated to the l$^-$ decay first, $3\pi^0$ decay later.

For this Reference, the CPT-conjugated process is $K_- \to \bar{K}^0$, separated out from the CP-conjugated $\bar{K}^0 \to K_-$ and the T-conjugated $K_- \to K^0$. The CPT-conjugated process is associated to the time-ordered decay channels

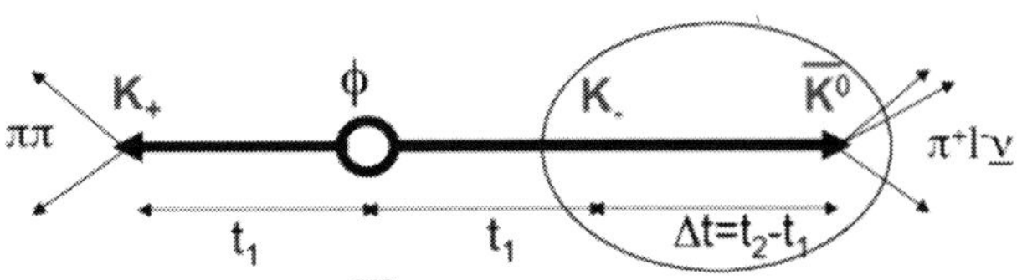

Figure 10- The CPT-transformed $K_- \to \bar{K}^0$ is associated to the 2π decay first, the l$^-$ decay later.

For a CPT-symmetry test one has a priori 4 possible comparisons of transitions as given in Table 2

CPT symmetry test

Reference		*CPT*-conjugate	
Transition	Decay products	Transition	Decay products
$K^0 \to K_+$	$(\ell^-, \pi\pi)$	$K_+ \to \bar{K}^0$	$(3\pi^0, \ell^-)$
$K^0 \to K_-$	$(\ell^-, 3\pi^0)$	$K_- \to \bar{K}^0$	$(\pi\pi, \ell^-)$
$\bar{K}^0 \to K_+$	$(\ell^+, \pi\pi)$	$K_+ \to K^0$	$(3\pi^0, \ell^+)$
$\bar{K}^0 \to K_-$	$(\ell^+, 3\pi^0)$	$K_- \to K^0$	$(\pi\pi, \ell^+)$

Table 2.- Reference transitions and their CPT-transformed transitions.

One can then define [33] the following ratios of time dependent probabilities

$$\begin{aligned} R_{1,\mathcal{CPT}}(\Delta t) &= P\left[K_+(0) \to \bar{K}^0(\Delta t)\right] / P\left[K^0(0) \to K_+(\Delta t)\right] \\ R_{2,\mathcal{CPT}}(\Delta t) &= P\left[K^0(0) \to K_-(\Delta t)\right] / P\left[K_-(0) \to \bar{K}^0(\Delta t)\right] \\ R_{3,\mathcal{CPT}}(\Delta t) &= P\left[K_+(0) \to K^0(\Delta t)\right] / P\left[\bar{K}^0(0) \to K_+(\Delta t)\right] \\ R_{4,\mathcal{CPT}}(\Delta t) &= P\left[\bar{K}^0(0) \to K_-(\Delta t)\right] / P\left[K_-(0) \to K^0(\Delta t)\right] \end{aligned} \quad (20)$$

Any deviation from $R_{i,CPT} = 1$ constitutes a direct evidence of a Violation of CPT-symmetry.

These theoretical ratios for meson state transitions can be connected to experimental ratios of Intensities for pairs of decay channels as indicated in Table 2.

For visualisation purposes, we plot [33] in Fig. 11 these 4 ratios for CPT-Violation using a "δ" parameter in the Hamiltonian matrix with values $\mathrm{Re}(\delta)=3.3 \cdot 10^{-4}$ and $\mathrm{Im}(\delta)=1.6 \cdot 10^{-5}$ (continous line -) and $\mathrm{Im}(\delta) = 0$ (broken line - -).

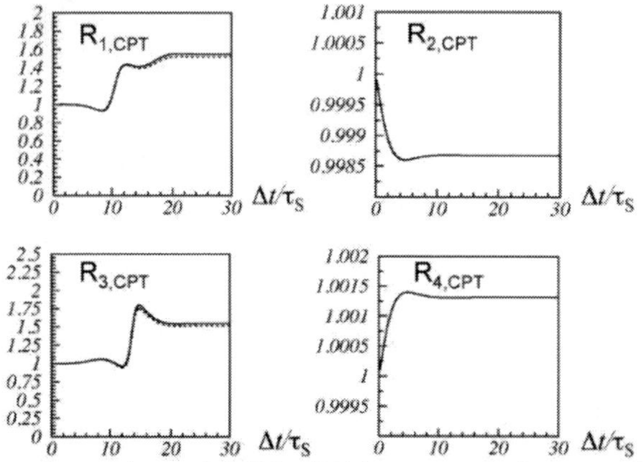

Figure 11.- The ratios of transition probabilities (20) between a Reference and its CPT- transformed for values of the δ parameter as explained in the text.

For the most interesting Δt region larger than the short lifetime of K_S τ_S, one has

$$R_{2,CPT}(\Delta t >> \tau_S) = 1 - 4\mathrm{Re}(\delta)$$

$$R_{4,CPT}(\Delta t >> \tau_S) = 1 + 4\mathrm{Re}(\delta) \quad (21)$$

One cannot imagine a better clearcut test of CPT symmetry in transitions as the one shown in Eqs. (21). This test is feasible at KLOE-2.

6. The ω-EFFECT

In presence of decoherence and CPT breaking induced by quantum gravity effects, the CPT operator is "ill-defined" and the definition of the particle-antiparticle states could be modified. This in turn could induce a weakening of the Entanglement [31] imposed by Bose statistics as an EPR correlation

$$\begin{aligned} |i\rangle &\propto \left(|K^0\rangle|\overline{K}^0\rangle - |\overline{K}^0\rangle|K^0\rangle\right) + \omega\left(|K^0\rangle|\overline{K}^0\rangle + |\overline{K}^0\rangle|K^0\rangle\right) \\ &\propto \left(|K_S\rangle|K_L\rangle - |K_L\rangle|K_S\rangle\right) + \omega\left(|K_S\rangle|K_S\rangle - |K_L\rangle|K_L\rangle\right) \end{aligned} \tag{22}$$

In some microscopic models of space-time foam [34] with stochastic fluctuations of defect recoils in the propagation of particles, the magnitude of the ω-parameter in Eq. (22) can reach values $\omega \sim 10^{-4} - 10^{-5}$.

The maximum sensitivity to ω is expected for a symmetric decay pair of the entangled state f1= f2= $\pi^+ \pi^-$, which is CP-violating from the dominant antisymmetric state of Eq. (22). As a consequence, one gets a fantastic enhancement of the ω-effect from the "wrong" component in Eq. (22) as ω/ε, where ε is the small CPV parameter in K-physics. In Fig. 12 we plot the expected double decay rate time-dependent Intensities for vanishing and non-vanishing values of ω accessible at KLOE-2.

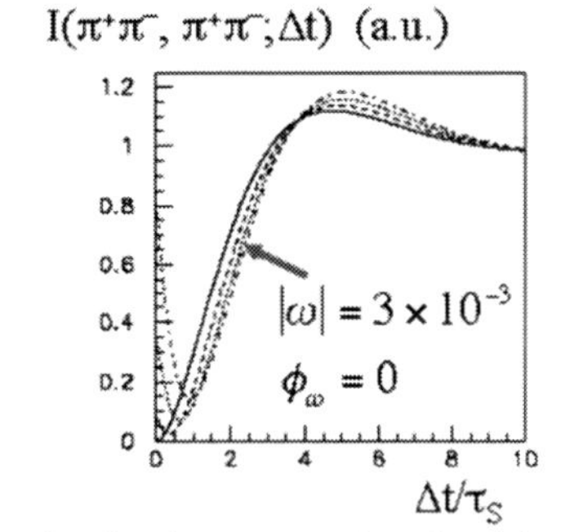

Figure 12.- Time-dependent Intensity for the symmetric pair of decay channels ($\pi^+\pi^-$, $\pi^+\pi^-$) for vanishing and non-vanishing ω-parameter.

The higher luminosity of present DAPHNE and a good time resolution at KLOE-2 allow to reach an interesting sensitivity to this most fascinating ω-effect.

7. Conclusions

I summarize the main Conclusions and Prospects in the studies of the Discrete Symmetries T, CP, CPT with Neutral Mesons:

- The fundamental role that Discrete Symmetry Breaking has played in the understanding of the Laws of Physics emphasizes the importance of looking for Direct Asymmetries of Separate T, CP, CPT in Transitions of Neutral Meson states. In dealing with the Antiunitary T and CPT, Entanglement is needed.
- The measurement of the semileptonic charge asymmetry from K_S by KLOE-2 at DAPHNE allows the separation of CPTV in the Mass Matrix of $K^0 - \overline{K}^0$.

- The Flavour-CP Transitions in Entangled $B^0 - \bar{B}^0$ have demonstrated Genuine Separate Asymmetries for T and CP with high statistical significance and, with a 2σ tension, compatibility with CPT invariance.
- KLOE-2 is in the best condition to accomplish a complete program of Genuine Separate Asymmetries for T, CP and CPT in Flavour-CP Transitions for $K^0 - \bar{K}^0$. Possible fake effects can be controlled within the same experiment.
- The best way to study the w-effect weakening Entanglement due to an ill-defined CPT operator is the CP-Violating Correlated Decay (π^+, π^-, π^+, π^-) at KLOE-2. This measurement has a distinctive signature from the dynamical CPTV in the Effective Hamiltonian for $K^0 - \bar{K}^0$.

Acknowledgements

This research has been supported by MINECO Project FPA 2014-54459-P, Generalitat Valenciana Project GV PROMETEO II 2013-017 and Severo Ochoa Excellence Centre Project SEV 2014-0398.

References

[1] F. Englert, R. Brout, Phys.Rev.Lett. **13**(1964) 321;
P.W. Higgs, Phys.Rev.Lett. **13**(1964)508

[2] ATLAS Collaboration, Phys. Lett. B **716**(2012)1;
CMS Collaboration, Phys.Lett. B **716**(2012)30

[3] T.D.Lee, C.-N. Yang, Phys.Rev. **104**(1956)254;
C.S.Wu et al., Phys.Rev. **105**(1957)1413

[4] R.L. Garwin et al., Phys. Rev. **105**(1957)1415

[5] R.P. Feynman, M. Gell-Mann, Phys.Rev. **109**(1958)193

[6] S.L. Glashow, Nucl. Phys. **22**(1961)579

[7] S. Weinberg, Phys. Rev. Lett. **19**(1967)1264

[8] A. Salam, Conf. Proc. C **680519**(1968)367

[9] Gargamelle Neutrino Collaboration, Phys. Lett. B **46** (1973)138

[10] J. H. Christenson et al., Phys. Rev. Lett. **13**(1964)138

[11] CPLEAR Collaboration, Phys. Rept. **374**(2003)165.

[12] BABAR Collaboration, Phys.Rev.Lett. **87**(2001)091801.

[13] Belle Collaboration, Phys.Rev.Lett. **87**(2001)091802.

[14] N.Cabibbo, Phys.Rev.Lett. **10**(1963)531.

[15] M. Kobayashi, T-Maskawa, Progr.Theor.Phys. **49**(1973)652.

[16] E.P.Wigner, Nachr.Ges.Wiss. Göttingen **32**(1932)35

[17] G.Lüders, Annals Phys. **2**(1957)1;
W.Pauli et al., "Niels Bohr and the development of Physics", Pergamon NY(1995)

[18] L. Wolfenstein, Phys.Rev.Lett. **83**(1999)911.

[19] M.C.Banuls, J.Bernabeu, Phys.Lett.B **464**(1999)117

[20] M.C. Banuls, J. Bernabeu, Nucl. Phys. B **590**(2000)19

[21] BABAR Collaboration, Phys.Rev.Lett. **109**(2012)211801

[22] J. Bernabeu et al., JHEP **1208**(2012)064

[23] E. Applebaum et al., Phys.Rev. D **89**(2014)076011

[24] J. Bernabeu et al., Phys.Lett. B **728**(2014)95

[25] V. Weisskopf, E.P. Wigner, Z. Phys. **63**(1930)54;
Z. Phys. **65**(1930)18

[26] CPLEAR Collaboration, Phys. Lett. B **444**(1998)52

[27] KTeV Collaboration, Phys. Rev. Lett. **88**(2002)181601

[28] KLOE Collaboration, Phys. Lett. B **636**(2006)173

[29] J. Bernabeu, F. Martinez-Vidal, Rev. Mod. Phys. **87**(2015)165

[30] J. Bernabeu et al., JHEP **1606** (2016)100

[31] J. Bernabeu et al., Phys. Rev. Lett. **92**(2004)131601

[32] J. Bernabeu et al., Nucl. Phys. B **868**(2013)102

[33] J. Bernabeu et al., JHEP **1510**(2015)139

[34] J. Bernabeu et al., Phys. Rev. D**74**(2006)045014

Models & Searches of CPT Violation: a personal, very partial, list

Nick E. Mavromatos[1,★]

[1] *King's College London, Department of Physics, Theoretical Particle Physics and Cosmology Group, Strand London WC2R 2LS, UK.*

Abstract. In this talk, first I motivate theoretically, and then I review the phenomenology of, some models entailing CPT Violation (CPTV). The latter is argued to be responsible for the observed matter-antimatter asymmetry in the Cosmos, and may owe its origin to either Lorentz-violating background geometries, whose effects are strong in early epochs of the Universe but very weak today, being temperature dependent in general, or to an ill-defined CPT generator in some quantum gravity models entailing decoherence of quantum matter as a result of quantum degrees of freedom in the gravity sector that are inaccessible to the low-energy observers. In particular, for the latter category of CPTV, I argue that entangled states of neutral mesons (Kaons or B-systems), of central relevance to KLOE-2 experiment, can provide smoking-gun sensitive tests or even falsify some of these models. If CPT is ill-defined one may also encounter violations of the spin-statistics theorem, with possible consequences for the Pauli Exclusion Principle, which I only briefly touch upon.

1 Introduction and Motivation

Invariance of a relativistic (*i.e.* Lorentz Invariant), local and unitary field theory Lagrangian under the combined transformations of Charge Conjugation (C), Parity (spatial reflexions, (P)) and reversal in Time (T), at any order, is guaranteed by the corresponding celebrated theorem [1]. This has important implications for the equality of masses and the absolute values of the various quantum numbers that characterise particles and antiparticles, and the equal amounts of matter and antimatter when they were created in the beginning of the Universe's evolution. On the other hand, today, there is an overwhelming dominance of matter over antimatter in the Cosmos, which calls for an explanation. Assuming the validity of the CPT theorem, A. Sakharov [2] has suggested that the observed matter-antimatter asymmetry in the Universe today is the result of *out of thermal equilibrium* processes in the early universe that violate C, CP and Baryon number (B) symmetries. The out of equilibrium assumption is a crucial one so that any asymmetry generated by the violation of C , B and CP symmetries in the expanding Universe is not washed out but remains to the current epoch, thereby explaining the observed dominance of matter over antimatter today.

Although all of Sakharov conditions are met qualitatively by the Standard Model (SM) of Particle Physics, unfortunately they are not valid quantitatively, meaning that the amount of CP violation within the SM is some ten orders of magnitude smaller than the required one to produce the observable

★e-mail: Nikolaos.Mavromatos@kcl.ac.uk

matter-antimatter asymmetry [3]. According to observations, the abundance of baryons over that of antibaryons is of order [4]

$$Y_{\Delta B} = \frac{n_B - n_{\bar{B}}}{n_\gamma} = (6.1 \pm 0.3) \times 10^{-10} \tag{1}$$

for temperatures $T \gtrsim 1$ GeV, where n_B is the number density of baryons, $n_{\bar{B}}$ is the number density of antibaryons and n_γ is the density of photons (proportional to the entropy density s of the Universe). This number was determined with accurate measurements of the Cosmic Microwave Background (CMB) radiation [5]. Out of equilibrium processes that generate a baryon asymmetry in the Universe are called collectively *Baryogenesis* [6]. Similarly, the generation of an asymmetry between leptons. and antileptons is known as *Leptogenesis* [7], and is expected to be of the same order of magnitude as $Y_{\Delta B}$. Since the amount of CP violation in the SM is not sufficient to produce (1), one should look for models beyond the standard model that contain new sources of CP violation, that can be tuned so as to produce sufficient Baryogenesis and Leptogenesis. There is a plethora of such models in the current literature, ranging from supersymmetric theories, to theories with extra dimensions of space, including strings. There is no experimental evidence as yet for the realisation of such extended models in Nature.

Interesting models [8, 9] of Baryon asymmetry in the Universe involve a two stage process, during which one generates first a lepton asymmetry in the early Universe, by appropriate CP-violating non-equilibrium processes in the early Universe, which is then communicated to the baryon sector, at temperatures lower than those of the electroweak phase transition ($T \lesssim 100$ GeV), by means of Baryon-minus-Lepton-(B-L)-number-conserving sphaleron processes within the SM sector of the theory. In this important scenario of Leptogenesis as the path to Baryogenesis, pioneered by Fukugita and Yanagida [8, 9], the lepton abundance is produced by the decay of heavy right-handed Majorana neutrinos (and so represents physics beyond the Standard Model (BSM)). The difference in the branching ratios of the channels of production of leptons and antileptons is equal to the imaginary part of the interference term of tree-level and one-loop diagrams for the decay processes. For the interference to generate a non-zero CP violating phase, at least two generations of right-handed neutrinos are needed [7, 8]. In fact, at least two right-handed neutrinos are *also* required in the see-saw mechanism [10] for the generation of light neutrino masses, as necessitated by the observed phenomena of neutrino flavour oscillations among the light neutrinos [11], which require at least two of the active neutrinos to have a mass. The model of Fukugita and Yanagida, therefore, connects an explanation of leptogenesis with the see-saw mechanism of light neutrino masses observed in Nature. The model thus represents an economical extension of SM, since, at least for the purpose of generating baryon asymmetry it does not require other particles apart from the massive right-handed neutrinos.

In the framework of ref. [8], the right handed neutrinos are very massive, and thus have decayed today, leaving no other trace apart from the see-saw type masses of the active neutrinos. Shaposhnikov and collaborators [12] have also made a proposal for a Minimal extension of the SM, termed νMSM, involving three generations of right-handed neutrinos, whose masses though are much lower than the corresponding ones in the model of [8]. In fact, in the νMSM the two heavier neutrinos are almost degenerate, with masses of order $O(1)$ GeV, while the lightest of the right-handed neutrinos may have masses of order of a few keV, and has a life time longer than the age of the Universe. In this sense, the model can provide a natural candidate for dark matter (DM) [1]. The relatively light masses of the right-handed neutrinos in the νMSM makes the Baryogenesis process in this model rather complicated [14],

[1]In this respect, we mention that an O(50) keV right-handed neutrino DM can play an important role in explaining galactic structures and thus resolving some of the tensions between ΛCDM model and observations at small (galactic) scales, especially if appropriate self-interactions among the keV-right-handed neutrinos are introduced [13]. The coincidence in the range of the allowed right-handed neutrino DM mass obtained in this approach, based on purely astrophysical reasons at galactic scales, with the one of the νMSM, induced by particle physics and DM cosmology reasons, is intriguing.

and certainly dissociated from the Leptogenesis path of [8]. From our point of view in this work we shall focus our attention to the model of [8], but taking into account the coupling of the fermions in the model to the gravitational background; this will have interesting consequences.

However, we shall not simply consider the Robertson-Walker Universe. Indeed, an alternative approach to baryon asymmetry, which avoids altogether the Sakharov condition on out of equilibrium processes, is the one in which from the beginning, the Universe has a matter-antimatter asymmetry, as a result of *violations* of the CPT symmetry. Keeping locality and unitarity intact, violation of Lorentz symmetry seems the most plausible possibility for evading the CPT theorem, and in fact this lead to a sort of anti-CPT theorem by Greenberg [15], claiming that the assumption of Lorentz symmetry plays somehow a more fundamental rôle than the other assumptions of the CPT theorem, with the conclusion that CPT violation *implies* necessarily a violation of Lorentz invariance. However, in his proof, Greenberg did assume the existence of a well-defined scattering transfer matrix, and in this sense his anti-CPT theorem is equivalent to the original one on CPT symmetry. Explicit counter examples to these claims, where violations of CPT occur in a Lorentz invariant way, *e.g.* in some non-local theories, have been provided [16]. Moreover there are Lorentz-invariant models with intrinsic decoherence [17], and it is well known that the presence of decoherence may lead [18] to an ill-defined nature of the CPT operator in the effective theory, obtained after tracing over the appropriate environmental degrees of freedom. Nevertheless, in our discussions in this talk, we shall restrict ourselves to spontaneous violation of Lorentz symmetry implying CPT Violation either through background geometries or through decoherence induced by a quantum gravity Lorentz-violating environment. In the latter case, the environmentally-induced intrinsic CPT violation may lead to interesting and unique effects in entangled states of mesons (ω-effect [19]), which we shall review at the end of talk.

The structure of the talk is the following: In section 2 we shall motivate CPT Violation (CPTV) theoretically, by considering early universe models, inspired from string theory, in which (a cosmic-temperature dependent) CPTV may generate the observed matter-antimatter asymmetry in the cosmos in relatively simple models. These are much simpler than current extensions of the Standard Models, where such asymmetry requires new sources of CP violation. In section 3 we shall discuss the present-day phneomenology of CPTV by concentrating on models inspired by the cosmologies discussed in the previous section. We discuss stringent bounds on *both*, geometrically-induced CPTV, due to Lorentz Violating backgrounds, within the local effective field theory framework of the Standard Model Extension (SME) [20] (subsection 3.1), and decoherence-induced CPTV, in which the generator of the CPT transformations is ill-defined, due to information loss carried by inaccessible (to a low-energy observer) degrees of freedom of quantum gravity. This latter type of CPTV lies beyond the effective local field theory framework, and leads to characteristic evidence in entangled quantum states of neutral mesons (ω-effect, subsection 3.2). Finally, instead of conclusions, in section 4, we discuss how the latter type of CPTV, may lead to (small) spin-statistics violations.

2 CPT Violation in Stringy Universes

We commence our discussion on Baryon asymmetry generation by considering *spontaneous violations* of Lorentz symmetry, which in turn induce CPT violation [2]. Such violations can be provided

[2]For early pioneering works on matter-antimatter asymmetry generated by CPT Violating backgrounds within the SME framework we refer the reader to [21]. In that work, it was argued that, under certain circumstances, certain CPT violating terms within an SME effective Lagrangian [20] can produce large baryon asymmetry, at Grand Unified temperatures, which is eventually diluted to the current value by sphaleron processes within the SM sector. An alternative source of CPT Violating interactions that could lead to matter-antimatter asymmetry is the coupling of the baryon (or B-L) number (anomalous) current to scalar curvature $\mathcal{R}$ of space-time through a CP violating interaction Lagrangian $\mathcal{L}$, that could occur, *e.g.* within some Supergravity theories [22, 23]: $\mathcal{L} = \frac{1}{M_\star^2} \int d^4x \sqrt{-g} \left(\partial_\mu \mathcal{R}\right) J^\mu$, where $M_\star$ is a cut-off in the effective field theory and J^μ could be

by the background geometry, for instance within the context of string-inspired models [25, 26], by the field strength of the spin-one antisymmetric tensor Kalb-Ramond field, $H_{\mu\nu\rho} = \partial_{[\mu} B_{\nu\rho]}$, which is totally antisymmetric in its three indices, and under certain circumstances is constant in a given frame (to be identified with the co-moving frame of the observer in an expanding Universe framework). $H_{\mu\nu\rho}$ plays the rôle of a constant totally antisymmetric torsion background field, which couples to fermions via an axial $\gamma^\mu\gamma^5$ coupling. In four space-time dimensions, $H_{\mu\nu\rho} \propto \epsilon_{\mu\nu\rho\sigma}\partial^\sigma b$, where $b(x)$ is a Kalb-Ramond pseudoscalar (axion-like) field. In cosmological models $b(x) = b(t)$ is only a function of the cosmic time t.

The constant torsion background coupling with the axial fermions implies that we should study the following model of the right-handed heavy neutrinos [26]:

$$\mathcal{L} = i\overline{N}\partial\!\!\!/ N - \frac{m}{2}(\overline{N^c}N + \overline{N}N^c) - \overline{N}B\!\!\!/\gamma^5 N - Y_k\overline{L}_k\tilde{\phi}N + h.c.\,, \tag{2}$$

where N is the heavy right-handed Majorana field, satisfying the Majorana condition $N^c = N$, with the superscript c denoting Dirac's charge conjugation, and L_k is the lepton SU(2) left-handed doublet field of the SM, with k a generation index. The adjoint of the Higgs field is defined by the relation $\tilde{\phi}_i = \varepsilon_{ij}\phi_j$. Due to our constant H-torsion situation, the axial vector background is $B^\mu = \dot{b}\delta^{0\mu}$, where the overdot denotes derivative with respect to the cosmic time t. This is understood in what follows. For our purposes, we shall restrict ourselves [26] to *only one generation* of right-handed neutrinos [3]. Notice that the model has a portal of communication of the right-handed neutrino sector with the SM sector via the Yukawa interactions with couplings Y_k. We do not specify the mechanism for generating a mass m for the heavy right-handed neutrino. The mass m is determined self consistently by the requirement of the model generating sufficient leprogenesis [26].

The total lepton number asymmetry generated in this model by the tree-level decays $N \to \ell^+\,\phi\,(\ell^-\,\overline{\phi})$ at an appropriately calculated freeze-out temperature T_D, in the presence of a constant axial background $B_0 \neq 0$, is estimated to be [26]:

$$\Delta L^{TOT} = (2r-1)n_N = \frac{2\Omega B^0}{\Omega^2 + (B^0)^2} n_N \simeq \frac{2B^0}{m}\, n_N, \quad \Omega \equiv \sqrt{(B^0)^2 + m^2}\,. \tag{3}$$

where n_N is the density of right-handed neutrinos N. The high-temperature expansion of n_N yields for small $B_0 \ll T_D \simeq m$ [26]

$$n_N(T_D) = e^{-\beta\overline{m}}\left(\frac{\overline{m}}{2\pi\beta}\right)^{\frac{3}{2}} + O(B_0^2)\Big|_{T\simeq T_D} \simeq 0.023\,\overline{m}^3. \tag{4}$$

The lepton asymmetry $\frac{\Delta L^{TOT}}{n_\gamma}$, where n_γ is the photon density, is expected to be of the same order of magnitude of the baryon asymmetry (1), given that in scenarios of baryogenesis via leptogenesis we are adopting here, the asymmetry in the lepton sector is communicated to the baryon sector via Baryon (B) and Lepton (L) number violating, B-L conserving, sphaleron processes in the SM sector [3]. On making the approximation $T_D \simeq m$, retaining only first order terms in $\frac{B_0}{\overline{m}}$, and recalling that the photon number density is [28] $n_\gamma \simeq \frac{2\zeta(3)}{\pi^2}\,T^3 \simeq 0.24\,T^3$, we obtain:

$$\frac{B_0}{\overline{m}} \simeq 5\cdot 10^{-8}\,, \quad \text{at} \quad T = T_D \simeq \overline{m} \sim 100\ \text{TeV}\,. \tag{5}$$

the current associated with $B - L$ (L being the lepton number). Recently some potential problems with such models have been pointed out in [24], regarding instabilities of theories with such interactions.

[3]This will be sufficient for leptogenesis in our model, but of course in such a case one should consider alternative ways [27] to seesaw mechanism [10] in order to give masses to the active light neutrinos of the SM sector (that are parts of the doublet L_k). In case one requires a seesaw type mass generation for neutrinos, then at least two right-handed neutrino generations are needed. In such a case, there are extra sources of CP violation, which contribute to the lepton asymmetry [8].

If other mechanisms contributed to the lepton asymmetry in the universe, or the Yukawa couplings assume smaller values, the minimum value of B_0 would be smaller than the one given here.

In [26] the (constant) background B^0 is assumed to undergo a phase transition at temperatures near $T_D = 100$ TeV, such that

$$B_0 = c_0 T^3 , \quad c_0 > 0 , \quad \text{for} \quad T \lesssim T_D . \tag{6}$$

The parameter c_0 is a phenomenological, and can be constrained by requiring that B^0 today must be at most equal to the experimental upper bounds [29] of the b_0 (temporal) axial Lorentz and CPT violating coefficient of the Standard Model Extension [20]. Taking into account that the temperature of the Universe today (from the CMB measurements) is $T_{\rm CMB} = 2.725$ K $= 0.2348$ meV, we obtain from (6) and (5) [26] $c_0 = 1$ MeV(100 TeV)$^{-3} = 10^{-42}$ meV^{-2}. implying a current value of B_0 of order

$$B_{0\,\rm today} = \mathcal{O}\big(10^{-44}\big)\,\text{meV} , \tag{7}$$

way too small for any experimental detection [29]. Moreover, in this case, the value of B^0 at Big-Bang-Nucleosynthesis (BBN) temperature of a few MeV [28], $B^0(T = T_{\rm BBN}) \simeq 10^{-18}$ MeV, is also very small, so there are negligible effects of the background on the formation of the material elements. However, in general, the critical temperature at which the phase transition for the demise of the background B_0 occurs could be significantly lower than the decoupling temperature for the right handed neutrinos, T_D; applying the experimental upper bounds on the magnitude of the background H-torsion field B^0 today, then, we can constrain the parameter c_0 in the cooling law (6).

A different scenario for CPT-induced baryon asymmetry was discussed in [30], which is based on string Universe models with bulk D-brane defects [31]. In these models our world is a (compactified) brane Universe, with three large spatial dimensions, propagating in a higher-dimensional space bulk punctured by (inhomogeneous in general) populations of D-particle "effectively" point-like defects [4]. The interesting feature of this second class of models is that there is an intrinsic CPT violation in such cases, due to unobserved (by a low-energy observer) degrees of freedom associated with the recoil of the D-brane defects on the brane Universe during their interaction with string matter. In these models, CPT violation is primarily associated with an ω-type effect [19], and there are different dispersion relations between particles and antiparticles as they propagate in the "medium" of D-brane defects [30], which can lead to matter-antimatter asymmetry already in thermal equilibrium.

Indeed, if $\vec{u}_\parallel$ denotes the recoil velocity vector (parallel to the brane world's longitudinal directions), one may parameterise the momentum transfer during the scattering of a particle excitation off a D-particle defect by the fraction parameter of the incident momentum r, which is in turn assumed stochastic, that is

$$u_i = \frac{g_s}{M_s}\Delta p_i \to g_s r_i \frac{p_i}{M_s} \quad \text{(no sum over i)} , \quad \ll r_i \gg = 0 , \quad \ll r_i r_j \gg = \Delta^2 \delta_{ij} . \tag{8}$$

where $\ll \cdots \gg$ denotes a statistical average over D-particle populations (as well as average over quantum fluctuations of D-particles) [30]. Δ is a phenomenological parameter to be bound by experiment.

The D-foam is predominantly transparent to electrically charged matter, on account of electric charge conservation. Hence, from the SM elementary constituents of matter on the brane, mainly neutrinos and photons are interactlng dominantly with the D-particles (but in general, neutral composite particles, such as neutral mesons, are also assumed to undergo non-trivial interactions with the

[4]In phenomenologically relevant type IIB string theories, the D-"particles" are themselves compactified branes wrapped around circles of small radii [32], so that, from the point of view of a low-energy (with respect to the string mass scale M_s) observer in the brane, they look "effectively" point-like.

D-foam medium, although their quark substructure may suppress them, when compared to those of neutrinos and photons. Lacking, at present, a complete theory of D-foam-matter interactions, the latter can only be studied phenomenologically). As shown in [30], the dispersion relation of the (active or right-handed) neutrino particles, of mass m, viewed as appropriate excitations of the open matter string, are modified due to *both* the metric distortion of space-time, as a result of the recoil of the D-particle, and the capture/splitting/re-emission process of the matter string by the D-particle (*cf.* (8)):

$$\begin{aligned} \ll E_\nu \gg &= \sqrt{p^2 + m_\nu^2}\left(1 + \frac{g_s^2}{2\,M_s^2}\Delta^2\,p^2\right) - \frac{g_s}{2\,M_s}\Delta^2\,p^2 \\ \ll \overline{E}_{\overline{\nu}} \gg &= \sqrt{p^2 + m_\nu^2}\left(1 + \frac{g_s^2}{2\,M_s^2}\Delta^2\,p^2\right) + \frac{g_s}{2\,M_s}\Delta^2\,p^2\,, \end{aligned} \tag{9}$$

where $\overline{E} > 0$ is the (positive) energy of the antiparticle excitation. The first terms on the right-hand-side of these equations are due to the propagation of neutrinos in recoil-distorted space-time geometry, while the second terms are due to the capture process of a particle by the D-particle defect.

There can thus be *local* CPTV in the sense that the effective dispersion relation between neutrinos and antineutrinos are different in D-foam backgrounds. This difference induces a difference between neutrino and antineutrino phase-space distribution functions, which in turn generates a matter-antimatter lepton asymmetry in the relevant densities

$$\ll n - \overline{n} \gg = g_{d.o.f.} \int \frac{d^3p}{(2\pi)^3} \ll [f(E) - f(\overline{E})] \gg\,, \tag{10}$$

where $g_{d.o.f.}$ denotes the number of degrees of freedom (d.o.f.) of relativistic neutrinos and $f(E)$ is the Fermi-Dirac distribution. The lepton asymmetry can be calculated from the integrated difference (9), for $\Delta^2 < 1$, but, in contrast to conventional point-like field theory models, where the upper limit of momentum integration can be extended to ∞, in D-foam models, due to (8), this is extended up to the value for which the D-particle recoil velocity approaches the (upper-limiting) value of the speed of light *in vacuo*, c=1 in our units, *i.e.* $p_{\rm max} \equiv |\vec{p}|_{\rm max} = \frac{M_s}{g_s\sqrt{\Delta^2}}$, where r is the stochastic variable satisfying (8). The resulting integrals in (10) then become:

$$\Delta n_\nu = \frac{g_{d.o.f.}}{2\pi^2}\,T^3 \int_0^{\frac{M_s}{T g_s \sqrt{\Delta^2}}} d\tilde{u} \left(\frac{1}{1 + e^{\tilde{u} - \tilde{u}^2 \frac{g_s \Delta^2 T}{2M_s}}} - \frac{1}{1 + e^{\tilde{u} + \tilde{u}^2 \frac{g_s \Delta^2 T}{2M_s}}} \right) \simeq \frac{g_{d.o.f.}}{\pi^2} T^4 \frac{\Delta^2\, g_s}{M_s} \tag{11}$$

The lepton asymmetry resulting from (11) freezes out at temperature T_d and is given by:

$$\Delta L(T < T_d) = \frac{\Delta n_\nu}{s} = \frac{2\Delta^2 g_s T_d}{M_s}. \tag{12}$$

From (12), we observe that for a freeze-out temperature or order $T_d \sim 10^{15}$ GeV in, say, Grand Unified Models, communicating the lepton asymmetry to baryon asymmetry via B-L preserving GUT interactions, the phenomenological value $\Delta L \sim 10^{-10}$ is attained for

$$\frac{M_s}{g_s} \sim 10^{25}\,\Delta^2\,\text{GeV}\,. \tag{13}$$

For $\Delta^2 \sim 10^{-6}$ a Planck size D-particle mass $M_s/g_s \sim 10^{19}$ GeV is required so that the D-foam provides the physically observed Lepton and, thus, Baryon Asymmetry. For the unnaturally small $\Delta^2 < 10^{-21}$ one arrives at $M_s/g_s \sim 10$ TeV. For $\Delta^2 \sim \mathcal{O}(1)$ transplanckian D-particle masses are required. We should stress that the above conclusions were based on specific assumptions on the freeze-out temperature.

An important remark is in order at this point. In case of right-handed Majorana neutrinos, which we discussed in the beginning of this section in connection with the model of [26], we note that they are their own antiparticles, and hence the effects of D-foam in inducing asymmetries in thermal equilibrium (10) between particles and antiparticles due to different dispersion relations do not concern them, but refer to light neutrinos that have a Dirac component in seesaw scenarios. Such asymmetries could co-exist with the mechanism of leptogenesis of [26], reviewed above, which however we would consider as the dominant one. In such models, the D-particle bulk foam could provide a way [26] for inducing the phase transition at a given era of the Universe, *e.g.* shortly after the decoupling of the right-handed neutrino species, that leads to the temperature-driven demise of the axial CPTV background B_0 (6), so that, in the current epoch, the value of B^0 is consistent with the stringent phenomenological bounds [29] and also with the BBN constraints.

An induced ω-effect. In addition to producing matter-antimatter asymmetries, the model of [30] entails an interesting ω-effect, due to decoherence associated with recoil degrees of freedom inaccessible by a low-energy observer. Indeed, consider the case of an (unnormalised) initial entangled quantum state $|i\rangle$ is given by

$$|\psi\rangle = |k,\uparrow\rangle^{(1)}\,|-k,\downarrow\rangle^{(2)} - |k,\downarrow\rangle^{(1)}\,|-k,\uparrow\rangle^{(2)} + \xi\,|k,\uparrow\rangle^{(1)}\,|-k,\uparrow\rangle^{(2)} + \xi'\,|k,\downarrow\rangle^{(1)}\,|-k,\downarrow\rangle^{(2)} \qquad (14)$$

where $\left|M_L\left(\overrightarrow{k}\right)\right\rangle = |k,\uparrow\rangle$ in an actual situation may represent a neutral meson (Kaon K^0 or B^0 meson) and we have taken the momentum $\overrightarrow{k}$ to have only a non-zero component k in the x-direction for brevity and concreteness; superscripts label the two separated detectors of the collinear meson pair, ξ and ξ' are complex constants and we have left the state $|\psi\rangle$ unnormalised.

The effect of space-time foam on the initial entangled state of two neutral mesons, say, has been estimated in [30] by applying non-degenerate perturbation theory to the states $|k,\uparrow\rangle^{(i)}$, $|k,\downarrow\rangle^{(i)}$, $i = 1, 2$, where the perturbation is associated with appropriate gravitational dressing of the states, due to the distortion of space-time as a result of recoil of the space-time defect, allowing for flip of the quantum numbers $\uparrow, \downarrow$ (viewed as "flavour"), during the interaction of the states with the D-particle. Assuming that the gravitationally perturbed states are still the momentum eigenstates, the dominant features of a possible ω-effect can be seen from a term $\widehat{H_I}$ in the single-particle interaction hamiltonian [30]

$$\widehat{H_I} = -(r_1\sigma_1 + r_2\sigma_2)\,\widehat{k}\,, \quad \text{with } \sigma^a\,,\ a = 1, 2, \text{ the corresponding Pauli matrices}\,, \qquad (15)$$

which is the leading order contribution in the small stochastic parameters $r_i \ll 1$ that satisfy (*cf.* 8),

$$\langle r_i\rangle = 0, \quad \left\langle r_i r_j\right\rangle = \Delta_i\,\delta_{ij}\,, \quad i, j = 1, 2\,. \qquad (16)$$

where we allowed for anisotropic momentum transfers in general, depending on the (left or right) direction $i = 1, 2$ of the neutral mesons after the decay of the initial particle in meson factories. In (16) we consider explicitly quantum averages, but statistical averages over D-particle populations should also be understood on top of the former.

In first order in perturbation theory the "gravitational dressing" of $|k,\downarrow\rangle^{(i)}$ leads to a state:

$$\left|k^{(i)},\downarrow\right\rangle^{(i)}_{QG} = \left|k^{(i)},\downarrow\right\rangle^{(i)} + \left|k^{(i)},\uparrow\right\rangle^{(i)}\alpha^{(i)}, \qquad \alpha^{(i)} = \frac{{}^{(i)}\left\langle\uparrow, k^{(i)}\right|\widehat{H_I}\left|k^{(i)},\downarrow\right\rangle^{(i)}}{E_2 - E_1} \qquad (17)$$

and correspondingly for $\left|k^{(i)},\uparrow\right\rangle^{(i)}$ the dressed state is obtained from (17) by exchanging $|\downarrow\rangle \leftrightarrow |\uparrow\rangle$ and $\alpha^{(i)} \to \beta^{(ii)}$ where

$$\beta^{(i)} = \frac{{}^{(i)}\left\langle \downarrow, k^{(i)}\right| \widehat{H_I} \left|k^{(i)},\uparrow\right\rangle^{(i)}}{E_1 - E_2} \tag{18}$$

Here the quantities $E_i = (m_i^2 + k^2)^{1/2}$ denote the energy eigenvalues, and $i = 1$ is associated with the up state and $i = 2$ with the down state. With this in mind the totally antisymmetric "gravitationally-dressed" state can be expressed in terms of the unperturbed single-particle states as:

$$\begin{aligned}
&|k,\uparrow\rangle^{(1)}_{QG}|-k,\downarrow\rangle^{(2)}_{QG} - |k,\downarrow\rangle^{(1)}_{QG}|-k,\uparrow\rangle^{(2)}_{QG} = |k,\uparrow\rangle^{(1)}|-k,\downarrow\rangle^{(2)} - |k,\downarrow\rangle^{(1)}|-k,\uparrow\rangle^{(2)}\\
&+ |k,\downarrow\rangle^{(1)}|-k,\downarrow\rangle^{(2)}\left(\beta^{(1)} - \beta^{(2)}\right) + |k,\uparrow\rangle^{(1)}|-k,\uparrow\rangle^{(2)}\left(\alpha^{(2)} - \alpha^{(1)}\right)\\
&+\beta^{(1)}\alpha^{(2)}|k,\downarrow\rangle^{(1)}|-k,\uparrow\rangle^{(2)} - \alpha^{(1)}\beta^{(2)}|k,\uparrow\rangle^{(1)}|-k,\downarrow\rangle^{(2)}
\end{aligned} \tag{19}$$

It should be noted that for $r_i \propto \delta_{i1}$ the generated ω-like effect corresponds to the case $\xi = \xi'$ in (14) since $\alpha^{(i)} = -\beta^{(i)}$, while the ω-effect of [19], specific to neutral mesons, corresponds to $r_i \propto \delta_{i2}$ (and the generation of $\xi = -\xi'$) since $\alpha^{(i)} = \beta^{(i)}$. In the density matrix these cases can be distinguished by the off-diagonal terms. These two cases are physically very different. In the case of Φ-factories [33], the former corresponds to non-definite strangeness in the initial state of the neutral Kaons (seen explicitly when written in terms of $K_0 - \overline{K}_0$), and hence strangeness nonconservation in the initial decay of the Φ-meson, while the latter conserves this quantum number. We remind the reader that in a stochastic quantum-gravity situation, strangeness, or, in that matter, the appropriate quantum number in the case of other neutral mesons, is not necessarily conserved, and this is reflected in the above-described general parametrisation of the interaction Hamiltonian (15) in "flavour" space.

We next remark that on averaging the density matrix over the random variables r_i, we observe that only terms of order $|\omega|^2$ will survive, with the order of $|\omega|^2$ being $|\omega|^2 = O\left(\frac{1}{(E_1-E_2)}(\langle\downarrow,k|H_I|k,\uparrow\rangle)^2\right) = O\left(\frac{\Delta_2 k^2}{(E_1-E_2)^2}\right) \sim \frac{\Delta_2 k^2}{(m_1-m_2)^2}$ for the physically interesting case in which the momenta are of order of the rest energies (*i.e.* masses of the constituent states). The variance Δ_2 (and also Δ_1) is of the order of the square of the momentum transfer during the scattering of the single particle state off a space-time-foam defect (*cf.* (9)), *i.e.* $\Delta_2 = g_s^2\,\frac{\Delta^2 k^2}{M_s^2}$, where M_s/g_s the D-particle defect rest mass (with M_s the string mass scale, and $g_s < 1$ the (perturbative) string coupling). As already mentioned, the parameter Δ^2 is at present a phenomenological parameter, which is proportional to the probability of interaction of the string matter with the D-particles, and therefore the total cross section of such processes. It cannot be further determined due to the lack (at present) of a complete theory of (string/brane) quantum gravity. Thus, we arrive at the following estimate of the order of ω in this model of foam [30]:

$$|\omega|^2 \sim g_s^2\,\frac{\Delta^2 k^4}{M_s^2(m_1 - m_2)^2}\,. \tag{20}$$

In the case of neutral mesons, which (ignoring other quantum numbers, such as strangeness or beuaty) are treated as identical bosons, if the CPT operator was well defined as a quantum mechanical operator, one should have $\xi = \xi' = 0$ [34]. On the other hand, in case of quantum gravity environments, entailing loss of information for a low-energy observer, as is the case of the D-foam quantum fluctuations degrees of freedom, discussed above, the quantum generator of the CPT transformations may not be well-defined for the low-energy observer [18], which, in the case of neutral mesons, implies an initial entangled state (14) (*cf.* also (19)) with a non-trivial but *small* contamination $\xi = \xi' = \omega \neq 0$, with $\omega = |\omega|\,e^{i\Omega}$ a complex parameter, so that the antiparticle state is *perturbatively defined*, but the Einstein-Podoskly-Rosen (EPR) correlations associated with the decay products of the initial entangled state (14) get modified. This is the ω-effect [19], whose phenomenology will be discussed in the next section. The reader should notice that the explicit ω-effect discussed in this section is Lorentz-violating, depending on the magnitude of spatial momenta of the entangled states, as a result of its origin, due to spatial recoil-velocity degrees of freedom of the D-particles, violating Lorentz symmetry *locally* in space and time [31].

3 Current Phenomenology of CPT Violation

In this section we shall discuss some aspects of the current era phenomenology of the two types of CPT Violation (CPTV) motivated by early Universe matter-antimatter asymmetry generation in the previous section. The first type of CPTV (subsection 3.1, below) is induced by Lorentz violating backgrounds, which can be studied within the framework of effective field theories, and in particular the Standard Model Extension (SME) [20] parametrization. The other CPTV type (subsection 3.2, below) goes beyond local effective field theories, and is linked to situations of environmental decoherence in which the CPT operator is not well-defined, leading to ω-type modifications [19] of the Einstein-Podolsky-Rosen (EPR) correlators of entangled boson (specifically, meson) states in appropriate meson factories (Φ-factories [33] or B-factories [35]).

3.1 CPT Violation within Standard Model Extension Framework

Let us commence our discussion with the phenomenology of CPTV within the SME framework [20]. There is a phethora of precision tests which are tabularted and regularly updated in ref. [29], where we refer the interested reader for details. For our purposes here we shall only concentrate on bounds of the axial background Lorentz and CPT Violating coefficient b_μ in the notation of [20, 29], and in particular its temporal component, which coincides with our axial background vector B_0 in (2). The first few terms in the fermion sector of the SME read [20, 29]:

$$\begin{aligned}
\mathcal{L}_{SME} &\ni \frac{1}{2} i\overline{\psi}\,\Gamma^\nu\,\partial_\nu\psi - \overline{\psi}\,\mathcal{M}\,\psi\,, \\
\mathcal{M} &\equiv m + a_\mu\gamma^\mu + b_\mu\gamma^5\gamma^\mu + \frac{1}{2}H^{\mu\nu}\sigma_{\mu\nu}\,, \quad \sigma_{\mu\nu} = \frac{1}{4}[\gamma_\mu, \gamma_\nu]\,, \\
\Gamma^\nu &\equiv \gamma^\nu + c^{\mu\nu}\gamma_\mu + d^{\mu\nu}\gamma_5\gamma_\mu + e^\nu + i f^\nu\gamma_5 + \frac{1}{2}g^{\lambda\mu\nu}\sigma_{\lambda\mu}\,. \qquad (21)
\end{aligned}$$

For our purposes we note that the terms proportional to a_μ and b_μ violate both Lorentz and CPT symmetries, unlike the $c_{\mu\nu}, d_{\mu\nu}$ and $H_{\mu\nu}$ terms that violate only Lorentz symmetries. Analogous observations can be made for the various terms inside the Γ^μ structure of (21). The main assumption behind the form of such operators is that an unknown physics at high energy scales could lead to a spontaneous breaking of Lorentz invariance by giving an expectation value to certain tensorial fields, which are not in the Standard Model (SM) spectrum. The interaction of these fields with operators composed from the SM fields, which are fully Lorentz-symmetric before the spontaneous breaking, will manifest itself as effective Lorentz-Violating (LV) terms, which below the scale of the LV condensation would have the schematic form: $O^{\rm SM}_{\mu\nu\ldots}\, C^{\mu\nu\ldots} \to O^{\rm SM}\,\langle C^{\mu\nu\ldots}\rangle$, where $C^{\mu\nu\ldots}$ is an external field that undergoes condensation and $O^{\rm SM}$ is a SM field operator that transforms properly under the Lorentz group. The classification of ref. [36] requires that the independent dimension-5 operators must be gauge invariant, Lorentz invariant after contraction with the background tensors $\langle C^{\mu_1\mu_2\cdots}\rangle$, not reducible to total derivatives or to lower-dimension operators by the use of equations of motion, and they should couple to an irreducible background tensor. Several experiments, of diverse origin, can be used in order to impose stringent constraints on the relevant SME coefficients [29], that range from searches for forbidden atomic transitions in precision experiments and studies of low-energy antiprotonic atoms and antimatter factories, to high-energy cosmic rays, nuclear spin precession and atomic and nuclear Electric Dipole Moments (EDM) measurements, as well as data on neutrino oscillations. In the following we shall discuss some phenomenological consequences of some of the above coefficients, especially in the context of neutral mesons, forbidden transitions in (anti)hydrogen molecules or atomic dipole moment measurements. Then we shall scale the appropriate bounds on b_μ back in time, in the context of our CPTV stringy Universe of section 2, in order to see whether (some of) our leptogenesis scenarios discussed above can be falsified.

We commence with SME tests in antiprotonic atoms [37], in particular antihydrogen ($\overline{\rm H}$) [38]. Motivated by the theoretical microscopic models of section 2, I shall restrict myself to constraining the b_μ coefficients of the SME (21) using spectroscopy, in particular looking for forbidden transitions, *e.g.* $1s \to 2s$. Within H spectrocopic measurements, the presence of a b_μ coefficient in the SME (21) leads to the relevant transition of the electron in the H atom. The sensitivity of the tests depend crucially whether the atoms are free or trapped in an external magnetic field. In the case of free H (and $\overline{\rm H}$), the frequency shift of the 1s-2s transition is a higher-loop quantum effect in the SME/Quantum-Electrodynamics (QED) lagrangian, and thus the effect is suppressed by the square of the fine structure constant, α^2: $\delta_{1s-2s}\nu^{\rm H} \simeq -\alpha^2\, b_3^e/8\pi$, *i.e.* the pertinent sensitivity of such experiments would be about five orders of magnitude smaller compared to tests involving the corresponding transitions in trapped H and $\overline{\rm H}$. However, in the latter tests, the corresponding frequency shifts are proportional to the difference $b_3^e - b_3^p$ of the third spatial component of b_μ between electrons (e) and protons(p) (in a frame where the direction of the external magnetic field is along the z axis). In view of the universal character of B^μ vectors due to background space-time geometries discussed in section 2, for this model the above difference would vanish. To cover ourselves against such cases, it is therefore imperative to either measure the sum of the coefficients $b_\mu^{e,p}$, or isolate them experimentally. The former can be achieved by examining hyperfine structure transitions in atomic (anti)matter. Indeed, within 1s transitions of H or $\overline{\rm H}$, one can determine the relevant energy shifts induced by b_μ [37]:

$$\Delta^H_{a\to b} \simeq (b^e_\mu + b^p_\mu)/\pi + \dots \qquad (22)$$

where the ... denote contributions from the rest of the SME coefficients (21), which are not written explicitly here. Hyperfine transitions within the 1S level of H can be measured with accuracies exceeding 1 mHz in masers. So transitions of this type in trapped H and $\overline{\rm H}$ are interesting candidates for performing tests of Lorentz or CPT symmetry, although to achieve resolutions of 1 mHz in trapped antihydrogen does not seem feasible in the foreseeable future.

Another possibility would be to measure [37] radio-frequency transitions between states within the triplet of hyperfine levels in H and $\overline{\rm H}$, in particular the so called $|d\rangle_1 \to |c\rangle_1$ transition at external magnetic fields of order B $\simeq$ 0.65 Tesla. The corresponding frequency shifts depend solely on b_3^p:

$$\Delta^{\rm H}_{c\to d} \simeq -b_3^p/\pi, \qquad \Delta^{\overline{\rm H}}_{c\to d} \simeq +b_3^p/\pi \qquad (23)$$

where we took into account that under the action of CPT operation, which exchanges H and $\overline{\rm H}$, the coefficient of the b_3^p changes sign. Thus, comparison of the above spectroscopic measurement between trapped H and $\overline{\rm H}$ would yield immediately a bound (or a measurement!) on b_3^p. If a frequency resolution of 1 mHz could be attained (which at present is far from being plausible), then, one could obtained $|b_3^p| \le 10^{-27}$ GeV. Still such bounds are about four orders of magnitude smaller that the ones coming from masers. We also note that, although, clock-comparison experiments are able to resolve spectral lines to about 1 μHz, nevertheless, isolating b_i^p is very complicated due to the complex structure of the nuclei involved.

The above experiments are sensitive only to spatial components of Lorentz-violating couplings. Sensitivity to timelike couplings, b^0, would require appropriate boosts. On the other hand, in the context of the model (2) of section 2, such experiments can bound the combinations $\gamma \vec{v}_3 B^0$, where $\vec{v}$ is the current-era relative velocity of us (as local observers) with respect to the CMB (or Friedmann-Robertson-Walker) frame. Currently, we can quote the following bounds on b_μ coefficients for electrons [29]:

$$b^0 \leq 0.02\,\text{eV}\,, \quad |\vec{b}| \leq 10^{-21}\ \text{eV}\,. \tag{24}$$

It should be mentioned that the stringent limits on $|\vec{b}|$ have been obtained in measurements using torsion pendulum containing macroscopic numbers of polarised electrons [39]. New interactions, such as the above-mentioned Lorentz- and CPT-violating ones are then searched for (and bounded) by looking at the corresponding effects on the electron spin. Such tests may also be performed in man-made antihydorgen [38] or other anti-atoms, with the aim of providing direct comparison of CPT properties and thus tests of CPT invariance.

Next we describe the situation governing the constraints on the relevant dimension-5 terms of the SME lagrangian coming from EDM. These are generically found to be of order [36] $\leq\ 10^{-25}\,e\,$cm. The overall expression for the total EDM, due to the CP Violating conventional QED terms and the CPT Violating terms due to the presence of an appropriate Lorentz-violating background vector n^μ, is obtained from the effective Lagrangian

$$\mathcal{L}_{\text{EDM}} = -i\frac{1}{2}d_{\text{CP}}\overline{\psi}\,\sigma^{\mu\nu}\,F_{\mu\nu}(A)\psi + d_{\text{CPT}}\overline{\psi}\,\gamma^\mu\,\gamma^5\,F_{\mu\nu}(A)\,n^\nu\,\psi\,, \tag{25}$$

where $F_{\mu\nu}$ is the Maxwell field strength. The currently null result on the neutron dipole moment imposes the constraint $d_{\text{CP}} + d_{\text{CPT}} = 0$. The lagrangian (25) should be completed with the a^μ and b^μ SME terms (21), as well as the appropriate dimension-5 operators from the QED sector of the SME [36]:

$$\mathcal{L}_5 = \sum_{\text{fermion species}} \left[c^\mu\,\overline{\psi}\gamma^\lambda F_{\lambda\mu}\psi + d^\mu\,\overline{\psi}\gamma^\lambda\,\gamma^5\,F_{\lambda\mu}\psi + g^\mu\overline{\psi}\gamma^\lambda\tilde{F}_{\lambda\mu}\psi + f^\mu\,\overline{\psi}\gamma^\lambda\,\gamma^5\tilde{F}_{\lambda\mu}\psi\right], \tag{26}$$

where $\tilde{F}_{\mu\nu}$ is the dual of the Maxwell tensor. The various terms in (26) have different transformation properties under the action of the discrete symmetries C, P and T, which, together with the corresponding terms of (21), are indicated in the table of fig. 1, on the assumption that the vector backgrounds are time-like and invariant under C,P and T reflections [36]. Experimentally [40], one can disentangle CP-odd from CPT-odd operators, because of different suppression scales. Specifically, the former require helicity flip and are thus represented by dimenion-six operators in the SME effective lagrangian, with suppression by the CP breaking scale of order $1/\Lambda^2_{\text{CP}}$. Such operators imply spin precession in a magnetic field relative to the direction of $\mathbf{B} \times \mathbf{v}$. On the other hand, the CPT-odd operators are of dimension 5, as they do not require helicity flip, *e.g.* in the quark sector such operators are of the form $\overline{q}_{R(L)}\gamma^\nu\,\gamma^5\,F_{\nu\mu}\,q_{R(L)}$, and $\overline{q}_L\gamma^\nu\,\gamma^5\,F^a_{\nu\mu}\,\tau^a\,q_L$, where τ^a, $a = 1, 2, 3$ are the SU(2) generators of the weak interaction standard model group, and $F_{\mu\nu}$ and $F^a_{\mu\nu}$ are the U(1) and SU(2) gauge field strengths respectively. These operators are suppressed linearly by the CPT-breaking scale, $1/\Lambda_{\text{CPT}}$.

EDMs have been bounded with high precision in several occasions [40]: **(i)** neutrons, with the bound $d_n\ <\ 3 \times 10^{-26}\,$e cm , **(ii)** diamagnetic atoms (such as Hg, Xe, ...), for which the corresponding EDMs are induced by the EDMs of the valence nucleons (for the case of mercury EDM, one has the (approximate) relation: $d_{\text{Hg}} \simeq -5 \times 10^{-4}(d_n + 0.1\,d_p) \sim -5 \times 10^{-4}\,d_n$, which implies that a signal consistent with CPT violation would occur, if non-zero d_n, d_{Hg} were to be found), and **(iii)** paramagnetic atoms (such as Tl, Cs, ...): their EDM are extremely suppressed as a result of the absence of a CPT-odd electron EDM.

Coefficient	Operator	C	P	T
a^0	$\overline{\psi}\gamma_0\psi$	$-$	$+$	$+$
b^0	$\overline{\psi}\gamma_0\gamma_5\psi$	$+$	$-$	$+$
c^0	$F_{\lambda 0}\overline{\psi}\gamma^\lambda\psi$	$+$	$+$	$-$
d^0	$F_{\lambda 0}\overline{\psi}\gamma^\lambda\gamma^5\psi$	$-$	$-$	$-$
f^0	$\widetilde{F}_{\lambda 0}\overline{\psi}\gamma^\lambda\gamma^5\psi$	$-$	$+$	$+$
g^0	$\widetilde{F}_{\lambda 0}\overline{\psi}\gamma^\lambda\psi$	$+$	$-$	$+$

Figure 1. Tranformation properties of the various terms in (26) under the action of the discrete symmetries C, P and T. A + (−) indicates an even (odd) function. From ref. [36].

In general, theoretical estimates of dimension-three operators induced by multiloop CP violating corrections in the standard model, imply the following bounds of the SME coefficients in (21) [40]

$$a^\mu, b^\mu \sim d^\mu \left(10^{-20} - 10^{-18}\right) \mathrm{GeV}^2 , \tag{27}$$

providing sensitivity to $d^\mu \leq 10^{-12}$ GeV^{-1} and thus $\Lambda_{\rm CPT} \sim \left(10^{11} - 10^{12}\right)$ GeV, if one takes into account the current bounds on b_μ [29] (*cf.* (24)).

Higher LV background tensors, *e.g.* terms in SME effective lagrangian of the form $\mathcal{D}_{\mu\nu\rho}\overline{e}\,\gamma^\rho\,\gamma^5\,e\,F_{\mu\nu}$, can also be bounded experimentally with high accuracy, by looking [40] for corrections to the spin precession frequency of the form $(\mathcal{D}^{i[0k]} + \mathcal{D}^{k[0i]})\,E_i\,B_k$, which changes sign under the reversal of the electric field E_i. The relative signal changes during the day as a result of the change of the Laboratory orientation relative to the tensor background. We also mention the interesting suggestion of ref. [41] on further tests of CPT symmetry due to the CPT-odd axial vector background b^μ, which has been of interest to us in section 2. According to this work, within the framework of LV extended electrodynamics, the Dirac equation for a bound electron in an external electromagnetic field has been considered, assuming the interaction with the background field b_μ. A Foldy-Wouthysen quasi-relativistic (1/c)-series expansion (truncated to order $1/c^2$) has been applied to obtain an effective Hamiltonian for the hydrogen atom and through this the relativistic Dirac eigenstates in a spherically-symmetric potential to second order in b^0. The b^0-induced CPT-odd corrections to the electromagnetic dipole moment operators of a bound electron have been calculated. Such corrections contribute to the anapole moment of the atomic orbital and may cause a specific asymmetry of the angular distribution of the radiation of a hydrogen atom, in particular the $2p_{1/2,1/2} \rightarrow 1s_{1/2,-1/2}$ (*cf.* fig. 2). The non-observation currently of such asymmetries leads to bounds of the magnitude of $|b^0|$: $|b^0| \leq 2\times 10^{-8}\, m_e\, c^2 \simeq 10^{-11}$ GeV, which are consistent with the general bounds (24) for the SME coefficient b_μ for electrons [29].

Finally we mention that, further tests of CPT invariance can be made by direct measurements of particle antiparticle mass and charge differences, which we are not going to discuss here. However, in the spirit of our cosmological model discussed in section 2, we do mention that, if the observed

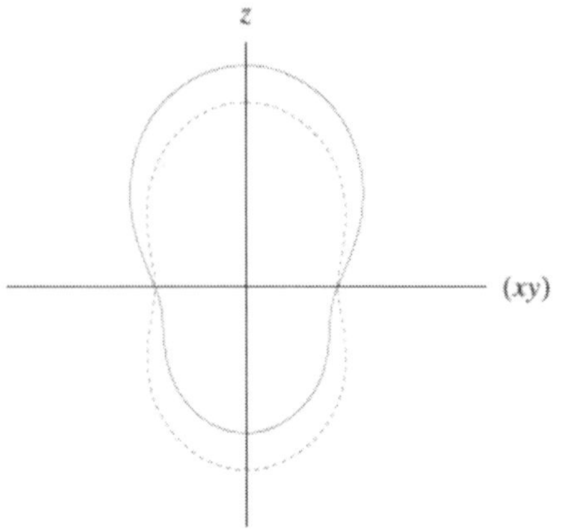

Figure 2. Anglular distribution for spontaneous radiation for the atomic transition $2p_{1/2,1/2} \rightarrow 1s_{1/2,-1/2}$ in the presence of a CPT-odd SME axial background vector b^μ. The dashed line indicates the standard electrodynamics $b^\mu = 0$ case. From ref. [41].

matter/antimatter asymmetry were due to a mass difference between particle and antiparticles, then, one may make the reasonable assumption that baryogenesis could be due to mass differences between quarks and antiquarks [42]. The latter nay depend linearly with temperature, $m_q(T) \sim gT$, as a consequence of known high-temperature properties of Quantum Chromodynamics (QCD). Furthermore, it is reasonable (although not strictly necessary) to assume that the quantk-antiquark differences today are bound by the current bound on proton-antiproton mass difference, which is of order 7×10^{-10} GeV, as provided in 2011 by the ASACUSA Collaboration [38]. Scaling back in temperature such differences, up to the respective decoupling temperature of the quarks, lead to baryon asymmetries that are much smaller than the observed one [42].

In this sense the model of [25] can still survive, given that, even if a $B^0 < 0.02$ eV is observed today, according to the current SME limits, the Universe may have undergone such a (or series of) phase transition at $T \sim 10^9$ GeV towards a smaller (or zero) H-torsion background. This is an (crude) example of how one can use current SME bounds to fit early universe cosmologies. In a similar spirit, the model of [26] on leptogenesis through CPT and CP violating decays of heavy right-handed neutrinos, discussed in section 2, based on the Lagrangian (2), with the assumption that the critical temperature of the phase transition that triggers the demise of the axial background B_0 is of the same order as the decoupling temperature of right-handed neutrinos, $T_D \sim 100$ TeV (5), lies comfortably within the limits (24), given the scaling (6) of the background $B^0 = b^0$ and its current value (7). Reversing the logic, we can assume the upper bound of B^0 today (24), scale back in time with the scaling (6) up to BBN temperatures, $T_{\rm BBN} \sim \mathcal{O}(1)$ MeV, and then constrain the coefficient c_0 by the requirement that the BBN conditions are not disturbed. Then. one can continue scaling back in the cosmic time, to check at which temperature range sufficient leptogenesis is produced, if at all. Such procedures require of course detailed models of baryogenesis via leptogenesis, which we reserve for future studies.

3.2 Intrinsic CPT Violation in Quantum Decoherence Models - the ω-effect

We now come to examine the current phenomenology of the ω-effect [19], which is associated with a second type of CPTV, already discussed in the context of the D-foam model in section 2, in which the quantum CPT operator is not well-defined. We stress again that this latter type of CPTV goes beyond local effective field theories, and its most sensitive bounds can be placed in experimental facilities involving entangled states of neutral mesons, such as neutral Kaon(Φ) factories [33, 43] or $B - \overline{B}$ meson factories [35].

In case of loss of information for a low energy observer, carried by d.o.f. inaccessible to him/her due to quantum gravity environments (*e.g.* the gravitational reocoil d.o.f. in the D-foam example of section 2), the quantum operator that generates CPT symmetry may not be well defined [18]. A proof of this statement is obtained by *recuctio ad absurdum*, that is by first assuming the existence of a well-defined *unitary* and *invertible* CPT operator acting on density matrices (antiunitary if acting on state vectors): Θ, such that (a bar above an operator denotes a quantity pertaining to anti-matter states, obtained via the action of the CPT transformation):

$$\rho_{\rm out}(t \to \infty) = \$\,\overline{\rho}_{\rm in}(t \to -\infty) \tag{28}$$

where the subscript "in" and "out" denotes asymptotic states, and the density matrix is defined as $\rho = {\rm tr}|\psi\rangle\langle\psi|$, where the trace operation "tr" is over quantum states inaccessible to a low energy observer. The "superscattering matrix" \$ is a linear operator acting on density matrices *without an inverse*, due to the existance of information loss in the problem. We may then write (from now one we ignore the time t arguments of the asymptotic ρ, for brevity):

$$\rho_{\rm out} = \$\,\rho_{\rm in} \Rightarrow \Theta\rho_{\rm in} = \$\,\Theta^{-1}\,\overline{\rho}_{\rm out} \Rightarrow \overline{\rho}_{\rm in} = \Theta^{-1}\,\$\,\Theta^{-1}\,\overline{\Theta}_{\rm out}\,. \tag{29}$$

However, since $\overline{\rho}_{\rm out} = \$\,\overline{\rho}_{\rm in}$, the last relation on the right-hand-side of (29) implies

$$\overline{\rho}_{\rm in} = \Theta^{-1}\,\$\,\Theta^{-1}\,\$\,\overline{\rho}_{\rm in}\,. \tag{30}$$

But this is *impossible*, as it would imply that the superscattering operator \$ has an inverse $\Theta^{-1}\,\$\,\Theta^{-1}$, that contradicts the initial assumption of information loss [5]. This is the so called *strong form of (intrinsic) CPT violation*, which would imply a *microscopic time arrow*.

Nevertheless in nature there could be [18] a *weak form* of CPT invariance, according to which the microscopic time arrow does not show in any scattering experiments. Indeed, in such a case the experimentalist would be able to prepare initial pure quantum mechanical state vectors, and there should a well defined transition probability $\mathcal{P}$ from the initial pure state $|\psi\rangle$ to the final state $|\phi\rangle$, such that

$$\mathcal{P}\big(\psi \to \phi\big) = \mathcal{P}\big(\theta^{-1}\,\phi \to \theta\,\psi\big) \tag{31}$$

where the (antiunitary) CPT operator θ acts on "in" and "out" Hilbert spaces $\mathcal{H}$ vectors now, θ : $\mathcal{H}_{\rm in} \to \mathcal{H}_{\rm out}$, and is such that

$$\Theta\rho = \theta\,\rho\,\theta^{\dagger}\,, \quad \theta^{\dagger} = -\theta^{-1}\,, \tag{32}$$

which in terms of the \$ matrix can be written as

$$\$^{\dagger} = \Theta^{-1}\,\$\,\Theta^{-1}\,. \tag{33}$$

Whether there exists such a situation of weak form of CPT invariance is in general an experimental question. The ω-effect [19] is one way to answer this question experimentally, and, as we have seen

[5] In a theory without information loss, of course, \$ would factorise to the Heisenberg scattering matrix S as $\$ = S\,S^{\dagger}$, and would have a well-defined inverse, $\$^{-1} = \$^{\dagger}$, in which case the CPT operator would be well defined satisfying $\Theta = \$\,\Theta^{-1}\,\$$.

in 2, it characterises the D-foam example. We next proceed to the phenomenology of this effect in entangled states of mesons, which, if observed, would constitute a "smoking-gun" evidence for such an *intrinsic* CPT violation.

We commence our discussion by briefly mentioning direct tests of Time reversal invariance, independent of CP and CPT violation, within the Lorentz invariant Standard Model theory, using entangled neutral mesons. Eventually we shall use those tests assuming the presence of a CPTV ω-effect in order to constrain the real and imaginary parts of the corresponding parameter. These tests have been initially proposed in [44], leading to the recent observation of *direct* T violation by the Ba-Bar collaboration [35], through the exchange of initial and final states in transitions that can only be connected by a T -symmetry transformation. For example, the transition $\overline{B}^0 \rightarrow B^-$ for the second B to decay, at time t_2, once the first B (entangled with the second) has been tagged at time t_1, is identified by reconstructing events in the time-ordered final states $(\ell^+ X, J/\psi\, K_s^0)$. The rate of this transition is then compared to that of the $B^- \rightarrow \overline{B}^0$ transition, that exchanges initial and final states, which is identified by the reconstruction of the final states $(J/\psi\, K_L^0, \ell^- X)$. Any observed difference between these two rates, would thus indicate direct observation of T violation, independent of CP properties [45]. This would also imply an independent test of CPT symmetry within the Standard Model. Similar tests of T violation in entangled Kaon Φ factories have also been suggested [46], by identifying the appropriate reactions that exchange initial and final states. However, if CPT is *intrinsically* violated, in the sense of being not well defined due to decoherence [18] induced by quantum gravity [47], the above-mentioned direct observation of T violation cannot constitute a test of decoherence-induced CPT breaking. This is because in such a case a distinct phenomenon, associated with the ill-defined nature of CPT operator, emerges, termed ω-effect [19]. If the ω effect were present, such direct T-violation tests using entangled states of B-mesons [48] would allow the experimenter to disentangle it from conventional CPT violating effects in the Hamiltonian, within the SME framework, and also to measure independently Imω and Reω. We shall comment briefly on this later in the section.

For the moment, let us concentrate first to the neutral Kaon system, where the ω-effects are dominant, as compared to other entangled neutral mesons, although conceptually our analysis applies equally [49] to entangled B-meson factories as well, such as those of [35]. In a quantum-gravity induced decohered situation, the Neutral mesons K^0 and $\overline{K}^0$ should *no longer* be treated as *identical particles*. As a consequence [19], the initial entangled state in Φ factories $|i>$, after the Φ-meson decay, assumes the form:

$$|i\rangle = \mathcal{N}\Big[\Big(|K_S(\vec{k}), K_L(-\vec{k})\rangle - |K_L(\vec{k}), K_S(-\vec{k})\rangle\Big) + \omega\Big(|K_S(\vec{k}), K_S(-\vec{k})\rangle - |K_L(\vec{k}), K_L(-\vec{k})\rangle\Big)\Big], \tag{34}$$

where $\omega = |\omega|e^{i\Omega}$ is a complex parameter, parametrizing the intrinsic CPTV modifications of the EPR correlations [19]. The ω-parameter controls the amount of contamination of the final C(odd) state by the "wrong" (C(even)) symmetry state. The appropriate observable (*c.f.* fig. 3) is the "intensity" $I(\Delta t) = \int_{\Delta t \equiv |t_1 - t_2|}^{\infty} |A(X, Y)|^2$, with $A(X, Y)$ the appropriate Φ decay amplitude [19], where one of the Kaon products decays to the final state X at t_1 and the other to the final state Y at time t_2 (with $t = 0$ the moment of the Φ decay).

It must be noted that in Kaon factories there is a particularly good channel, the one with bi-pion states $\pi^+\pi^-$ as final decay products, which *enhances the sensitivity* to the ω-effect by three orders of magnitude. This is due to the fact that the relevant terms [19] in the intensity $I(\Delta t)$ (c.f. fig. 3) contain the combination $\omega/|\eta_{+-}|$, where η_{+-} is the relevant CP-violating amplitude for the $\pi^+\pi^-$ states, which is of order 10^{-3}. The KLOE experiment bounds of the ω parameter are [43]:

$$\mathrm{Re}(\omega) = \left(-1.6^{+3.0}_{-2.1\,\mathrm{stat}} \pm 0.4_{\mathrm{syst}}\right) \times 10^{-4}\,, \quad \mathrm{Im}(\omega) = \left(-1.7^{+3.3}_{-3.0\,\mathrm{stat}} \pm 1.2_{\mathrm{syst}}\right) \times 10^{-4}\,. \tag{35}$$

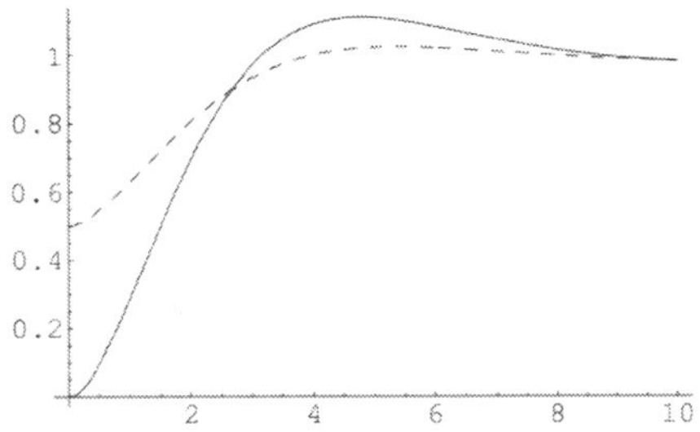

Figure 3. A characteristic case of the intensity $I(\Delta t)$ (vertical axis) as a function of Δt (horizontal axis), with $|\omega| = 0$ (solid line) vs $I(\Delta t)$ (dashed line) with $|\omega| = |\eta_{+-}|$, $\Omega = \phi_{+-} - 0.16\pi$, for definiteness [19].

At least an order of magnitude improvement is expected for upgraded facilities such as KLOE-2 at (the upgraded) DAΦNE-2 [43].

This sensitivity is not far from certain optimistic models of space time foam leading to ω-like effects [30]. Indeed, let us comparing these bounds to the D-foam case (20). First, we recall that successful leptogenesis from this class of models requires heavy D-particles masses (13). Assuming for definiteness D-particle masses of order of the Planck mass, $M_s/g_s \sim M_P$, we recall that the variance $\Delta^2 \sim O(10^{-6})$ (13) in order to have phenomenolgically acceptable leptogenesis. In the case of neutral kaons, with momenta of the order of the rest energies ($\sim$ 1 GeV) and mass differences $m_1 - m_2 \simeq 10^{-15}$ GeV, we observe from (20) that $|\omega| \sim 10^{-4}|\Delta|$ (whilst for B-mesons we have $|\omega| \sim 10^{-6}|\Delta|$). For $1 > \Delta \geq 10^{-3}$ these values for ω are not far below the sensitivity of current facilities, such as KLOE-2 at DAΦNE, and thus such leptogenesis CPTV models may be constrained experimentally in the foreseeable future.

In B-factories one can look for similar ω-like effects. Although in this case there is no particularly good channel to lead to enhancement of the sensitivity, as in the Φ-factories, nevertheless one gains in statistics, and hence interesting limits may also be obtained [49]. The presence of a quantum-gravity induced ω-effect in B systems is associated with a theoretical limitation on flavour tagging, namely the fact that in the absence of such effects the knowledge that one of the two-mesons in a meson factory decays at a given time through a flavour-specific channel determines unambiguously the flavour of the other meson at the same time. This is not true if intrinsic CPT Violation is present. One of the relevant observables [49] is given by the CP-violating semi-leptonic decay charge asymmetry (in equal-sign dilepton channel), with the first decay $B \to X\ell^\pm$ being time-separated from the second decay $B \to X'\ell^\pm$ by an interval Δt. In the absence of ω-effects, the intensity at equal decay times vanishes, $I_{\rm sl}(\ell^\pm, \ell^\pm, \Delta t = 0) = 0$, whilst in the presence of a complex $\omega = |\omega|e^{i\Omega}$, $I_{\rm sl}(\ell^\pm, \ell^\pm, \Delta t = 0) \sim |\omega|^2$. In such a case, the asymmetry observable exhibits a peak, whose position depends on $|\omega|$, while the shape of the curve itself depends on the phase Ω [49]. The analysis of [49], using the above charge asymmetry method and comparing with currently available experimental data, leads to the following bounds:

$$-0.0084 \leq \mathrm{Re}(\omega) \leq 0.0100\,, \quad \text{at}\, 95\%\ \text{C.L.}\,. \qquad (36)$$

Such tests for intrinsic CPT violation may be performed simultaneously with the above-mentioned observations of direct T violation, as they are completely independent. Quite recently, we have embarked [48] on a detailed study of ω-effects in B_d-system, using the experimental procedure suggested in [44], and implemented in [35], for tests of T violation in etnangled meson systems independent of CP violation. We have identified how to probe the complex ω parameter in the entangled B_d-system using Flavour(f)-CP(g) eigenstate decay channels: the connection between the Intensities for the two time-ordered decays (f, g) and (g, f) is lost in the presence of a non-zero ω. Appropriate observables have been constructed allowing independent experimental determinations of Re(ω) and Im(ω), disentangled from CPT violation in the evolution Hamiltonian, Re(θ) and Im(θ), which parametrise CPTV within the SME local effective field theory frameworks. The general analysis of [48] has found

$$\begin{aligned} \mathrm{Im}\,(\omega) &= \pm(6.40 \pm 2.80) \times 10^{-2}\,, \quad \mathrm{Re}\,(\omega) = (1.09 \pm 1.60) \times 10^{-2}\,, \\ \mathrm{Im}\,(\theta) &= \pm(6.11 \pm 3.45) \times 10^{-2}\,, \quad \mathrm{Re}\,(\theta) = (0.99 \pm 1.98) \times 10^{-2}\,, \end{aligned} \tag{37}$$

where the 2.4 σ deviations from Im(ω) = 0 and Re(θ) = 0 are interpreted as upper bounds. These 2σ tensions have been shown to be uncorrelated [48].

At this juncture, I would like to point out that an observation of the ω-effect in both the Φ and B-factories could also provide an independent test of Lorentz symmetry properties of the intrinsic CPT Violation, namely whether the effect respects Lorentz symmetry. This is because, although the Φ particle in neutral Kaon factories is produced at rest, the corresponding Υ state in B-factories is boosted, and hence there is a frame change between the two experiments. If the quantum gravity ω-effect is Lorentz violating, as is the case of the models considered in section 2 [30], then a difference in the value of ω between the two experiments should be expected.

Finally, since Lorentz Violation has been mentioned, I also point out that bounds of the LV SME coefficients a^μ (*cf.* eq. (21)) can be placed by measurements in the entangled Kaon Φ factories [43]. In particular by adopting the relevant SME terms to the quark sector, relevant for Kaon physics, one can bound differences $\Delta a^\mu = a^\mu_{q_1} - a^\mu_{q_2}$, where q_i, i =, 2 denote appropriate quark states. The current experimental limits for the coefficients Δ^μ_a are: from the KTeV Collaboration $\Delta_X, \Delta a_Z < 9.2 \times 10^{-22}$ GeV, while from the the KLOE Collaboration in the DaΦNE Φ factory [43] are less competitive but with the advantage that entangled meson factories have sensitivity to all four coefficients Δa^μ, in particular: $\Delta a^0 = (0.4 \pm 1.8) \times 10^{-17}$ GeV, from KLOE, with expected sensitivity at KLOE-2 in upgraded DAΦNE facilties for $\Delta a^{X,Y,Z} = \mathcal{O}(10^{-18})$ GeV. Unfortunately, entangled meson factories have only sensitivity to differences Δa^μ rather than absolute coefficients a^μ. Of course, if gravity acts universally for all quark species, such differences may be zero.

4 Instead of Conclusions : Intrinsic CPTV and Spin-Statistics Theorem

Before closing the talk I would like to make some remarks on another potential effect of quantum-gravity-decoherence-induced CPTV, namely tiny violations of the Spin-Statistics theorem [1] and in particular Pauli exclusion principle and related searches [50]. The *Spin-Statistics Theorem* states that the wave function of a system of identical integer-spin particles has the same value when the positions of any two particles are swapped, while the wave function of a system of identical half-integer spin particles changes sign when two particles are swapped. Particles with wave functions symmetric under exchange are called bosons and those with wave functions antisymmetric under exchange are called fermions. The theorem was first proposed by Fierz in 1939, by Pauli in a more systematic formulation in 1940, and in a rigorous mathematical formalism, using quantum field theory path integrals, by Schwinger in 1950, where the underlying mathematical assumptions were made clear:

- (1) The theory has a Lorentz and CPT invariant Lagrangian and relativistic causality.
- (2) The vacuum is Lorentz-invariant (can be weakened).
- (3) The particle is a localized excitation. Microscopically, it is not attached to a string or domain wall.
- (4) The particle is propagating (has a non-infinite mass).

- (5) The particle is a real excitation, meaning that states containing this particle have a positive-definite norm and has positive energy.

An important consequence of the spin-statistics theorem is that the wavefunction of two identical fermions is zero, hence two identical fermions (i.e. with all quantum numbers the same) cannot occupy the same state, which is the celebrated *Pauli exclusion Principle* (PEP), that was postulated by Pauli in 1925, without knowledge of the spin-statistics theorem at the time.

We should remark at this point that when a violation of the spin-statistics theorem appears, it is one or more of the above assumptions that they are violated. For instance, spinless anticommuting fields, which could exist in condensed matter models are not relativistic invariant; ghost fields in gauge theories are spinless fermions but they have negative norm. The 2+1 dimensional Chern-Simons theory has anyons (fractional spin excitations) but in such a case the wave function of the planar system splits between the bulk and the boundary, and hence is somehow delocalised. One remark concerns quarks: Despite being attached to a confining string, QCD quarks can have a spin-statistics relation proven at short distances (ultraviolet limit) due to asymptotic freedom.

We have mentioned above that CPT and Lorentz invariance are crucial assumptions for the validity of the spin-statistics theorem. Although spontaneous violation of Lorentz and CPT symmetry may be tolerated (see assumption (2), which may be relaxed), nevertheless if the CPT is ill defined, and there is a strong (intrinsic) form of CPT violation, as is the case of quantum-gravity induced decoherence (or D-foam situations, as in section 2), then there may be tiny violations of the spin-statistics theorem. In this respect, we recall that it was the Bose-Statistics requirement of the neutral mesons that resulted in the antisymmetric initial state (34) when $\omega = 0$. In D-foam situations [31], for instance, the matter excitation is dressed by open strings stretched between the defect and the brane world, and in this sense, assumption (3) of the theorem is violated, together with assumption (1), due to the ill-defined nature of the CPT operator. In such cases, there are hidden degrees of freedom in a particle state, and thus an evasion, in case of fermions, of the PEP may be understood by the fact that the otherwise looking identical quantum states entering the PEP formulation, actually differ by hidden quantum numbers. Thus we believe that searches for intrinsic CPT violation could also be complemented by searches of PEP. The current searches of PEP [50] are mainly for charged particles (electrons), for which, as we have discussed previously [32], the effects of the D-particle foam are expected to be strongly suppressed. Nevertheless, this is only one model and experiment should be independent of any theoretical assumptions, hence searches of PEP are equally encouraged as searches of intrinsic CPT violation.

Acknowledgements

I thank the organisers of KLOE-2 2016 *Workshop on e^+-e^- collisions physics at 1 GeV* (Frascati Lab. INFN (Italy), October 26-28 2016) for their kind invitation to give a talk and for stimulating discussions. The work of N.E.M. is partially supported by STFC (UK) via the research grant ST/L000326/1.

References

[1] R. F. Streater and A. S. Wightman, *PCT, spin and statistics, and all that* (Princeton, USA: Princeton Univ. Press) and references therein.

[2] A. D. Sakharov , Pisma Zh. Eksp. Teor. Fiz. **5**, 32 (1967) [JETP Lett. **5**, 24 (1967)] [Sov. Phys. Usp. **34**, 392 (1991)] [Usp. Fiz. Nauk **161**, 61 (1991)].

[3] V. A. Kuzmin, V. A. Rubakov and M. E. Shaposhnikov , Phys. Lett. B **155**, 36 (1985); M. B. Gavela, P. Hernandez, J. Orloff and O. Pene , Mod. Phys. Lett. A **9**, 795 (1994)

doi:10.1142/S0217732394000629 [hep-ph/9312215]; M. B. Gavela, P. Hernandez, J. Orloff, O. Pene and C. Quimbay, Nucl. Phys. B **430**, 382 (1994) doi:10.1016/0550-3213(94)00410-2 [hep-ph/9406289].

[4] For a review see: W. Buchmuller, R. D. Peccei and T. Yanagida, Ann. Rev. Nucl. Part. Sci. **55**, 311 (2005) [hep-ph/0502169], and references therein.

[5] D. N. Spergel *et al.* [WMAP Collaboration] , Astrophys. J. Suppl. **148**, 175 (2003) [astro-ph/0302209]. P. A. R. Ade *et al.* [Planck Collaboration] , Astron. Astrophys. **571**, A16 (2014) [arXiv:1303.5076 [astro-ph.CO]].

[6] R. N. Mohapatra and P. B. Pal, World Sci. Lect. Notes Phys. **60**, 1 (1998) [World Sci. Lect. Notes Phys. **72**, 1 (2004)]. M. Trodden, Rev. Mod. Phys. **71**, 1463 (1999) [hep-ph/9803479]; W. Buchmuller, arXiv:0710.5857 [hep-ph].

[7] M. C. Chen, *TASI 2006 Lectures on Leptogenesis* (2007), hep-ph/0703087 [HEP-PH]. A. Pilaftsis, J. Phys. Conf. Ser. **171**, 012017 (2009) [arXiv:0904.1182 [hep-ph]]; J. Phys. Conf. Ser. **447**, 012007 (2013).

[8] M. Fukugita and T. Yanagida, Phys. Lett. B **174**, 45 (1986).

[9] M. A. Luty, Phys. Rev. D **45**, 455 (1992).

[10] P. Minkowski, Phys. Lett. B **67**, 421 (1977); M. Gell-Mann, P. Ramond and R. Slansky, in *Supergravity*, eds. D.Z. Freedman and P. van Nieuwenhuizen (North-Holland, Amsterdam, 1979); T. Yanagida, in Proc. of the *Workshop on the Unified Theory and the Baryon Number in the Universe* (Tsukuba, Japan, eds. O. Sawada and A. Sugamoto, 1979); R. N. Mohapatra and G. Senjanovic, Phys. Rev. Lett. **44**, 912 (1980); J. Schechter and J. W. F. Valle, Phys. Rev. D **22**, 2227 (1980); G. Lazarides, Q. Shafi and C. Wetterich , Nucl. Phys. B **181**, 287 (1981); doi:10.1016/0550-3213(81)90354-0

[11] M. C. Gonzalez-Garcia and M. Maltoni, Phys. Rept. **460**, 1 (2008) [arXiv:0704.1800 [hep-ph]].

[12] T. Asaka, S. Blanchet and M. Shaposhnikov, Phys. Lett. B **631**, 151 (2005); doi:10.1016/j.physletb.2005.09.070 [hep-ph/0503065]. For cosmological/astrophysical applications of this model see: A. Boyarsky, O. Ruchayskiy and M. Shaposhnikov, Ann. Rev. Nucl. Part. Sci. **59**, 191 (2009) doi:10.1146/annurev.nucl.010909.083654 [arXiv:0901.0011 [hep-ph]]; M. Shaposhnikov, Prog. Theor. Phys. **122**, 185 (2009). doi:10.1143/PTP.122.185

[13] C. R. Argüelles, N. E. Mavromatos, J. A. Rueda and R. Ruffini, JCAP **1604**, no. 04, 038 (2016) doi:10.1088/1475-7516/2016/04/038 [arXiv:1502.00136 [astro-ph.GA]]; N. E. Mavromatos, C. R. Argüelles, R. Ruffini and J. A. Rueda, Int. J. Mod. Phys. D **26**, no. 03, 1730007 (2016); doi:10.1142/S0218271817300075, and references therein.

[14] L. Canetti, M. Drewes, T. Frossard and M. Shaposhnikov, Phys. Rev. D **87**, 093006 (2013) doi:10.1103/PhysRevD.87.093006 [arXiv:1208.4607 [hep-ph]].

[15] O. W. Greenberg, Phys. Rev. Lett. **89**, 231602 (2002) doi:10.1103/PhysRevLett.89.231602 [hep-ph/0201258].

[16] M. Chaichian, K. Fujikawa and A. Tureanu, Phys. Lett. B **712**, 115 (2012) doi:10.1016/j.physletb.2012.04.048 [arXiv:1203.0267 [hep-th]]; Eur. Phys. J. C **73**, no. 3, 2349 (2012) doi:10.1140/epjc/s10052-013-2349-2 [arXiv:1205.0152 [hep-th]].

[17] G. J. Milburn, New J. Phys. **8**, 96 (2006) doi:10.1088/1367-2630/8/6/096 [gr-qc/0308021].

[18] R. M. Wald, Phys. Rev. D **21**, 2742 (1980) doi:10.1103/PhysRevD.21.2742

[19] J. Bernabeu, N. E. Mavromatos and J. Papavassiliou, Phys. Rev. Lett. **92**, 131601 (2004) doi:10.1103/PhysRevLett.92.131601 [hep-ph/0310180]; E. Alvarez, J. Bernabeu, N. E. Mavromatos, M. Nebot and J. Papavassiliou, Phys. Lett. B **607**, 197 (2005) doi:10.1016/j.physletb.2004.12.032 [hep-ph/0410409]; J. Bernabeu, N. E. Mavro-

matos, J. Papavassiliou and A. Waldron-Lauda, Nucl. Phys. B **744**, 180 (2006) doi:10.1016/j.nuclphysb.2006.03.028 [hep-ph/0506025].

[20] D. Colladay and V. A. Kostelecky, Phys. Rev. D **58**, 116002 (1998) [hep-ph/9809521].

[21] O. Bertolami, D. Colladay, V. A. Kostelecky and R. Potting, Phys. Lett. B **395**, 178 (1997) doi:10.1016/S0370-2693(97)00062-2 [hep-ph/9612437].

[22] H. Davoudiasl, R. Kitano, G. D. Kribs, H. Murayama and P. J. Steinhardt, Phys. Rev. Lett. **93**, 201301 (2004) doi:10.1103/PhysRevLett.93.201301 [hep-ph/0403019].

[23] B. Mukhopadhyay, Mod. Phys. Lett. A **20**, 2145 (2005) doi:10.1142/S0217732305017640 [astro-ph/0505460]. U. Debnath, B. Mukhopadhyay and N. Dadhich, Mod. Phys. Lett. A **21**, 399 (2006) doi:10.1142/S0217732306019542 [hep-ph/0510351]. G. Lambiase and S. Mohanty, Phys. Rev. D **84**, 023509 (2011) doi:10.1103/PhysRevD.84.023509 [arXiv:1107.1213 [hep-ph]]. G. Lambiase, Phys. Lett. B **642**, 9 (2006) doi:10.1016/j.physletb.2006.09.037 [hep-ph/0612212]. S. Capozziello and G. Lambiase, Frascati Phys. Ser. **58**, 17 (2014) [arXiv:1409.3370 [gr-qc]]. J. I. McDonald and G. M. Shore, Phys. Lett. B **751**, 469 (2015) doi:10.1016/j.physletb.2015.10.075 [arXiv:1508.04119 [hep-ph]]. Phys. Lett. B **766**, 162 (2017) doi:10.1016/j.physletb.2016.11.057 [arXiv:1604.08213 [hep-ph]].

[24] E. V. Arbuzova and A. D. Dolgov, arXiv:1702.07477 [gr-qc]; arXiv:1612.06206 [gr-qc].

[25] J. Ellis, N. E. Mavromatos and S. Sarkar, Phys. Lett. B **725**, 407 (2013) doi:10.1016/j.physletb.2013.07.016 [arXiv:1304.5433 [gr-qc]].

[26] M. de Cesare, N. E. Mavromatos and S. Sarkar, Eur. Phys. J. C **75**, no. 10, 514 (2015) doi:10.1140/epjc/s10052-015-3731-z [arXiv:1412.7077 [hep-ph]].

[27] N. E. Mavromatos and A. Pilaftsis, Phys. Rev. D **86**, 124038 (2012) doi:10.1103/PhysRevD.86.124038 [arXiv:1209.6387 [hep-ph]].

[28] E. W. Kolb and M. S. Turner, *The Early Universe*, Front. Phys. **69**, 1-547.

[29] V. A. Kostelecky and N. Russell, *Data Tables for Lorentz and CPT Violation*, Rev. Mod. Phys. **83**, 11 (2011) doi:10.1103/RevModPhys.83.11 [arXiv:0801.0287 [hep-ph]].

[30] J. Bernabeu, N. E. Mavromatos and S. Sarkar, Phys. Rev. D **74**, 045014 (2006) doi:10.1103/PhysRevD.74.045014 [hep-th/0606137]; N. E. Mavromatos and S. Sarkar, Eur. Phys. J. C **73**, no. 3, 2359(2013) doi:10.1140/epjc/s10052-013-2359-0 [arXiv:1211.0968 [hep-ph]].

[31] J. R. Ellis, N. E. Mavromatos and M. Westmuckett , Phys. Rev. D **70**, 044036 (2004) doi:10.1103/PhysRevD.70.044036 [gr-qc/0405066]; *ibid.* Phys. Rev. D **71**, 106006 (2005) doi:10.1103/PhysRevD.71.106006 [gr-qc/0501060]; J. R. Ellis, N. E. Mavromatos, D. V. Nanopoulos and M. Westmuckett, Int. J. Mod. Phys. A **21**, 1379 (2006) doi:10.1142/S0217751X06028990 [gr-qc/0508105]; N. E. Mavromatos, V. A. Mitsou, S. Sarkar and A. Vergou, Eur. Phys. J. C **72**, 1956 (2012) doi:10.1140/epjc/s10052-012-1956-7 [arXiv:1012.4094 [hep-ph]]; N. E. Mavromatos, M. Sakellariadou and M. F. Yusaf, JCAP **1303**, 015 (2013) doi:10.1088/1475-7516/2013/03/015 [arXiv:1211.1726 [hep-th]]; T. Elghozi, N. E. Mavromatos, M. Sakellariadou and M. F. Yusaf, JCAP **1602**, no. 02, 060 (2016) doi:10.1088/1475-7516/2016/02/060 [arXiv:1512.03331 [hep-th]].

[32] T. Li, N. E. Mavromatos, D. V. Nanopoulos and D. Xie, Phys. Lett. B **679**, 407 (2009) doi:10.1016/j.physletb.2009.07.062 [arXiv:0903.1303 [hep-th]].

[33] A. Di Domenico, *Search for CPT and Lorentz-Symmetry Violation in Entangled Neutral Kaons,* arXiv:1608.00241 [hep-ph]; A. De Santis [KLOE-2 Collaboration and DAFNE Team], *DAΦNE and KLOE-2*, Phys. Scripta T **166**, 014015 (2015) doi:10.1088/0031-8949/2015/T166/014015 [arXiv:1503.06002 [physics.ins-det]] and references therein.

[34] J. Bernabeu, F. J. Botella and J. Roldan, FTUV/89-35, IFIC/89-11, Proc. Blois Conference on CP Violations, pp. 389-400 (1989)

[35] J. P. Lees *et al.* [BaBar Collaboration], Phys. Rev. Lett. **109**, 211801 (2012) [arXiv:1207.5832 [hep-ex]]; G. Eigen [BaBar Collaboration], *Recent BABAR Results*, J. Phys. Conf. Ser. **631**, no. 1, 012033 (2015) doi:10.1088/1742-6596/631/1/012033 [arXiv:1503.02867 [hep-ex]] and references therein.

[36] P. A. Bolokhov and M. Pospelov, Phys. Rev. D **77**, 025022 (2008) [hep-ph/0703291 [HEP-PH]], and references therein.

[37] N. Russell, hep-ph/0209251; M. Hori and J. Walz, Prog. Part. Nucl. Phys. **72**, 206 (2013) doi:10.1016/j.ppnp.2013.02.004 [arXiv:1304.3721 [physics.atom-ph]], and references therein.

[38] M. Hori *et al.* [ASACUSA Collaboration], Nature **475**, 484 (2011); doi:10.1038/nature10260; M. Hori *et al.* [ASACUSA Collaboration], Science **354**, no. 6312, 610 (2016) doi:10.1126/science.aaf6702

[39] B. R. Heckel, E. G. Adelberger, C. E. Cramer, T. S. Cook, S. Schlamminger and U. Schmidt, Phys. Rev. D **78**, 092006 (2008) [arXiv:0808.2673 [hep-ex]].

[40] P. A. Bolokhov, M. Pospelov and M. Romalis, Phys. Rev. D **78**, 057702 (2008) [hep-ph/0609153] and references therein.

[41] O. G. Kharlanov and V. C. .Zhukovsky, J. Math. Phys. **48**, 092302 (2007) [arXiv:0705.3306 [hep-th]].

[42] A.D. Dolgov, Phys. Atom. Nucl., **73**, 588 (2010), arXiv: 0903.4318, and references therein.

[43] F. Ambrosino *et al.* [KLOE Collaboration], Phys. Lett. B **642**, 315 (2006); A. Di Domenico, arXiv:1608.00241 [hep-ph].

[44] M. C. Banuls and J. Bernabeu, Nucl. Phys. B **590**, 19 (2000) [hep-ph/0005323]; J. Bernabeu, M. C. Banuls and F. Martinez-Vidal, PoS HEP **2001**, 057 (2001) [hep-ph/0111073]; J. Bernabeu, F. Martinez-Vidal and P. Villanueva-Perez, JHEP **1208**, 064 (2012) [arXiv:1203.0171 [hep-ph]]; See also: E. Alvarez and A. Szynkman, Mod. Phys. Lett. A **23**, 2085 (2008) [hep-ph/0611370].

[45] J. Bernabeu, F. J. Botella and M. Nebot, JHEP **1606**, 100 (2016) doi:10.1007/JHEP06(2016)100 [arXiv:1605.03925 [hep-ph]].

[46] J. Bernabeu, A. Di Domenico and P. Villanueva-Perez, Nucl. Phys. B **868**, 102 (2013) [arXiv:1208.0773 [hep-ph]].

[47] J. R. Ellis, J. S. Hagelin, D. V. Nanopoulos and M. Srednicki, Nucl. Phys. B **241**, 381 (1984); J. R. Ellis, J. L. Lopez, N. E. Mavromatos and D. V. Nanopoulos, Phys. Rev. D **53**, 3846 (1996); P. Huet and M. E. Peskin, Nucl. Phys. B **434**, 3 (1995); F. Benatti and R. Floreanini, Nucl. Phys. B **488**, 335 (1997).

[48] J. Bernabeu, F. J. Botella, N. E. Mavromatos and M. Nebot, arXiv:1612.05652 [hep-ph].

[49] E. Alvarez, J. Bernabeu, N. E. Mavromatos, M. Nebot and J. Papavassiliou, Phys. Lett. B **607**, 197 (2005); E. Alvarez, J. Bernabeu and M. Nebot, JHEP **0611**, 087 (2006).

[50] J. Marton *et al.* , arXiv:1703.01615 [quant-ph]; A. Pichler *et al.*, J. Phys. Conf. Ser. **718**, no. 5, 052030 (2016) doi:10.1088/1742-6596/718/5/052030 [arXiv:1602.00898 [physics.ins-det]]; C. Curceanu *et al.*, Found. Phys. **46**, 263 (2016) doi:10.1007/D10701-015-9923-4, 10.1007/s10701-015-9923-4 [arXiv:1601.06617 [quant-ph]]. H. Shi *et al.* [VIP Collaboration], J. Phys. Conf. Ser. **718**, no. 4, 042055 (2016) doi:10.1088/1742-6596/718/4/042055 [arXiv:1601.05828 [physics.ins-det]]. C. Curceanu *et al.*, J. Phys. Conf. Ser. **631**, no. 1, 012068 (2015) doi:10.1088/1742-6596/631/1/012068.

Can a spontaneous collapse in flavour oscillations be tested at KLOE?

Kyrylo Simonov[1,*] and *Beatrix C.* Hiesmayr[1,**]

[1] *University of Vienna, Faculty of Physics, Boltzmanngasse 5, 1090 Vienna, Austria.*

Abstract. Why do we never see a table in a superposition of here and there? This problem gets a solution by so called collapse models assuming the collapse as a genuinely physical process. Here we consider two specific collapse models and apply them to systems at high energies, i.e. flavour oscillating neutral meson systems. We find on one hand a potentially new interpretation of the decay rates introduced by hand in the standard formalism and on the other hand that these systems at high energies constrain by experimental data the possible collapse scenarios.

1 Introduction: The Measurement Problem

Quantum mechanics is an exceedingly successful theory which can explain a plethora of experimental results covering physical phenomena on different energy scales. However, in its standard formulation quantum mechanics meets some conceptual problems. For instance, formally a macroscopic object can be in a superposition as the famous example of the Schrödinger's cat. The interaction of a seemingly macroscopic system such as a measurement apparatus with a seemingly microscopic system, a quantum system, is postulated to force a reduction/collapse of the wave function of the quantum system, i.e. breaking up the superposition. Standard quantum theory neither provides a mechanism how this collapse takes place nor reveals whether it is a real physical process. We have two completely different types of dynamics at hand, i.e.

- a deterministic unitary time evolution ruled by the Schrödinger equation and
- a stochastic and non-unitary reduction process of the wave function caused by a measurement obeying Born's rule.

Obviously, several questions arise: What defines a system to be microscopic versus macroscopic? The Copenhagen interpretation uses this concept but never defines it. Is the collapse procedure a real physical process? Does the collapse manifest at high energies differently than for systems at lower energies? Do collapse models allow for a different interpretation of the measurable dynamics of flavour oscillating mesons? In turn, how do these systems at high energies restrict the plethora of collapse models?

Dynamical reduction models (or collapse models) claim to solve this issue by modeling the collapse as a genuinely physical process. The aim of this contribution is to investigate collapse models

* e-mail: Kyrylo.Simonov@univie.ac.at
** e-mail: Beatrix.Hiesmayr@univie.ac.at

for oscillation systems at high energies such as the K-mesons produced by DAΦNE. The paper is organised by giving a short introduction to collapse models and then applying two popular models to neutral meson systems followed by interpreting the results.

2 Collapse models: A Solution to the Measurement Problem?

Dynamical reduction models provide one of the possible solutions to the measurement problem. In general they propose a new universal dynamics which introduces the collapse of the wave function as an objective physical process. In order to recover the predictions of quantum mechanics this dynamics should fulfill the following properties

- non-linearity: The new dynamics should break superpositions on a macroscopic level, particularly, during a measurement.
- stochasticity: The new dynamics should produce the quantum probabilities obeying Born's rule.
- no superluminal signaling: The new dynamics should not be in conflict with the special relativity.

Let us mention that it is by no means trivial that a mathematically consistent framework exists with the above properties and simultaneously agrees with all observed data. The first dynamical reduction model was the GRW model introduced in 1985 by Ghirardi, Rimini and Weber [1]. It proposes a simple mechanism for the collapse for a system of N particles through spontaneous localisations, i.e.

1. Each particle undergoes a sudden localisation at a random time t (mathematically modelled by a multiplication with a peaked Gaussian function).
2. Between such localisations the state of the system evolves due to the Schrödinger equation.

In this way the GRW model, and collapse models in general, introduce two new fundamental constants, the localisation rate λ_{GRW}, which fixes the mean rate of the localisation process, and the coherence length r_C, which fixes the width of the Gaussian function, i.e. the localisation effect. These two constants allow to divide the world into a microscopic regime, where the Schrödinger equation rules the dynamics mainly, and a macroscopic regime where superposition breaks (Schrödinger's cat is alive). Consequently, taking collapse models seriously there are two new natural constants ruling our world.

Around 1989 two more realistic dynamical reduction models appeared on the market. These are the QMUPL model (Quantum Mechanics with Universal Position Localisation) [2] and the CSL model (Continuous Spontaneous Localisation) [3–5]. In contrast to the GRW model these models introduce the spontaneous collapse as a continuous process. The dynamics is obtained by adding to the Schrödinger equation non-linear and stochastic terms. For a given Hamiltonian $\hat{H}$ and state $|\psi_t\rangle$ at time t the dynamics is defined by [6]

$$d|\psi_t\rangle \quad = \quad \left[-i\hat{H}\,dt + \sqrt{\lambda}\sum_{i=1}^{N}(\hat{A}_i - \langle\hat{A}_i\rangle_t)dW_{i,t} - \frac{\lambda}{2}\sum_{i=1}^{N}(\hat{A}_i - \langle\hat{A}_i\rangle_t)^2 dt\right]|\psi_t\rangle, \qquad (1)$$

where $\hbar = 1$ is taken, $\langle\hat{A}_i\rangle_t := \langle\psi_t|\hat{A}_i|\psi_t\rangle$ is the quantum mechanical expectation value, and λ sets the strength of the collapse (connected to the rate and coherence length). Here $\hat{A}_i$ represent a set of N self-adjoint commuting operators introducing the collapse, $W_{i,t}$ represent a set of N independent standard Wiener processes, one for each collapse operator $\hat{A}_i$, and lead to a white noise field $w_{i,t} = \frac{d}{dt}W_{i,t}$ which plays a crucial role in the model. The two collapse models differ mainly in the choice of the collapse operators $\hat{A}_i$.

Both models have maintained a high interest over the last years being intensively investigated from the fundamental point of view as well as in experiments. Particularly, the QMUPL model has been investigated by analysing the spontaneous radiation emission from a free charged particle [7, 8] and X-ray emission by an isolated slab of Germanium [9–11]. For the CSL model experiments with optomechanical cavities [12, 13] and mechanical oscillators [14] were proposed, contributions from cold-atom experiments [15] were also recently obtained.

In the high energy regime the first computations for neutral meson and neutrino systems can be found in Refs. [16, 17]. These predictions were compared to specific decoherence models [18–20], where the decoherence strength was computed by KLOE data [21–23] and by data from B-factories [24–26]. Last but not least it was shown that the collapse models can contribute to an effective cosmological constant [27]. Thus the dynamical reduction models reveal a rich spectrum of potential experimental tests.

3 Neutral Mesons: A Laboratory for Testing the Collapse Models

Flavour oscillating systems, in particular the neutral K-mesons, have proven to be an exceptional laboratory for exploring quantum foundations, for instance a puzzling connection between the violation of the $\mathcal{CP}$ symmetry and the violation of a Bell inequality [28] has been found (experimentally feasible with the KLOE detector). Neutral mesons are massive systems with two different mass eigenstates $|M_L\rangle$ and $|M_H\rangle$ which decay with distinct decay rates. The difference between decay rates for K-mesons is rather huge, in strong contrast to the other mesons. In what follows we neglect the tiny violation of the $\mathcal{CP}$ symmetry (which means that $\langle M_H|M_L\rangle = 0$). A neutral meson consists of quark-antiquark pair, and both the particle state $|M^0\rangle$ and the antiparticle state $|\bar{M}^0\rangle$ can decay into the same final states. Therefore, neutral mesons have to be considered as a two-state system and their dynamics is then described by the following effective Schrödinger equation due to the Wigner–Weisskopf approximation

$$\frac{d}{dt}\,|\psi_t\rangle = -i\,\hat{H}\,|\psi_t\rangle \implies |\psi_t\rangle = a(t)\,|M^0\rangle + b(t)\,|\bar{M}^0\rangle, \tag{2}$$

where the effective Hamiltonian $\hat{H} = \hat{M} + \frac{i}{2}\hat{\Gamma}$ is non-hermitian, and $\hat{M}$ is the hermitian mass operator, and $\hat{\Gamma}$ is a hermitian operator which describes the decay. The diagonalised Hamiltonian defines the mass eigenstates

$$\hat{H}\,|M_n\rangle = \left(m_n + \frac{i}{2}\Gamma_n\right)|M_n\rangle, \tag{3}$$

where Γ_L, Γ_H are the corresponding decay rates (with $c = 1$). The flavour eigenstates are related to the mass eigenstates by

$$|M^0\rangle = \tfrac{1}{\sqrt{2}}\big(|M_H\rangle + |M_L\rangle\big), \tag{4a}$$

$$|\bar{M}^0\rangle = \tfrac{1}{\sqrt{2}}\big(|M_H\rangle - |M_L\rangle\big). \tag{4b}$$

The general collapse scenario, Eq. (1), has been shown to be in our case equivalent to [29, 30]

$$i\frac{d}{dt}|\psi_t\rangle = \left[\hat{H} - \sqrt{\lambda}\sum_{i=1}^{N}\hat{A}_i w_{i,t}\right]|\psi_t\rangle := \left[\hat{H} + \hat{N}(t)\right]|\psi_t\rangle, \tag{5}$$

which has a form of a standard Schrödinger equation with a random perturbation $\hat{N}(t)$. Solving this equation (5) with the help of Dyson series up to second order we obtained after a cumbersome computation the following probabilities [31, 32] for the propagation of the mass eigenstates for the QMUPL and CSL models

$$P^{QMUPL}_{M_{\mu=L/H}\to M_{\nu=L/H}}(t) = \delta_{\mu\nu}\left(1-\Lambda_\mu^{QMUPL}\cdot t+3\cdot\frac{1}{2}(\Lambda_\mu^{QMUPL})^2\cdot t^2\right)\cdot e^{-\Gamma_\mu t}, \tag{6}$$

$$P^{CSL}_{M_{\mu=L/H}\to M_{\nu=L/H}}(t) = \delta_{\mu\nu}\left(1-\Gamma_\mu^{CSL}\cdot t+\frac{1}{2}(\Gamma_\mu^{CSL})^2\cdot t^2\right)\cdot e^{-\Gamma_\mu t}\approx\delta_{\mu\nu}\,e^{-(\Gamma_\mu^{CSL}+\Gamma_\mu)t}, \tag{7}$$

and for the propagation of the flavour states

$$\begin{aligned}
P^{QMUPL}_{M^0\to M^0/\bar{M}^0} &= \frac{1}{4}\Bigg\{\sum_{i=H,L}e^{-\Gamma_i t}\left(1-\Lambda_i^{QMUPL}\cdot t+3\cdot\frac{1}{2}(\Lambda_i^{QMUPL})^2\cdot t^2\right)\\
&\pm 2\cos(\Delta m t)\,e^{-\frac{\Gamma_H+\Gamma_L}{2}t}\cdot\Bigg(1-\frac{\alpha\lambda}{2}\left[\frac{\Delta m^2}{m_0^2}(1-\theta(0))+\frac{m_H m_L}{m_0^2}(1-2\theta(0))\right]\cdot t\\
&+3\cdot\frac{1}{2}\left(\frac{\alpha\lambda}{2}\left[\frac{\Delta m^2}{m_0^2}(1-\theta(0))+\frac{m_H m_L}{m_0^2}(1-2\theta(0))\right]\right)^2\cdot t^2\Bigg)\Bigg\},
\end{aligned} \tag{8}$$

$$\begin{aligned}
P^{CSL}_{M^0\to M^0/\bar{M}^0} &= \frac{1}{4}\Bigg\{\sum_{i=H,L}e^{-\Gamma_i t}\left(1-\Gamma_i^{CSL}\cdot t+\frac{1}{2}(\Gamma_i^{CSL})^2\cdot t^2\right)\\
&\pm 2\cos(\Delta m t)\,e^{-\frac{\Gamma_H+\Gamma_L}{2}t}\cdot\Bigg(1-\frac{\gamma}{(\sqrt{4\pi}r_C)^d}\left[\frac{\Delta m^2}{m_0^2}(1-\theta(0))+\frac{m_H m_L}{m_0^2}(1-2\theta(0))\right]\cdot t\\
&+\frac{1}{2}\left(\frac{\gamma}{(\sqrt{4\pi}r_C)^d}\left[\frac{\Delta m^2}{m_0^2}(1-\theta(0))+\frac{m_H m_L}{m_0^2}(1-2\theta(0))\right]\right)^2\cdot t^2\Bigg)\Bigg\}\\
&\approx \frac{e^{-(\Gamma_L+\Gamma_L^{CSL})t}+e^{-(\Gamma_H+\Gamma_H^{CSL})t}}{4}\cdot\left\{1\pm\frac{\cos(\Delta m t)}{\cosh(\frac{(\Gamma_L+\Gamma_L^{CSL})-(\Gamma_H+\Gamma_H^{CSL})}{2}\cdot t)}\cdot e^{-\frac{\gamma}{(\sqrt{4\pi}r_C)^d}\frac{(\Delta m)^2}{2m_0^2}t}\right\},
\end{aligned} \tag{9}$$

where we used the abbreviations

$$\Lambda_\mu^{QMUPL}=\frac{\alpha\lambda}{2}\cdot\frac{m_\mu^2}{m_0^2}\cdot\left(1-2\theta(0)\right) \tag{10}$$

and

$$\Gamma_\mu^{CSL}=\frac{\gamma}{(\sqrt{4\pi}r_C)^d}\cdot\frac{m_\mu^2}{m_0^2}\cdot\left(1-2\theta(0)\right). \tag{11}$$

Here d is the dimensionality, $\lambda=\frac{\lambda_{GRW}}{2r_C^2}$ is the collapse strength in the QMUPL model, $\gamma=\lambda_{GRW}\cdot(\sqrt{4\pi}r_C)^d$ is the collapse strength in the CSL model, m_0 is a reference mass which is taken usually to be the nucleon mass (we will use the rest mass of the corresponding neutral meson), $\Delta m=m_H-m_L$ is the mass difference, $\theta(0)$ is the Heaviside function at zero, and $\sqrt{\alpha}$ represents a width of the wave packet that is assumed to be the initial state of the meson for the QMUPL model.

The obtained probabilities reveal several interesting results, namely:

- For the QMUPL model the probabilities have a non-exponential contribution which makes the collapse effect in principle observable.

	$\Gamma_L^{\exp}\,[s^{-1}]$	$\Gamma_H^{\exp}\,[s^{-1}]$	$\Delta m^{\exp}\,[\hbar s^{-1}]$	$m_L\,[\hbar s^{-1}]$	$m_H\,[\hbar s^{-1}]$
K-mesons	$1.117 \cdot 10^{10}$	$1.955 \cdot 10^{7}$	$0.529 \cdot 10^{10}$	$2.311 \cdot 10^{8}$	$5.524 \cdot 10^{9}$
D-mesons	$2.454 \cdot 10^{12}$	$2.423 \cdot 10^{12}$	$0.950 \cdot 10^{10}$	$1.468 \cdot 10^{12}$	$1.477 \cdot 10^{12}$
B_d-mesons	$6.582 \cdot 10^{11}$	$6.576 \cdot 10^{11}$	$0.510 \cdot 10^{12}$	$1.020 \cdot 10^{15}$	$1.020 \cdot 10^{15}$
B_s-mesons	$7.072 \cdot 10^{11}$	$6.158 \cdot 10^{11}$	$1.776 \cdot 10^{13}$	$2.477 \cdot 10^{14}$	$2.655 \cdot 10^{14}$

Table 1. Experimental values of the decay rates, the mass difference and the computed values of the absolute masses for K-, D-, B_d- and B_s-mesons. The table is taken from [32].

- For the CSL model the probabilities can be expanded to exponentials, allowing for new interpretations.
- The collapse dynamics contributes to the transition probabilities by two independent effects, a damping of the oscillation and a contribution to the decay rates.
- While the damping of the oscillation is proportional to the experimentally measurable squared mass difference, the effective decay constants are proportional to the squared absolute masses (which do not show up in the standard quantum mechanical framework as measurable quantities).

In general, the obtained probabilities are strongly sensitive to the very nature of the noise field which results in the freedom to choose the value of the Heaviside function at zero $\theta(0) \in [0, 1]$. Furthermore, a fully new interpretation of the dynamics can be obtained [32]. The decay rates can be fully derived as an effect of the collapse. Note that in any standard framework handling the meson dynamics only the mass difference $m_H - m_L$ appears, never the absolute masses. This can be achieved by reversing the mass ratio $\frac{m_\mu}{m_0}$ in the decay constants Γ_μ^{CSL} (which is more reasonable from a collapse model perspective) and taking $\theta(0) \in [0, 1/2\}$ to save their positivity. Considering such a scenario allows to calculate the absolute masses of the mesons using the decay rates and mass difference measured in experiments via

$$\frac{\Gamma_L^{CSL} - \Gamma_H^{CSL}}{\Gamma_L^{CSL} + \Gamma_H^{CSL}} = \frac{m_H^2 - m_L^2}{m_H^2 + m_L^2} = 1 - \frac{m_L^2}{m_L^2 + m_L \Delta m + \frac{1}{2}(\Delta m)^2}, \tag{12}$$

and the obtained absolute masses of some neutral mesons are listed in the Table 1. Thus the decay rates are functions of the absolute masses and do not need an additional introduction as in the standard approach by making the Hamiltonian non-hermitian or treating the decay as a decoherence effect [35, 36].

From the perspective of collapse models neutral mesons allow us to derive the CSL collapse rate in return

$$\lambda^{CSL} := \Gamma_\mu^{\exp} \cdot \frac{m_\mu^2}{m_0^2} \frac{1}{1 - 2\theta(0)} = \frac{1}{(\sqrt{\Gamma_L^{-1}} - \sqrt{\Gamma_H^{-1}})^2} \frac{(\Delta m)^2}{m_0^2} \frac{1}{1 - 2\theta(0)}. \tag{13}$$

The predicted values of the collapse rate are plotted in Fig. 1 for the different types of mesons and can be compared in Fig. 2 with the experimental data of different physical systems (summarised recently in Ref. [33]).

The plots show that the values for all the mesons except for K-mesons are of the order of the collapse rate assumed by Adler, $\lambda_{Adler}^{CSL} = 10^{-(8\pm2)} s^{-1}$ [34]. K-mesons tend to be closer to the original value of the collapse rate proposed by Ghirardi, Rimini and Weber, $\lambda_{GRW}^{CSL} = 10^{-16} s^{-1}$ [1], but require a value of $\theta(0)$ close to $\frac{1}{2}$.

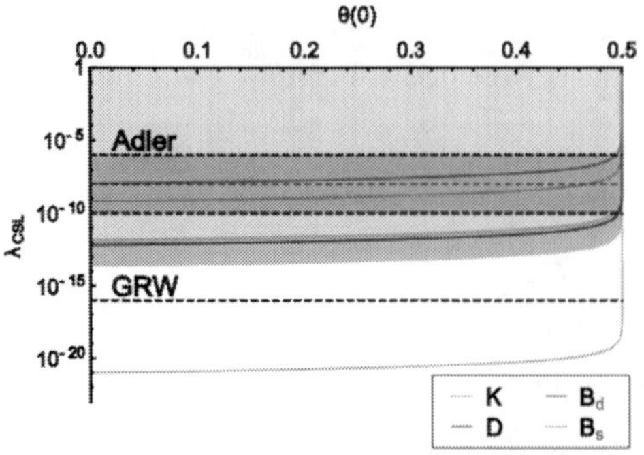

Figure 1. Values of the collapse rate including experimental errors as a function of $\theta(0)$ based on the values $\Gamma_H^{\text{exp}}, \Gamma_L^{\text{exp}}, \Delta m^{\text{exp}}$ known from experiments for the different types of neutral mesons. The plot is taken from [32].

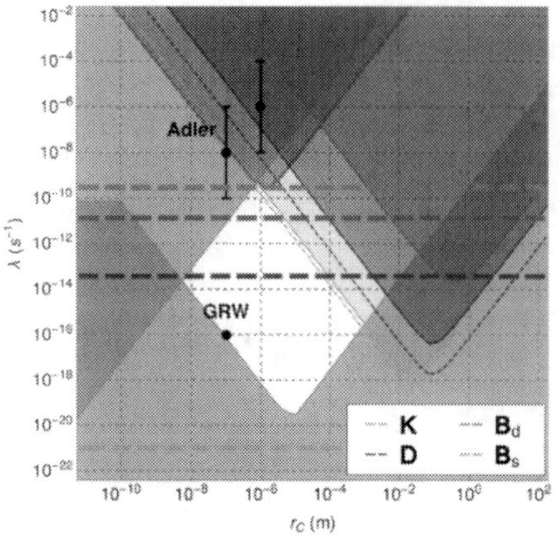

Figure 2. Bounds on the natural constants of the CSL model based on LIGO, LISA Pathfinder and AURIGA experiments (blue, green, and red lines), ultracold cantilever experiments (purple line), X-ray experiments (light blue line) and theoretical estimations (grey line) (taken from Ref. [33]). The thick dashed lines refer to the lower bounds due to our computations of λ^{CSL}, Eq. (13), for the neutral meson system. Note that r_C is not bounded.

4 Conclusions

Dynamical reduction models provide a possible solution to the measurement problem of standard quantum mechanics by introducing a physical mechanism of the collapse of the wave function. We have investigated two popular collapse models and derived their respective deviations from the standard dynamics of flavour oscillations which are intensively studied at accelerator facilities. We have found that in principle such changes are observable and the dynamics can change the oscillation behaviour and under certain constraints explain the decay behaviour of neutral mesons completely. We have also illustrated how the meson system restricts the plethora of collapse models considerably.

Precision data from experiments such as KLOE will allow further to restrict or even rule out certain collapse scenarios when one extents the dynamics to two entangled pairs of mesons or/and includes $\mathcal{CP}$ violation.

Acknowledgements: Both authors acknowledge gratefully the Austrian Science Fund (FWF-P26783).

References

[1] G. C. Ghirardi, A. Rimini and T. Weber, Phys. Rev. D **34**, 470 (1986).
[2] L. Diósi, Phys. Rev. A **40**, 1165 (1989).
[3] P. Pearle, Phys. Rev. A **39**, 2277 (1989).
[4] G. C. Ghirardi, P. Pearle and A. Rimini, Phys. Rev. A **42**, 78 (1990).
[5] G. C. Ghirardi, R. Grassi and F. Benatti, Found. Phys. **25**, 5 (1995).
[6] A. Bassi, D. Dürr and G. Hinrichs, Phys. Rev. Lett. **111**, 210401 (2011).

[7] A. Bassi and D. Dürr, J. Phys. A **42**, 485302 (2009).
[8] A. Bassi and S. Donadi, Phys. Lett. A **378**, 761 (2014).
[9] C. Curceanu, B. C. Hiesmayr and K. Piscicchia, J. Adv. Phys. **4**, 263 (2015).
[10] C. Curceanu et al., Found. Phys. **46**, 263 (2016).
[11] K. Piscicchia et al., arXiv:1501.04462 (2015).
[12] M. Bahrami et al., Phys. Rev. Lett. **112**, 210404 (2014).
[13] S. Nimmrichter, K. Hornberger and K. Hammerer, Phys. Rev. Lett. **113**, 020405 (2014).
[14] L. Diósi, Phys. Rev. Lett. **114**, 050403 (2015).
[15] M. Bilardello et al., arXiv:1605.01891 (2016).
[16] S. Donadi et al., Found. Phys. **43**, 813 (2013).
[17] M. Bahrami et al., Sci. Rep. **3**, 1952 (2013).
[18] R. A. Bertlmann, W. Grimus, B. C. Hiesmayr, Phys. Rev. D **60**, 114032 (1999).
[19] B. C. Hiesmayr, Found. Phys. Lett. **14**, 231 (2001).
[20] R. A. Bertlmann, K. Durstberger, B. C. Hiesmayr, Phys. Rev. A **68**, 012111 (2003).
[21] F. Ambrosino et al. (KLOE collaboration), Phys. Lett. B **642**, 315 (2006).
[22] A. Di Domenico et al. (KLOE Collaboration), J. Phys.: Conf. Ser. **171**, 012008 (2008).
[23] A. Di Domenico et al. (KLOE Collaboration), Found. Phys. **40**, 852 (2010).
[24] A. Go et al. (Belle Collaboration), Phys. Rev. Lett. **99**, 131802 (2007).
[25] G. Richter, *"Stability of Nonlocal Quantum Correlations in Neutral B-Meson Systems"* (PhD Thesis, Technische Universität Wien, 2008).
[26] B. D. Yabsley, *"Quantum entanglement at the psi(3770) and Upsilon(4S)"* (Flavour Physics & CP Violation Conference, Taipei, 2008).
[27] T. Josset, A. Perez and D. Sudarsky, arXiv:1604.04183 (2016).
[28] B. C. Hiesmayr et al., Eur. Phys. J. C **72**, 1856 (2012).
[29] S. L. Adler and A. Bassi, J. Phys. A **40**, 15083 (2007).
[30] S. Donadi, *"Electromagnetic Radiation Emission and Flavour Oscillations in Collapse Models"* (PhD Thesis, Università degli studi di Trieste, 2012).
[31] K. Simonov and B. C. Hiesmayr, Phys. Lett. A **380**, 1253 (2016).
[32] K. Simonov and B. C. Hiesmayr, Phys. Rev. A **94**, 052128 (2016).
[33] M. Carlesso, A. Bassi, P. Falferi and A. Vinante, Phys. Rev. D **94**, 124306 (2016).
[34] S. L. Adler, J. Phys. A: Math. Theor. **40**, 2935 (2007).
[35] R. A. Bertlmann, W. Grimus, B. C. Hiesmayr, Phys. Rev. A **73**, 054101 (2006).
[36] J. Bernabéu, N. E. Mavromatos and P. Villanueva-Pérez, Phys. Lett. B **724**, 269 (2013).

Overview of KLOE results on kaon physics and KLOE–2 perspectives

Eryk Czerwiński[1,⋆] on behalf of the KLOE-2 Collaboration

[1] *Faculty of Physics, Astronomy and Applied Computer Science, Jagiellonian University, 30-348 Kraków, Poland*

Abstract. Kaon physics program of KLOE is being continued at the upgraded KLOE-2 system, therefore an overview of KLOE results in this field is shortly presented together with prospects for KLOE-2.

1 Introduction

The National Laboratory in Frascati (LNF-INFN, Italy) hosts the e^+e^- collider known as ϕ-factory DAΦNE [1–3]. The collider was designed to operate at the peak of ϕ resonance ($\sqrt{s} = m_\phi \approx 1019\ MeV$) producing ϕ mesons almost at rest ($\beta_\phi \approx 0.015$) since electrons and positrons collide with small transverse momenta. A ϕ meson decays mostly into kaon pairs (49% into K^+K^- and 34% into $K_S K_L$), which makes a ϕ-factory the natural place for kaon physics studies. From 2001 to 2006 KLOE has collected 2.5 fb^{-1} of integrated luminosity [4]. The detector consists of a cylindrical drift chamber [5] and an electromagnetic calorimeter [6] surrounded by a superconducting coil which produces an axial magnetic field parallel to the beam axis.

At KLOE, reconstruction of the K_S decay close to interaction region (clean selection of $K_S \to \pi^+\pi^-$, BR=69%) allows to tag a K_L presence which makes KLOE an excellent place for K_L decay measurements. Additionally a detection of the K_L hit in the calorimeter module tags the presence of the K_S, which makes KLOE a unique place to study pure K_S beams e.g. the measurement of $BR(K_S \to \pi^0\pi^0\pi^0)$. A study of both kaons decays in single event is also possible. Since both of them are produced in a pure quantum state ($J^{PC} = 1^{--}$), it is possible to study e.g. quantum interference effects.

At KLOE, the physics program was related to the good accuracy of reconstruction of K_L decays in large fiducial volume, while at KLOE-2 an increased interest will be focused on the physics close to the interaction point as rare K_S decays, $K_S - K_L$ interference, multi-lepton events, as well as η, η' and $K^\pm$ decays. More details about whole KLOE-2 physics program can be found in Ref. [7]. The main modification of KLOE system from the point of view of kaon physics is installation of a light-material Inner Tracker detector based on the Cylindrical GEM technology to improve charged vertex reconstruction and to increase the acceptance for low transverse momentum tracks [8–11].

⋆e-mail: eryk.czerwinski@uj.edu.pl

Parameter	Measurement	Ref.
$BR(K_{e3}) = 0.4008(15)$ $BR(K_{\mu3}) = 0.2699(14)$	based on 13×10^6 K_L decays tagged by $K_S \to \pi^+\pi^-$	[12]
τ_L=50.92(30) ns	fit the time dependence over $0.4\tau_L$ of 8.5×10^6 $K_L \to 3\pi^0$ decays tagged by $K_S \to \pi^+\pi^-$	[13]
$\lambda'_+ = (25.5 \pm 1.8) \times 10^{-3}$ $\lambda''_+ = (1.4 \pm 0.8) \times 10^{-3}$	based on 2×10^6 $K_{L,e3}$ decays tagged by $K_S \to \pi^+\pi^-$	[14]
$BR(K_S \to \pi e\nu) = 7.046(91) \times 10^{-4}$	from tagged K_S beam 1.2×10^8 events (20% of full data sample)	[14]
$\lambda'_+ = (25.6 \pm 1.5_{stat} \pm 0.9_{syst}) \times 10^{-3}$ $\lambda''_+ = (1.5 \pm 0.7_{stat} \pm 0.4_{syst}) \times 10^{-3}$ $\lambda_0 = (15.4 \pm 1.8_{stat} \pm 1.3_{syst}) \times 10^{-3}$	based on 1.8×10^6 $K_{L,\mu3}$ decays tagged by $K_S \to \pi^+\pi^-$ and from combined with with $K_{L,e3}$ data	[15]

Table 1. Summary of KLOE results from neutral kaons useful for V_{us} measurement.

2 CKM matrix

The measurement of the CKM mixing matrix element V_{us} and information about lepton universality is provided at KLOE by precise measurements of semileptonic kaon decay rates. The ratio $\Gamma(K \to \mu\nu)/\Gamma(\pi \to \mu\nu)$ from the leptonic kaon decays is an independent measurement of $|V_{us}|^2$ / $|V_{ud}|^2$, which, together with the result of $|V_{ud}|$ from the superallowed $0^+ \to 0^+$ nuclear β transitions and V_{ub} from semileptonic B decays provide the most precise test of the CKM matrix unitarity using relation $|V_{ud}|^2 + |V_{us}|^2 + |V_{ub}|^2 = 1$. At the present level of accuracy $|V_{ub}|^2$ can be neglected. This allows the universality of lepton and quark weak couplings to be tested.

The kaon semileptonic decay rate is given by:

$$\Gamma(K_{l3}) = \frac{C_K^2 G_F^2 M_K^5}{192\pi^3} S_{EW} |V_{us}|^2 |f_+(0)|^2 I_{K,l}(\lambda)(1 + 2\Delta_K^{SU(2)} + 2\Delta_{K,l}^{EM}) \quad (1)$$

where $K = K^0, K^\pm$, $l = e, \mu$ and C_K is the Clebsch-Gordan coefficient, equal to 1/2 and 1 for $K^\pm$ and K^0, respectively. The experimental input is the decay width $\Gamma(K_{l3})$ obtained from the kaon lifetime and the semileptonic BRs, inclusive of radiation, while the theoretical inputs are: the form factor $f_+(0) \equiv f_+^{K^0\pi^-}(0)$ evaluated at zero momentum transfer, the long-distance electromagnetic corrections $\Delta_{K,l}^{EM}$, the universal short-distance electroweak correction $S_{EW} = 1.0232$ and the SU(2)-breaking $\Delta_K^{SU(2)}$. One or more slope parameters λ, measured from the decay spectra, are used to describe the form factor dependence on the momentum transfer which enters in the phase space integral $I_{K,l}(\lambda)$. KLOE has measured all relevant inputs for charged and neutral kaons: BR's, lifetimes (K_L, $K^\pm$), form factors.

The inputs for $|V_{us}|$ from the neutral kaons measurements at KLOE are gathered in table 1. The world average error on $V_{us} f_+(0)$ value determination is 0.19% [16], while taking into account KLOE results only it is 0.28% [17]. Combining these results together with the 5 fb^{-1} data foreseen at KLOE-2, we can improve the accuracy on the measurement of the K_L lifetime and K_{Se3} branching ratio. Statistical uncertainties on BRs and lifetimes have been obtained scaling present KLOE statistics to 5 fb^{-1} total integrated luminosity, while systematic errors have been estimated conservatively. The expected accuracies are presented in table 2.

Mode	$V_{us} f_+(0)$	World average [16] % err	BR	τ	Δ	I	KLOE-2 prospects with 5 fb^{-1} % err	BR	τ	Δ	I
K_{Le3}	0.2163(6)	0.26	0.09	0.20	0.11	0.05	0.20	0.09	0.13	0.11	0.06
$K_{L\mu3}$	0.2166(6)	0.28	0.15	0.18	0.11	0.06	0.24	0.15	0.13	0.11	0.08
$K_{S\,e3}$	0.2155(13)	0.61	0.60	0.02	0.11	0.05	0.32	0.30	0.03	0.11	0.06
$K^{\pm}_{e3}$	0.2172(8)	0.36	0.27	0.06	0.23	0.05	0.48	0.25	0.05	0.40	0.06
$K^{\pm}_{\mu3}$	0.2170(11)	0.51	0.45	0.06	0.23	0.06	0.48	0.27	0.05	0.39	0.08
Average	0.2165(4)	0.19					0.14				

Table 2. Comparison of world average value of $V_{us} f_+(0)$ [16] with KLOE-2 prospects [7].

The present fractional experimental uncertainty on $V_{\rm us}f_+(0)$ equal to 0.19% can be reduced to 0.14%, using KLOE and KLOE-2 statistics and combining with world average. This result together with more precise measurements of $f_+(0)$ and $V_{\rm ud}$ would allow to obtain precision at the level of a 10^{-4} in the test on the unitarity relation.

3 Charge asymmetry of K_S

The charge asymmetry for semileptonic decays of neutral kaons can be defined with K_S and K_L semileptonic decay widths as follows:

$$A_{S,L} = \frac{\Gamma(K_{S,L} \to \pi^- e^+ \nu) - \Gamma(K_{S,L} \to \pi^+ e^- \bar{\nu})}{\Gamma(K_{S,L} \to \pi^- e^+ \nu) + \Gamma(K_{S,L} \to \pi^+ e^- \bar{\nu})} = 2\left[Re\left(\epsilon_{S,L}\right) - Re(y) \pm Re(x_-)\right] \tag{2}$$

to the first order in parameters ϵ_S, ϵ_L which can be expressed in terms of the CP and CPT violation parameters ϵ_K and δ_K:

$$\epsilon_{L/S} = \epsilon_K \mp \delta_K. \tag{3}$$

The semileptonic kaon decays ($K \to \pi e \nu$) can be described with the following decay amplitudes:

$$\begin{aligned} \langle \pi^- e^+ \nu | H_{weak} | K^0 \rangle &= \mathcal{A}_+, \\ \langle \pi^+ e^- \bar{\nu} | H_{weak} | \bar{K^0} \rangle &= \bar{\mathcal{A}}_-, \\ \langle \pi^+ e^- \bar{\nu} | H_{weak} | K^0 \rangle &= \mathcal{A}_-, \\ \langle \pi^- e^+ \nu | H_{weak} | \bar{K^0} \rangle &= \bar{\mathcal{A}}_+, \end{aligned} \tag{4}$$

where H_{weak} stands for weak Hamiltonian. One introduces the following useful notation:

$$\begin{aligned} x &= \frac{\bar{\mathcal{A}}_+}{\mathcal{A}_+}, \quad \bar{x} = \left(\frac{\mathcal{A}_-}{\bar{\mathcal{A}}_-}\right)^*, \quad y = \frac{\bar{\mathcal{A}}_-^* - \mathcal{A}_+}{\bar{\mathcal{A}}_-^* + \mathcal{A}_+}, \\ x_\pm &= \frac{x \pm \bar{x}^*}{2} = \frac{1}{2}\left[\frac{\bar{\mathcal{A}}_+}{\mathcal{A}_+} \pm \left(\frac{\mathcal{A}_-}{\bar{\mathcal{A}}_-}\right)^*\right]. \end{aligned} \tag{5}$$

Sum and difference of the A_S and A_L allow to search for the CPT symmetry violation, either in the decay amplitudes through the parameter y or in the mass matrix through the parameter δ_K:

$$\begin{aligned} A_S + A_L &= 4Re(\epsilon) - 4Re\,(y), \\ A_S - A_L &= 4Re(\delta_K) + 4Re\,(x_-). \end{aligned} \tag{6}$$

The charge asymmetry was measured by the KTeV experiment for long-lived kaon [18] and by the KLOE - for the short-lived one [14]:

$$\begin{aligned} A_L &= (3.322 \pm 0.058_{stat} \pm 0.047_{syst}) \times 10^{-3}, \\ A_S &= (1.5 \pm 9.6_{stat} \pm 2.9_{syst}) \times 10^{-3}. \end{aligned} \tag{7}$$

These results allow for the most precise tests of the CPT symmetry in semileptonic decays of neutral kaons, although the accuracy on A_L is more than two orders of magnitude better then on the A_S. Additionally, the total error on A_S is dominated by the available statistics.

Presently the KLOE data set of 1.7 fb^{-1} is being reanalysed aiming at the determination of the semileptonic asymmetry for K_S. As the analyzed data sample is 4 times larger, an twofold reduction of the statistical error is expected. Moreover, the KLOE-2 experiment arises prospects of measuring A_S with a statistical uncertainty at the level of 3×10^{-3}.

4 Rare K_S decays

The mechanisms of the CP violation can be investigated by searches of the CP-prohibited rare decays $K_S \to 3\pi^0$. The channel $K_S \to 3\pi^0$ is a pure CP-violating process whereas $K_S \to \pi^+\pi^-\pi^0$ violates CP for I=1 or 3 in the final state. This CP violation is usually expressed in terms of the following amplitude ratios:

$$\begin{aligned} \eta_{+-0} &= \frac{\langle \pi^+\pi^-\pi^0 | H | K_S \rangle}{\langle \pi^+\pi^-\pi^0 | H | K_L \rangle} = \epsilon_K + \epsilon'_{+-0}, \\ \eta_{000} &= \frac{\langle \pi^0\pi^0\pi^0 | H | K_S \rangle}{\langle \pi^0\pi^0\pi^0 | H | K_L \rangle} = \epsilon_K + \epsilon'_{000}, \end{aligned} \tag{8}$$

where ϵ'_{+-0} and ϵ'_{000} describe CP violation in decay.

The KLOE experiment set the most precise upper limit on BR($K_S \to 3\pi^0$) using 1.7 fb^{-1} data-set [19]:

$$\mathrm{BR}(K_S \to 3\pi^0) < 2.6 \times 10^{-8}, \tag{9}$$

based on a kinematic fit, testing of signal and background hypotheses and kinematical differences between $K_S \to 2\pi^0$ and $K_S \to 3\pi^0$. This result allows for constraining $|\epsilon'_{000}| \leq 0.0088$ at 90% CL. The sensitivity on BR($K_S \to 3\pi^0$) can be improved to 10^{-8} with 5 fb^{-1} data-set to be collected by the KLOE-2 experiment.

Presently, the direct search of $K_S \to \pi^+\pi^-\pi^0$ is being performed on the same KLOE data set. The best value of the branching ratio for this process is an average of three indirect measurements with uncertainty at the level of 30% [20], while the ongoing analysis of KLOE data shows a possibility to obtain the relative uncertainty below 20% in a direct measurement.

5 Quantum Mechanics and CPT symmetry tests

Neutral kaons produced from decay of ϕ meson are in an entangled states. If both kaons decay into identical final states, e.g. $\pi^+\pi^-$ the decay rate of the system is proportional to:

$$I(\pi^+\pi^-, \pi^+\pi^-, \Delta t) \propto e^{-\Gamma_L \Delta t} + e^{-\Gamma_S \Delta t} - 2e^{-\frac{\Gamma_L + \Gamma_S}{2}\Delta t} cos(\Delta m \Delta t), \tag{10}$$

where Δm is the mass difference between K_L and K_S, Δt is the difference of decay times of two kaons. Presence of the interference term $2e^{-\frac{\Gamma_L + \Gamma_S}{2}\Delta t} cos(\Delta m \Delta t)$ implies that the kaons cannot decay at the same time, and it is an example of correlation of the type pointed out for the first time by Einstein, Podolsky and Rosen [21].

Both, T and CPT symmetry tests at KLOE-2 require identification of $K_S \to \pi\pi$, $K_L \to \pi^{\pm}\ell^{\mp}\nu$ and $K_S \to \pi^{\pm}\ell^{\mp}\nu$, $K_L \to 3\pi^0$ events. Reconstruction of $K_L \to 3\pi^0 \to 6\gamma$ decay has to be performed based on the calorimeter measurement only. The new reconstruction method is similar to Global Positioning System (GPS). For each cluster a sphere centered at its position is created. This sphere with unknown radius represents possible origin points of γ. The point of $K_L \to 3\pi^0 \to 6\gamma$ is identified as intersection of 6 such spheres (a common origin point of all photons).

5.1 Decoherence

It is physically possible that the coherent state of two kaons emerging from the phi decay can factorize and loose coherence [22]. The amount of deviation from prediction of quantum mechanics can be parametrized as the decoherence parameter ζ suppresing the interference in the following way:

$$I(\pi^+\pi^-, \pi^+\pi^-, \Delta t) \propto e^{-\Gamma_L \Delta t} + e^{-\Gamma_S \Delta t} - 2(1-\zeta_{SL}) e^{-\frac{\Gamma_L + \Gamma_S}{2}\Delta t} cos(\Delta m \Delta t) \,. \tag{11}$$

The usual quantum mechanics case corresponds to $\zeta = 0$, while $\zeta = 1$ implies the total decoherence. In general ζ depends on the basis in which the initial state is expressed. KLOE measured the decoherence parameter based on $\approx$1.5 fb^{-1} data sample [23]. The experimental points of signal events $\phi \to K_S K_L \to \pi^+\pi^-\pi^+\pi^-$ were fitted with Eq. 11 modified with parameters expressing decoherence. The fit was performed taking into account resolution and detection efficiency, the background from coherent and incoherent K_S regeneration on the beam pipe wall, and the small contamination from the non-resonant $e^+e^- \to \pi^+\pi^-\pi^+\pi^-$ channel. The obtained results show no deviation from quantum mechanics within the accuracy [23]:

$$\zeta_{SL} = (0.3 \pm 1.8_{stat} \pm 0.6_{syst}) \cdot 10^{-2} \,,$$
$$\zeta_{0\bar{0}} = (1.4 \pm 9.5_{stat} \pm 3.8_{syst}) \cdot 10^{-7} \,. \tag{12}$$

Increased statistics at KLOE-2 together with the usage of the inner tracker would allow to reach factor four of improvement in sensitivity.

5.2 CPT and Lorentz invariance test

In the framework of the Standard Model Extension (SME) [24–26], the δ_K parameter depends on the kaon four-momentum in the following way:

$$\delta_K \approx i\, \sin\phi_{SW} e^{i\phi_{SW}} \gamma_K (\Delta a_0 - \vec{\beta}_K \cdot \Delta\vec{a})/\Delta m, \tag{13}$$

where γ_K and $\vec{\beta}_K$ are the boost factor and velocity of the kaon in the observer rest frame, respectively, ϕ_{SW} = arctan$(2\Delta m/\Delta\Gamma)$ is the superweak phase with Δm and $\Delta\Gamma$ the differences of mass and width between K_S and K_L, respectively, and Δa_μ are four *CPT* and Lorentz violating coefficients. The experimental observable at fixed time difference $\Delta\tau = \tau_1 - \tau_2$ is therefore the following [22]:

$$I_{f_1 f_2}(\Delta\tau) \propto e^{-\Gamma|\Delta\tau|} \left[|\eta_1|^2 e^{\frac{1}{2}\Delta\Gamma\Delta\tau} + |\eta_2|^2 e^{\frac{1}{2}\Delta\Gamma\Delta\tau} - 2\mathrm{Re}\left(\eta_1 \eta_2^* e^{-i\Delta m \Delta\tau}\right) \right] \tag{14}$$

where $\eta_j = \langle f_j | T | K_L \rangle / \langle f_j | T | K_S \rangle \simeq \epsilon_K - \delta_K(\vec{p}_j, t_s)$, f_1 and f_2 denote kaon final states, $\Gamma = \Gamma_S + \Gamma_L$. In the reported measurement $f_1 = f_2 = \pi^+\pi^-$ and due to the fully destructive quantum interference at $\Delta\tau = 0$ the distribution (14) is sensitive to the η_1/η_2 ratio. This analysis is based on looking for a possible dependence of $I_{f_1 f_2}(\Delta\tau)$ on a sidereal time.

The data sample of 1.7 fb^{-1} total integrated luminosity was analyzed to select $\phi \to K_S K_L \to \pi^+\pi^-\pi^+\pi^-$ events selected using the invariant mass of pairs of particles, two-body kinematics and missing momentum and energy. The total contamination from background events was established from the Monte Carlo simulation and amounts to 1.5% coming mostly from kaons regenerated without momentum deflection. Selected events were transformed to the sidereal frame and K mesons were labeled accordingly to their momentum direction (forward, backward) with respect to the momentum of the ϕ meson. The two data samples are further divided into four 6 h long sidereal time periods, so finally the fit of distribution (14) is performed to eight data subsets. It results with the first measurement of all four parameters in the kaon sector and at present the most precise measurement of these parameters in the quark sector of SME [27]:

$$\begin{aligned}
\Delta a_0 &= (-6.0 \pm 7.7_{stat} \pm 3.1_{syst}) \times 10^{-18}\ \text{GeV},\\
\Delta a_x &= (\ \ 0.9 \pm 1.5_{stat} \pm 0.6_{syst}) \times 10^{-18}\ \text{GeV},\\
\Delta a_y &= (-2.0 \pm 1.5_{stat} \pm 0.5_{syst}) \times 10^{-18}\ \text{GeV},\\
\Delta a_z &= (\ \ 3.1 \pm 1.7_{stat} \pm 0.5_{syst}) \times 10^{-18}\ \text{GeV}.
\end{aligned}$$

The ongoing data-taking campaign of KLOE-2 at upgraded DAΦNE collider will improve statistical uncertainties due to higher luminosity of the collider and at the same time systematic uncertainties due to the installation of the Inner Tracker detector [8, 10, 11]. The Inner Tracker will improve resolution on the vertex position and acceptance for tracks with low transverse momentum. Expected accuracies on Δa_0, $\Delta a_{X,Y}$ and Δa_Z are (5.2, 1.3 and 2.2)×10^{-18}, respectively.

5.3 CPT and T symmetry test in transition

A direct test of the T symmetry is possible at the KLOE-2 detector with neutral kaons [28]. These states can be defined as states with definite flavour of CP:

$$\begin{aligned}
\mathcal{S}\,|\mathrm{K}^0\rangle &= +1\,|\mathrm{K}^0\rangle\,,\\
\mathcal{S}\,|\bar{\mathrm{K}}^0\rangle &= -1\,|\bar{\mathrm{K}}^0\rangle\,,
\end{aligned} \tag{15}$$

$$\begin{aligned}
|\mathrm{K}_+\rangle &= \frac{1}{\sqrt{2}}\left[|\mathrm{K}^0\rangle + |\bar{\mathrm{K}}^0\rangle\right] \ \text{withCP} = +1,\\
|\mathrm{K}_-\rangle &= \frac{1}{\sqrt{2}}\left[|\mathrm{K}^0\rangle - |\bar{\mathrm{K}}^0\rangle\right] \ \text{withCP} = -1,
\end{aligned} \tag{16}$$

and identified through observation of their decay products. At the moment of decay of the first kaon the state of the second, still-living kaon is immediately known (due to quantum entanglement). Identification of its state at the moment of decay after time Δt allows to observe a transition between the strangeness and CP-definite states. For each of the possible transitions a time-dependent ratio of probabilities can be defined as an observable of the T symmetry test. Two of those ratios can be measured at KLOE-2:

$$R_2(\Delta t) = \frac{P[\mathrm{K}^0(0) \to \mathrm{K}_-(\Delta t)]}{P[\mathrm{K}_-(0) \to \mathrm{K}^0(\Delta t)]} \sim \frac{\mathrm{I}(\ell^-, 3\pi^0; \Delta t)}{\mathrm{I}(\pi\pi, \ell^+; \Delta t)}, \tag{17}$$

$$R_4(\Delta t) = \frac{P[\bar{\mathrm{K}}^0(0) \to \mathrm{K}_-(\Delta t)]}{P[\mathrm{K}_-(0) \to \bar{\mathrm{K}}^0(\Delta t)]} \sim \frac{\mathrm{I}(\ell^+, 3\pi^0; \Delta t)}{\mathrm{I}(\pi\pi, \ell^-; \Delta t)}. \tag{18}$$

with a sensitivity at the level of 10^{-3}. Asymptotically, their deviation from unity should be proportional to real part of δ_K which is a T violating parameter of the neutral kaon system [28].

Constructing time-dependent ratios sensitive to CPT violation is also possible [29]:

$$R_{2,CPT}(\Delta t) = \frac{P[K^0(0) \to K_-(\Delta t)]}{P[K_-(0) \to K^0(\Delta t)]}, \tag{19}$$

$$R_{4,CPT}(\Delta t) = \frac{P[\bar{K}^0(0) \to K_-(\Delta t)]}{P[K_-(0) \to \bar{K}^0(\Delta t)]}. \tag{20}$$

leading to a double ratio as CPT violation observable in the $\Delta t \gg \tau_S$ limit is equal to $1 - 8\Re\delta_K - 8\Re x_-$. Again, possible deviation from unity would inform about CPT noninvariance [29]. So far, this parameter was not measured, but it is one of the KLOE-2 milestones.

6 Summary

In the recent years a long list of physics results from neutral kaon was provided by the KLOE experiment: measurement of V_{us}, study of CP and CPT discrete symmetries, search for decoherence of entangled kaons. Presently the KLOE-2 detector is collecting new data sample for broader and more precise results.

Acknowledgments

We warmly thank our former KLOE colleagues for the access to the data collected during the KLOE data taking campaign. We thank the DAΦNE team for their efforts in maintaining low background running conditions and their collaboration during all data taking. We want to thank our technical staff: G.F. Fortugno and F. Sborzacchi for their dedication in ensuring efficient operation of the KLOE computing facilities; M. Anelli for his continuous attention to the gas system and detector safety; A. Balla, M. Gatta, G. Corradi and G. Papalino for electronics maintenance; C. Piscitelli for his help during major maintenance periods. This work was supported in part by the Polish National Science Centre through the Grants No. 2013/08/M/ST2/00323, 2013/11/B/ST2/04245, 2014/14/E/ST2/00262, 2014/12/S/ST2/00459.

References

[1] A. Gallo *et al.*, Conf. Proc. C060626 (2006) 604, SLAC-PUB-12093.

[2] M. Zobov *et al.*, Phys. Rev. Lett. **104**, 174801 (2010).

[3] C. Milardi *et al.*, JINST **7**, 2012,T03002.

[4] F. Bossi, E. De Lucia, J. Lee-Franzini, S. Miscetti, M. Palutan and KLOE Collaboration, Rivista del Nuovo Cimento Vol.31 (2008) N.10

[5] M. Adinolfi, *et al.*, Nucl. Instrum. Meth. A **461** (2001) 25-28

[6] M. Adinolfi *et al.*, Nucl. Instrum. Meth. A **482** (2002) 364

[7] G. Amelino-Camelia *et al.* (KLOE–2 Collaboration), Eur. Phys. J. C **68** (2010) 619-681

[8] M. Alfonsi *et al.*, *Nucl. Instr. & Meth. A* **617** (2010) 151

[9] A. Balla *et al.*, Nucl. Instrum. Meth. A **845** (2017) 266

[10] KLOE-2 Collaboration, F. Archilli *et al.*, LNF-10/3(P) INFN-LNF, arXiv:1002.2572

[11] A. Balla *et al.*, Nucl. Instr. & Meth. A **604** (2009) 23

[12] KLOE Collaboration, F. Ambrosino et al., *Phys. Lett.* B **632** (2006) 76

[13] F. Ambrosino *et al.* [KLOE Collaboration], Phys. Lett. B **626** (2005) 15

[14] KLOE Collaboration, Phys. Lett. B636 (2006), p. 173-182
[15] F. Ambrosino *et al.* [KLOE Collaboration], JHEP **0712** (2007) 105
[16] M. Moulson, arXiv:1411.5252 [hep-ex].
[17] F. Ambrosino *et al.* [KLOE Collaboration], JHEP **0804** (2008) 059
[18] KTeV Collaboration, Phys. Rev. Lett. 88 (2002), p. 181601-181606
[19] D. Babusci *et al.* [KLOE Collaboration], Phys. Lett. B **723** (2013) 54
[20] C. Patrignani *et al.* [Particle Data Group], Chin. Phys. C **40** (2016) no.10, 100001.
[21] A. Einstein, B. Podolsky, N. Rosen, Physical Review **47** (1935) 777.
[22] *Handbook on neutral kaon interferometry at a Phi-factory*, editor: A. Di Domenico, Frascati Physics Series **43** (2007)
[23] A. Di Domenico *et al.* [KLOE Collaboration], J. Phys. Conf. Ser. **171** (2009) 012008
[24] V. A. Kostelecký, Phys. Rev. Let. **80** (1998) 1818
[25] V. A. Kostelecký, Phys. Rev. D **61** (1999) 016002
[26] V. A. Kostelecký, Phys. Rev. D **64** (2001) 076001
[27] D. Babusci *et al.* [KLOE-2 Collaboration], Phys. Lett. B **730** (2014) 89
[28] J. Bernabeu, A. Di Domenico and P. Villanueva-Perez, Nucl. Phys. B **868** (2013) 102
[29] J. Bernabeu, A. Di Domenico and P. Villanueva-Perez, JHEP **1510** (2015) 139

Decoherence and discrete symmetries in deformed relativistic kinematics

Michele Arzano[1,⋆]

[1] *Dipartimento di Fisica and INFN, "Sapienza" University of Rome, P.le A. Moro 2, 00185 Roma, EU*

Abstract. Models of deformed Poincaré symmetries based on group valued momenta have long been studied as effective modifications of relativistic kinematics possibly capturing quantum gravity effects. In this contribution we show how they naturally lead to a generalized quantum time evolution of the type proposed to model fundamental decoherence for quantum systems in the presence of an evaporating black hole. The same structures which determine such generalized evolution also lead to a modification of the action of discrete symmetries and of the CPT operator. These features can in principle be used to put phenomenological constraints on models of deformed relativistic symmetries using precision measurements of neutral kaons.

1 Introduction

Ordinary time evolution in a closed quantum system preserves purity: given a time evolution operator S the density matrices of the system before, ρ_{in}, and after, ρ_{fin}, the evolution are related by $\rho_{fin} = S\rho_{in}S^{\dagger}$. The operator S is unitary, $SS^{\dagger} = 1$, and, if the initial state is pure i.e. $\mathrm{Tr}\,\rho_{in}^2 = 1$, then $\mathrm{Tr}\,\rho_{fin}^2 = 1$ and thus *purity is eternal*.

The discovery of black hole quantum radiance [1] raised a conceptual puzzle concerning the fate of quantum states after the complete evaporation of the black hole. As Hawking first suggested [2, 3], the full disappearance of a black hole via quantum evaporation, could allow for the evolution of a pure quantum state into a mixed one in contrast with the well established picture recalled above. Hawking's first attitude was to accept *de facto* this possibility and postulate that in the presence of quantum gravitational effects the unitary evolution operator S should be replaced by a "superscattering" operator $\$$ evolving states according to $\rho_{fin} = \$\rho_{in} \neq S\rho_{in}S^{\dagger}$ so that starting from a pure state $\mathrm{Tr}\,\rho_{in}^2 = 1$ one could in principle evolve to a mixed state $\mathrm{Tr}\,\rho_{fin}^2 \leq 1$

Possible phenomenological signatures of such generalized quantum evolution in neutral kaon systems and for neutron interferometry were first discussed by Ellis, Hagelin, Nanopoulos and Srednicki in [4]. Their approach relied on the analysis of the dynamics associated to $\$$ in terms of a differential equation for ρ

$$\dot{\rho} = \not{H}\rho \neq -i[H,\rho]\,, \tag{1}$$

generalizing the usual Von Neumann evolution equation.

⋆e-mail: michele.arzano@roma1.infn.it

Building up on the results of [4] Banks, Peskin and Susskind [5] looked for a general form of the evolution term $\not{H}\rho$ assuming that hermiticity and trace of the density matrix, $\rho = \rho^\dagger$ and $\mathrm{Tr}\rho = 1$, are preserved by the evolution. The resulting evolution equation they found

$$\dot{\rho} = -i[H,\rho] - \frac{1}{2}h_{\alpha\beta}\left(Q^\alpha Q^\beta \rho + \rho\, Q^\beta Q^\alpha - 2Q^\alpha \rho\, Q^\beta\right), \tag{2}$$

had been already discovered few years before in the literature on open quantum systems and was known as *Lindblad equation* [6]. In (2) $h_{\alpha\beta}$ is a hermitian matrix of constants and Q^α form a basis of hermitian matrices. In the present contribution we will show how models of deformed kinematics based on Lie group-valued momenta naturally allow for deformed evolution equations of Lindblad type and are thus amenable to phenomenological constraints from precision measurements of fundamental decoherence. Since the seminal work [4] the neutral kaon system has been considered as one of the most efficient candidates to test fundamental decoherence (see e.g. [7] and [8] for a recent review). Over the years it has become clear that these systems provide also a natural testbed for departures from CPT invariance (see Prof. Bernabeu's contribution to these proceedings) and, in this light, in the final part of this contribution we will discuss recent results concerning the formulation of deformed discrete symmetries.

The theoretical framework in which group-valued momenta emerge most naturally is in the study of the kinematics of point particles coupled to gravity in 2 + 1-dimensional space-time. In the next Section we will begin with a brief description how how deformed kinematics arises in such setting.

2 Group valued momenta from three-dimensional gravity

A well known feature of general relativity in three space-time dimensions is that it does not admit local degrees of freedom [9, 10]. The geometry of a solution of the theory representing a point particle at rest is described by the *conical* metric

$$ds^2 = -dt^2 + dr^2 + (1 - 4Gm)^2 r^2 d\varphi^2\,, \tag{3}$$

where G is the three dimensional Newton's constant (of dimension of inverse mass) and m is the mass of the particle [10].

Such conical geometry is characterized by a deficit angle $\alpha = 8\pi Gm$ proportional to the particle's mass and the tip of the cone, a singular point of the geometry, is identified with the location of the particle. In other words due to the topological nature of the theory a localized point particle is represented by a conical space-time defect. This suggests that the kinematical properties of the particle must be encoded in inherently "non-local" quantities capturing their topological nature. Indeed as shown above the mass of the particle m is proportional to the deficit angle α which can be "measured" by calculating the holonomy of the flat connection along a circular path around the defect. Such holonomy is a group element belonging to $SL(2,\mathbb{R})$, the gauge group of the theory, and it is indeed a rotation by α which we denote by h_α. The *rest energy* of the particle is thus represented by a $SL(2,\mathbb{R})$ group element (see [11] and [12] for more technical details). According to this picture one would intuitively expect that the general three-momentum of a moving particle will be given by another $SL(2,\mathbb{R})$ group element obtained by "boosting" the rest momentum h_α. This is indeed the case and it can be shown [12, 13] that the three-momentum of a point particle of mass m is given by the group element obtained by boosting with a generic $g \in SL(2,\mathbb{R})$, via group conjugation, the rest three-momentum h_α:

$$h = g h_\alpha g^{-1}\,. \tag{4}$$

In order to find the actual components of such momentum one needs to introduce a parametrization of the group $SL(2,\mathbb{R})$. A typical choice is to write h using a basis of real two by two matrices given by the identity $\mathbb{1}$ and three linearly independent traceless matrices γ_μ:

$$h = u\mathbb{1} + \frac{p^\mu}{\kappa}\gamma_\mu \tag{5}$$

where $\kappa = (4\pi G)^{-1}$ is a constant with dimension of energy obtained from Newton's constant. The unit determinant condition on h leads to the condition $u^2 + p^2/\kappa^2 = 1$ where p^2 is the square of the three vector p^μ according to the usual Minkwoski metric with signature $(+,-,-)$. This is nothing but the equation defining three-dimensional anti-de Sitter space, the manifold of $SL(2,\mathbb{R})$, via embedding in a four-dimensional Minkowski space.

The non-abelian Lie group structure of momentum space has dramatic consequences at the quantum level for what concerns the action of symmetry generators on the Hilbert space of the theory. In order to appreciate this let us first recall some basic concepts in elementary quantum theory.

3 Hopf algebras and quantum mechanics

In the usual formulation of relativistic quantum mechanics one-particle states belong to a Hilbert space $\mathcal{H}$ which carries a unitary irreducible representation of the Poincaré group. Accordingly there is a well defined action of the symmetry generators belonging to the Poincaré algebra on these states. As it is well known, we use such action to label the states, for example in terms of their eigenvalue with respect to translation generators

$$P_\mu|k\rangle = k_\mu|k\rangle \tag{6}$$

where k_μ are on-shell components of the energy-momentum vector. The notion of dual representation of a Lie algebra allows us to define an action of the elements of the Poincaré algebra and in particular of P_μ on the dual Hilbert space $\mathcal{H}^*$ spanned by bras $\langle k|$ as

$$P_\mu\langle k| \equiv -k_\mu\langle k|\,, \tag{7}$$

notice that $P_\mu\langle k| = \langle k|(-k_\mu) = \langle k|(-P_\mu)$ thus we can define an operation on P_μ called the *antipode* such that

$$P_\mu\langle k| \equiv \langle k|\,S(P_\mu)\,, \tag{8}$$

and thus $S(P_\mu) = -P_\mu$.

Analogously one normally defines the action of the symmetry generators on multi-particle states, elements of tensor products of $\mathcal{H}$ in terms of a Leibniz rule. For example for a two-particle state $|k_1\rangle \otimes |k_2\rangle \in \mathcal{H} \otimes \mathcal{H}$ we have

$$P_\mu(|k_1\rangle \otimes |k_2\rangle) = P_\mu|k_1\rangle \otimes |k_2\rangle + |k_1\rangle \otimes P_\mu|k_2\rangle\,, \tag{9}$$

and again we can define an operation on P_μ, known as the *co-product*, which reproduces such action $\Delta P_\mu|k_1\rangle \otimes |k_2\rangle \equiv P_\mu(|k_1\rangle \otimes |k_2\rangle)$ and thus

$$\Delta P_\mu = P_\mu \otimes \mathbb{1} + \mathbb{1} \otimes P_\mu\,. \tag{10}$$

The operations of antipode and co-product are borrowed from the theory of Hopf algebras but, as we just showed, they are "built in" in our everyday quantum mechanics.

Two observations are crucial for our purposes at this point. The first is that these basic notions suffice to derive the action of symmetry generators, and in particular of P_μ, on a generic operator/observable of the theory. Let us take for example the projection operator $\pi_k = |k\rangle\langle k|$ on the state $|k\rangle$. This is a simple yet fundamental operator since it can be employed to express general operator in terms of their one-particle matrix elements [14]. Using the rules recalled above we have

$$P_\mu(\pi_k) = P_\mu(|k\rangle\langle k|) = P_\mu(|k\rangle)\langle k| + |k\rangle P_\mu(\langle k|) = P_\mu|k\rangle\langle k| - |k\rangle\langle k|P_\mu = [P_\mu, \pi_k]\,, \tag{11}$$

in other words the usual co-product and antipode for P_μ lead to the familiar *adjoint action* of symmetry generators on operators. The second observation is that the framework of Hopf algebras allows for non-trivial modifications of antipode and co-product which will lead to a deformed adjoint action. Models of relativistic kinematics in which momenta belong to a non-abelian Lie group provide a prototypical example of a modified action of translation generators on states and operators.

Indeed let us consider one-particle states labelled by elements of a non-abelian Lie group $|\pi\rangle$. One can define translation generators P_μ whose eigenvalues on these sates are coordinates functions $\mathcal{P}_\mu(\pi)$ on the group manifold

$$P_\mu\,|\pi\rangle = \mathcal{P}_\mu(\pi)\,|\pi\rangle\,. \tag{12}$$

We see that already at the level of dual representation or bra states, the non-trivial group operations enter the game. Indeed one must have that

$$P_\mu\langle\pi| = \mathcal{P}_\mu(\pi^{-1})\langle\pi| \equiv \langle\pi|\,S(P_\mu) \tag{13}$$

in other words the action on bra states (encoded in the antipode $S(P_\mu)$) leads to eigenvalues which read the coordinates of the *inverse* group element $\mathcal{P}_\mu(\pi^{-1})$. Similarly the action on multiparticle states is non-trivial

$$P_\mu(|\pi_1\rangle\otimes|\pi_2\rangle) = \mathcal{P}_\mu\,(\pi_1\cdot\pi_2)|\pi_1\rangle\otimes|\pi_2\rangle \equiv \Delta P_\mu\,|\pi_1\rangle\otimes|\pi_2\rangle \tag{14}$$

reflecting a non-abelian addition rule $\oplus$ for the momentum eigenvalues defined as follows

$$\mathcal{P}_\mu(\pi_1\cdot\pi_2) \equiv \mathcal{P}_\mu(\pi_1)\oplus\mathcal{P}_\mu(\pi_2) \neq \mathcal{P}_\mu(\pi_2\cdot\pi_1)\,, \tag{15}$$

where $\cdot$ denotes the (non-abelian) group composition. Notice that the non-trivial inverse operation (13) dictated by the antipode is required in order for the following consistency relation to hold

$$\mathcal{P}_\mu(\pi)\oplus\mathcal{P}_\mu(\pi^{-1}) = \mathcal{P}_\mu(\pi\cdot\pi^{-1}) = \mathcal{P}_\mu(\mathbb{1}) = 0\,. \tag{16}$$

Thus for a Lie group momentum space we have momentum eigenstates of the corresponding quantum system which combine in a non-trivial way reflecting the non-abelian structure of the group. These new rules reflect a deformed action of translation generators on the Hilbert space which are natural in the theory of Hopf algebra generalizations of Lie algebras and Lie groups (see e.g. [15] and references therein).

4 Deformed translations and Lindblad evolution in three dimensions

A concrete example of such non-trivial structures is given by the action of translation generators in the case when momentum space is the $SL(2,\mathbb{R})$ Lie group [16]. One can define translation generators [17] whose eigenvalues are given by embedding coordinates p^μ introduced above. In this case the coproduct and antipode read

$$\Delta P_\mu = P_\mu\otimes\mathbb{1} + \mathbb{1}\otimes P_\mu + \frac{1}{\kappa}\,\epsilon_{\mu\nu\sigma}P^\nu\otimes P^\sigma + O\left(\frac{1}{\kappa^2}\right), \qquad S(P_\mu) = -P_\mu\,, \tag{17}$$

where for simplicity we gave only the leading order in $1/\kappa$ expression for the coproduct while, for this choice of translation generators, the antipode is trivial at all orders in $1/\kappa$. As we illustrated in the previous section coproduct and antipode determine the action of symmetry generators on generic operators. For symmetry generators belonging to a non-trivial Hopf algebra, as for the case under study, the usual adjoint action on an operator ρ generalizes to the "quantum" adjoint action [18]

$$\mathrm{ad}_{P_\mu}(\rho) = (\mathrm{id} \otimes S)\Delta(P_\mu) \diamond \rho \tag{18}$$

where $(a \otimes b) \diamond c \equiv a\,c\,b$ and id is the identity operator. It is immediate to check that such action reduces to the ordinary commutator for trivial coproduct and antipode. This general expression can be read off the action (11) in the case when the operator ρ is the density matrix describing a pure state and easily generalizes to a generic quantum state.

Let us now focus on the action of the time translation generator P_0 in order to derive the infinitesimal time evolution equation for the density matrix. Plugging (17) into (18) we obtain

$$\mathrm{ad}_{P_0}(\rho) = [P_0, \rho] - \frac{1}{\kappa}\,\epsilon_{0ij} P^i \rho\, P^j\,, \tag{19}$$

leading to a three-dimensional Lindlblad equation

$$\dot{\rho} = -i[P_0, \rho] - \frac{1}{2} h_{ij} \left(P^i P^j \rho + \rho\, P^j P^i - 2P^j \rho\, P^i\right) \tag{20}$$

where the "decoherence" matrix h is given by

$$h = \frac{i}{\kappa} \begin{pmatrix} 0 & 0 & 0 \\ 0 & 0 & 1 \\ 0 & -1 & 0 \end{pmatrix}. \tag{21}$$

This simple three-dimensional example shows that the non-trivial action of translation generators imposed by the non-abelian nature of the group-valued momenta can naturally accommodate departures from the Von Neumann evolution equation of ordinary quantum mechanics. It is worth noticing that, starting from the seminal work [5], quantum evolution equations of the Lindblad type have always been discussed as the most general linear evolution equations that can allow for the evolution of pure states into mixed states. A main conceptual issue raised by these approaches was the lack of a more fundamental model form which these types of equations could be derived in a low energy limit [19]. Here we showed that, rather suggestively, a Lindblad equation can be obtained from leading order (remember that the $1/\kappa$ parameter is proportional to the inverse Planck mass) Hopf algebra deformations of relativistic symmetries emerging in three-dimensional gravity.

Below we illustrate how these structures can be generalized to the case of four space-time dimensions.

5 Deformed symmetries and quantum evolution in four dimensions

Historically Hopf algebra deformations of relativistic symmetries where first derived for the four dimensional Poincaré algebra. In [20], through a rather involved application of the procedure of quantum deformations of Lie algebras the authors obtained what is now known as the κ-Poincaré algebra. As for the three dimensional case discussed in the previous sections the deformation parameter κ carries dimension of energy and can be thought as a UV scale, possibly Planckian if one argues that such deformations emerge from the flat space limit of quantum gravity [21, 22]. It was only later realized that these deformations of relativistic symmetries can be interpreted in terms of group valued momenta very much in the spirit of the three-dimensional example discussed above. As was first suggested in [23] and later thoroughly explored in [24, 25] the full structure of the κ-Poincaré algebra can be obtained in terms of the properties of a momentum space described by the *abelian nilpotent* group $AN(3)$. Such group can be obtained from the Iwasawa decomposition of the five-dimensional Lorentz group $SO(4,1) \simeq SO(3,1)AN(3)$ [24] and this very structure ensures an action of the four-dimensional Lorentz group $SO(3,1)$ on the group valued momenta. Geometrically the $AN(3)$ group spans a submanifold of the four-dimensional de Sitter space dS_4 defined in an embedding five-dimensional Minkowski space as the subspace

$$-p_0^2 + p_1^2 + p_2^2 + p_3^2 + p_4^2 = \kappa^2\,; \qquad p_0 + p_4 > 0\,. \tag{22}$$

Notice how the deformation parameter κ determines the curvature of the momentum manifold.

In full analogy with the three dimensional case one associates translation generators[1] P_μ to the embedding coordinates p_μ. Their co-products and antipodes at leading order in κ are given by

$$\begin{aligned}
\Delta(P_0) &= P_0 \otimes \mathbb{1} + \mathbb{1} \otimes P_0 + \frac{1}{\kappa} P_m \otimes P_m\,,\\
\Delta(P_i) &= P_i \otimes \mathbb{1} + \mathbb{1} \otimes P_i + \frac{1}{\kappa} P_i \otimes P_0\,,\\
S(P_0) &= -P_0 + \frac{1}{\kappa}\vec{P}^2\,,\\
S(P_i) &= -P_i + \frac{1}{\kappa} P_i P_0\,.
\end{aligned}$$

This choice of translation generators is known as the "classical basis" of the κ-Poincaré algebra [29] since the action of the Lorentz generators on P_μ and the corresponding coordinates p_μ is undeformed and the mass Casimir of the algebra is the usual one $P_0^2 - \vec{P}^2 = const$. In other words at the algebra level the structures remain the same as in the ordinary Poincaré algebra and all the non-trivial structures are found at the "co-algebra" level i.e. in the form of the co-products and antipodes. These however will suffice to alter the action of translation generators on operators and, as in the previous section, on the density matrix ρ. Indeed a straightforward calculation shows that the time evolution equation is now deformed to a *non-symmetric* Lindblad equation [16]

$$\dot{\rho} = -i[P_0, \rho] + \frac{i}{\kappa} P_m \rho P_m - \frac{i}{\kappa} \rho \vec{P}^2 \tag{23}$$

where the $\vec{P}^2\rho$ term is missing. Such apparently odd form of the evolution equation is required by the non-trivial form of the antipode for P_0 (unlike the three dimensional example where it was undeformed). Indeed the antipode enters the expression of the hermitian adjoint of the adjoint action of P_0 via

$$(\mathrm{ad}_{P_0}(\cdot))^\dagger \equiv \mathrm{ad}_{S(P_0)}(\cdot) \tag{24}$$

and one can easily verify that, once this property is taken into account, (23) indeed preserve hermiticty of the density matrix ρ as it must be for a proper quantum evolution equation.

We have thus showed that it is possible to realize a generalized quantum evolution in four space-time dimensions when translation generators are deformed according to a particular realization of the κ-Poincaré algebra. In [19] it was claimed that a viable Lindblad-like evolution equation (an equation

[1] Choosing different parametrizations of $AN(3)$ leads, in general, to different choices of translation generators corresponding to dfferent "bases" of the κ-Poincaré algebra (See e.g. [26, 27] and [28] for a possible physical interpretation.)

in which the operators appearing in the decoherence term commute with P_0) would necessary conflict with Lorentz covariance. Is our evolution equation (23) covariant under the action of Lorentz boosts? It turns out that also in this case the role of a non-trivial antipode is fundamental. Indeed it can be shown that, as for P_0, the adjoint action of boosts on an operator is deformed according to

$$\mathrm{ad}_{N_i}(\rho) = [N_i, \rho] + \frac{1}{\kappa}[P_0, \rho]N_i + \frac{1}{\kappa}\epsilon^{ijm}[P_j, \rho]M_m\,, \tag{25}$$

and the adjoint actions of N_i and P_0 satisfy

$$\mathrm{ad}_{\mathrm{ad}N_i(P_0)}(\cdot) = \mathrm{ad}_{N_i}(\mathrm{ad}_{P_0})(\cdot) - \mathrm{ad}_{P_0}(\mathrm{ad}_{N_i})(\cdot) \tag{26}$$

thus ensuring that (23) satisfies a deformed notion of covariance. In this sense the non-trivial notion of covariance encoded in the Hopf algebraic structure of κ-deformations allows one to evade the no-go results of [19].

Besides investigating the actual physical viability of an evolution equation like (23) (checking, for example, whether it preserves positivity of the density matrix and further understanding the interplay between hermiticity and antipode from a quantum mechanical point of view), the most exciting development would be to phenomenologically constraint such models of deformed quantum evolution. As it stands, experimental searches for departures from ordinary time evolution in the form a Linbdlad equation have been ongoing for decades and have reached a striking sensitivity especially in precision measurements of neutral kaons and of the kaon-antikaon system, K^0-$\bar{K}^0$ [30]. It turns out that in order to properly appreciate the effects that a deformed evolution like (23) might have on the K^0-$\bar{K}^0$ system it will be crucial to have a clearer picture of the various ways the new features of deformed symmetries might alter the standard description of their states and evolution. In particular the role of discrete symmetries is central in this context and indeed the K^0-$\bar{K}^0$ system provides also a sensitive experimental probe for departures from CPT symmetry [30]. To this end in the final part of this contribution we will illustrate a first attempt at the formulation of discrete symmetry in a κ-deformed setting.

6 Deformed discrete symmetries

We report here on our recent proposal [31] to construct discrete symmetries associated to deformed symmetries based on group valued momenta and in particular for κ-Poincaré. The strategy followed in [31] has been to look for consistency with basic physical requirements that such discrete transformations should satisfy.

We start from parity and time-reversal transformations. The main physical requirement for the action of a parity transformation on momenta is that the total linear momentum of a one-particle state and its parity image *must vanish*. Following the discussion of Section 3 we see that such requirement leads to the parity image of the space translation generators to transform via their antipode

$$\mathbb{P} : P_i \to S(P)_i\,. \tag{27}$$

In order to define the action of the parity operator $\mathbb{P}$ on the remaining symmetry generators we supplement the input from the physical requirement from (27) with two algebraic consistency conditions. The first is that antipode mapping of one generator requires antipode mapping of all the other symmetry generators in order to preserve the structure of the commutators of the algebra. The second is a "correspondence" requirement that in the limit $\kappa \to \infty$, in which one obtains the usual Poincaré algebra, the usual action of discrete transformations on symmetry generators should be recovered. These conditions determine the following parity transformations on the elements of the κ-Poncaré algebra:

$$\mathbb{P}(P_i) = S(P)_i = -P_i + \frac{P_0 P_i}{\kappa} + O\left(\frac{1}{\kappa^2}\right); \quad \mathbb{P}(P_0) = -S(P)_0 = P_0 - \frac{\mathbf{P}^2}{\kappa} + O\left(\frac{1}{\kappa^2}\right)$$

$$\mathbb{P}(M_i) = -S(M)_i = M_i; \quad \mathbb{P}(N_i) = S(N)_i = -N_i + \frac{1}{\kappa}\left(-P_0 N_i + \epsilon_{ijk} P_j M_k\right) + O\left(\frac{1}{\kappa^2}\right). \tag{28}$$

Notice how the generator of rotations M_i remains untouched by parity, as in the standard case, while the generators of boosts acquire energy-momentum dependent terms. One can proceed in a similar fashion for the time reversal transformation operator but now requiring that in the limit $\kappa \to \infty$ the sign of the generator of rotations M_i should be changed. The resulting time reversal of symmetry generators is given by

$$\mathbb{T}(P_i) = S(P)_i, \quad \mathbb{T}(P_0) = -S(P)_0$$

$$\mathbb{T}(M_i) = S(M)_i, \quad \mathbb{T}(N_i) = -S(N)_i \tag{29}$$

in which the only difference from (28) is the change in sign in the time reversed rotation and boost generators.

To complete the picture we should construct an appropriate charge conjugation operator. To this extent let us recall that for a complex quantum scalar field anti-particle states belong to the *complex conjugate* space $\bar{\mathcal{H}}$ associated to the usual one-particle Hilbert space $\mathcal{H}$ [32]. The ordinary charge conjugation operator $\mathbb{C}$ provides a map bewteen these two spaces

$$\mathbb{C} : \phi(k) \in \mathcal{H} \to \bar{\phi}(-k) \in \bar{\mathcal{H}} \tag{30}$$

where $\phi(k)$ and $\bar{\phi}(k)$ are representatives of particle and antiparticle states in terms functions of on-shell momenta. The crucial point to note is that the complex conjugate space $\bar{\mathcal{H}}$ is isomorphic to the dual Hilbert space $\mathcal{H}^*$ [31] and thus symmetry generators must act on antiparticle states via their antipodes. Imposing that in the $\kappa \to \infty$ one recovers the ordinary charge conjugation map we obtain

$$\mathbb{C}(P_i) = -S(P)_i, \quad \mathbb{C}(P_0) = -S(P)_0$$

$$\mathbb{C}(M_i) = -S(M)_i, \quad \mathbb{C}(N_i) = -S(N)_i\,. \tag{31}$$

We thus see that all three discrete symmetries act on boosts and translations in a non-trivial way with the leading order departures from their usual actions suppressed by the inverse UV scale κ. Combining these relations together we can finally write down the explicit action of the deformed CPT operator on the generators of the κ-Poincaré algebra

$$\mathbb{CPT}(P_i) = P_i - \frac{P_0 P_i}{\kappa} + O\left(\frac{1}{\kappa^2}\right), \quad \mathbb{CPT}(P_0) = P_0 - \frac{\mathbf{P}^2}{\kappa} + O\left(\frac{1}{\kappa^2}\right) \tag{32}$$

$$\mathbb{CPT}(M_i) = -M_i, \quad \mathbb{CPT}(N_i) = -N_i + \frac{1}{\kappa}\left(-P_0 N_i + 3P_i + \epsilon_{ijk} P_j M_k\right) + O\left(\frac{1}{\kappa^2}\right). \tag{33}$$

In summary we have showed that symmetry deformations involving a non-trivial antipode imply a deformed action of discrete symmetries and of the CPT operator on symmetry generators. Let us notice that in the framework we illustrated the same structures which determine departures from ordinary CPT symmetry are also responsible for a generalized quantum evolution allowing, in principle, a map of pure states into mixed states. We should point out that this scenario differs from models in which possible fundamental decoherence is linked to violations of CPT symmetry according to a known theorem by Wald [33, 34]. In our framework CPT symmetry is not broken, we do have a well defined action of the CPT operator but this is deformed in the same way the action of translations and boosts on states and operators is deformed.

Of course, as for the generalized quantum evolution discussed in Section 5, much more needs to be understood for these new discrete symmetries starting from their action on fields of generic spin and the corresponding quantum states. Also in this case it will be very interesting to explore the possible phenomenological signatures of these departures from ordinary discrete transformations. The neutral kaon system, being already an extremely sensitive probe of for CPT violations, will be the main target of our future investigations.

7 Discussion

We illustrated how deformed symmetries associated to group valued momenta naturally predict departures from ordinary quantum evolution and CPT at energy scales of the order of the deformation parameter κ determined by the curvature of the momentum group manifold. What we showed are just the first steps illustrating two novel theoretical developments in the field of symmetry deformations and much still remains to be done to fully appreciate the actual physics beyond deformed quantum evolution and discrete symmetries. Certainly the most remarkable aspect of this new window on fundamental decoherence and departures from CPT symmetry is the potential to phenomenologically constrain the model with precision measurements of neutral kaons. Over the past fifteen years virtually all of the efforts concerning the phenomenology of models based on departures from standard relativistic kinematics have focused on establishing bounds from the highest energy astrophysical observation (see e.g. [35, 36]). The theoretical developments reported here open the way to new potential experimental constraints for κ-deformed symmetries, one of the most studied models of deformed relativistic symmetries.

References

[1] S. W. Hawking, Commun. Math. Phys. **43**, 199 (1975) [Erratum-ibid. **46**, 206 (1976)].
[2] S. W. Hawking, Phys. Rev. D **14**, 2460 (1976).
[3] S. W. Hawking, Commun. Math. Phys. **87**, 395 (1982).
[4] J. R. Ellis, J. S. Hagelin, D. V. Nanopoulos and M. Srednicki, Nucl. Phys. B **241**, 381 (1984).
[5] T. Banks, L. Susskind and M. E. Peskin, Nucl. Phys. B **244**, 125 (1984).
[6] G. Lindblad, Commun. Math. Phys. **48**, 119 (1976).
[7] J. R. Ellis, J. L. Lopez, N. E. Mavromatos and D. V. Nanopoulos, Phys. Rev. D **53**, 3846 (1996) [hep-ph/9505340].
[8] A. Di Domenico, arXiv:1608.00241 [hep-ph].
[9] A. Staruszkiewicz, Acta Phys. Polon. **24**, 735 (1963).
[10] S. Deser, R. Jackiw and G. 't Hooft, Annals Phys. **152**, 220 (1984).
[11] H. J. Matschull and M. Welling, Class. Quant. Grav. **15**, 2981 (1998) [arXiv:gr-qc/9708054].
[12] B. J. Schroers, PoS QG **-PH**, 035 (2007) [arXiv:0710.5844 [gr-qc]].
[13] M. Arzano, D. Latini and M. Lotito, SIGMA **10**, 079 (2014) [arXiv:1403.3038 [gr-qc]].
[14] M. Arzano, Phys. Rev. D **77**, 025013 (2008) [arXiv:0710.1083 [hep-th]].
[15] M. Arzano, arXiv:1212.1097 [hep-th].
[16] M. Arzano, Phys. Rev. D **90**, no. 2, 024016 (2014) [arXiv:1403.6457 [hep-th]].

[17] M. Arzano, J. Kowalski-Glikman and T. Trzesniewski, Class. Quant. Grav. **31**, no. 3, 035013 (2014) [arXiv:1305.6220 [hep-th]].
[18] H. Ruegg and V. N. Tolstoi, Lett. Math. Phys. **32**, 85 (1994) [hep-th/9406146].
[19] M. Srednicki, Nucl. Phys. B **410**, 143 (1993) [hep-th/9206056].
[20] J. Lukierski, A. Nowicki and H. Ruegg, Phys. Lett. B **293**, 344 (1992).
[21] G. Amelino-Camelia, L. Smolin and A. Starodubtsev, Class. Quant. Grav. **21**, 3095 (2004) [hep-th/0306134].
[22] J. Kowalski-Glikman, hep-th/0312140.
[23] S. Majid and H. Ruegg, Phys. Lett. B **334**, 348 (1994) [hep-th/9405107].
[24] J. Kowalski-Glikman and S. Nowak, hep-th/0411154.
[25] J. Kowalski-Glikman and S. Nowak, Class. Quant. Grav. **20**, 4799 (2003) doi:10.1088/0264-9381/20/22/006 [hep-th/0304101].
[26] J. Lukierski, H. Ruegg and V. N. Tolstoi, IC-94-250, C94-02-14.1.
[27] J. Kowalski-Glikman and S. Nowak, Phys. Lett. B **539**, 126 (2002) [hep-th/0203040].
[28] M. Arzano, Phys. Rev. D **83**, 025025 (2011) [arXiv:1009.1097 [hep-th]].
[29] A. Borowiec and A. Pachol, J. Phys. A **43**, 045203 (2010) [arXiv:0903.5251 [hep-th]].
[30] A. Di Domenico *et al.* [KLOE Collaboration], Found. Phys. **40**, 852 (2010).
[31] M. Arzano and J. Kowalski-Glikman, Phys. Lett. B **760**, 69 (2016) [arXiv:1605.01181 [hep-th]].
[32] Robert Geroch, *Quantum Field Theory: 1971 Lecture Notes*, Minkowski Institute Press, Montreal 2013
[33] N. E. Mavromatos, PoS KAON (2008) 041 [arXiv:0707.3422 [hep-ph]].
[34] R. M. Wald, Phys. Rev. D **21**, 2742 (1980).
[35] G. Amelino-Camelia and T. Piran, Phys. Rev. D **64**, 036005 (2001) [astro-ph/0008107].
[36] G. Amelino-Camelia, Living Rev. Rel. **16**, 5 (2013) [arXiv:0806.0339 [gr-qc]].

Status and prospects for the PADME experiment at LNF

Paola Gianotti[1,*]
on behalf of the PADME Collaboration**

[1] *Laboratori Nazionali di Frascati dell'INFN, via E. Fermi 40, I-00044 Frascati, Italy*

Abstract. The PADME collaboration aims to search for signals of a light dark photon A' using the beam of the LNF LINAC. The experiment, approved by INFN at the end of 2015, foresees to detect A', produced in the annihilation of positrons on a thin fixed target, by searching for missing mass signals. The detector construction is ongoing and should be completed within the end of 2017 in order to allow the collection of about 10^{13} positrons on target which are necessary to get a 10^{-3} sensitivity on the mixing parameter ε up to a dark photon mass of 23.7 MeV$/c^2$.

1 Introduction

Direct observations of dark matter signals are necessary to shed light on the cosmological and astrophysical evidences, in order to understand the nature of such unknown matter that at present has manifested itself only via gravitational interactions.

One of the present theoretical hypothesis foresees that the new dark particles are not directly connected with the Standard Model (SM) gauge fields, but only through mediators or "portals", linking our world with the new "secluded" sector [1]. A simple model that tries to explain dark matter, just adds an additional $U(1)$ symmetry, with its corresponding vector boson A'. All SM particles are neutral under this symmetry, and the new field couples to the charged particles of the SM with an effective charge εq, where q is the electromagnetic charge, induced by the kinetic mixing of the dark photon with the SM photon. If this dark photon A' would have a mass in the range from 1 to 100 MeV$/c^2$ and a coupling $\varepsilon \sim 10^{-3}$, it could also explain the discrepancy between theory and observation for the muon anomalous magnetic moment [2]. This possibility has been recently disproved under the assumption that the A' decays to SM particles only (for a recent review see [3]), but if the A' decays to dark sector particles, almost all of the available experimental constraints can be evaded and the dark photon is still a valuable explanation for the muon $(g-2)$ anomaly.

To test this hypothesis, PADME (**P**ositron **A**nnihilation into **D**ark **M**atter **E**xperiment) will search for invisible decays of the A' at the LNF LINAC. The experiment is designed to detect dark photons

*e-mail: paola.gianotti@lnf.infn.it

**G. Chiodini (INFN Lecce), A.P. Caricato, M. De Feudis, M. Martino, G. Maruccio, A. Monteduro, S. Spagnolo (INFN Lecce and Dip. Matematica e Fisica Università del Salento), R. Bedogni, B. Buonomo, F. Bossi, P. Gianotti, R. De Sangro, G. Finocchiaro, L.G. Foggetta, A. Ghigo, M. Palutan, G. Piperno, B. Sciascia, T. Spadaro (INFN Laboratori Nazionali di Frascati), G.V. Georgiev, V. Kozhuharov (University of Sofia and INFN Laboratori Nazionali di Frascati), U. Dosselli (INFN Padova), F. Ferrarotto, E. Leonardi, P. Valente (INFN Roma1), S. Fiore (INFN Roma1 and ENEA), G.C.Organtini, M.Raggi (INFN Roma1 and Dip. di Fisica Università "Sapienza" di Roma) produced in positron on fixed target annihilation $e^+e^- \to \gamma A'$ by measuring the final state missing mass.

The collaboration aims to complete the design and construction of the detector by the end of 2017 and to collect $\sim 10^{13}$ positrons of 550 MeV energy on target by the end of 2018. This will allow to reach a sensitivity $\varepsilon \sim 10^{-3}$ on the coupling, for dark photon masses up to 23.7 MeV$/c^2$.

In the following sections more details on the detector design and on the experiment schedule are given.

2 The PADME detector

The goal of the PADME experiment is to search for dark photons produced in the annihilation process of the positron beam of the DAΦNE LINAC with a thin carbon target and then identified using the missing mass technique [4]. The ordinary recoil photon is detected by a finely segmented electromagnetic calorimeter that also measures the particle four-momentum with high precision. This approach allows to identify the dark photon in case it is a stable particle, or if it decays to invisible particles, and also in case of visible decays, by disregarding the decay products of the dark photon. With the present foreseen positron energy of the LNF LINAC of 550 MeV, masses up to 23.7 MeV/c^2 could be explored. This region is currently unconstrained by direct measurement, and PADME has the unique opportunity to be the first dedicated experiment to search directly for the dark photon invisible decays in a model-independent way.

The basic elements of the PADME experiments are:

- a high intensity and low divergence positron beam, impinging on a thin, active target, capable of monitoring the beam spot;
- a vacuum chamber to avoid spurious positron interactions;
- a magnet to deflect the beam of positrons emerging from the target, with the additional task of measuring the momentum of the interacting positrons, thus improving the rejection of the Bremsstrahlung background;
- a finely-segmented, high-resolution e.m. calorimeter, to measure the momentum of the single SM photons.

A schematic view of the PADME layout can be seen in fig. 1.

The processes that will mainly take place in the beam-target interaction are Bremsstrahlung and $e^+e^- \to \gamma\gamma(\gamma)$. In order to cut out these background events, two extra components are crucial: a fast Small Angle Calorimeter (SAC) and a charged particle veto system. Fig. 2 shows the results of Montecarlo simulations performed to check background rejection capability requiring only one cluster in the main calorimeter (with energy in a range optimized depending on $m_{A'}$), no hits in the vetoes, and no photons with energy > 50 MeV in the SAC.

2.1 The positron beam

The PADME positron beam will be extracted from the warm LNF LINAC of the DAΦNE complex using the BTF (Beam Test Facility) line [5]. The typical size of the beam spot at the BTF is below 1 cm^2 RMS, while the energy spread is at the level of 1%. At present, this line is tuned to provides pulses at 510 MeV with a maximum length of 40 ns and a repetition rate up to 50 Hz. This is far from the requirements of the experiment where the interaction rate is ultimately dictated by the pile-up probability in the downstream calorimeters. The ideal PADME beam would be continuous or quasi-continuous, with an average time distance between two positrons larger than the characteristic

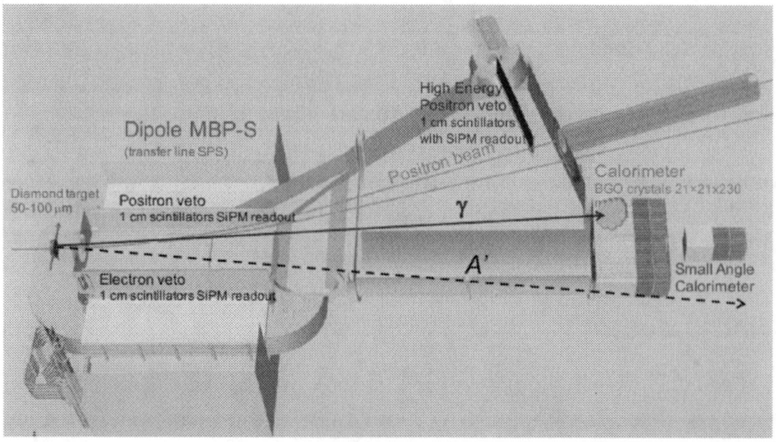

Figure 1: PADME detector layout. From left to right: the active target, the positron/electron vetoes inside the magnetic dipole, the high energy e^+ veto near the non-interacting beam exit, the e.m. calorimeters. The distance from the target to the first calorimeter is roughly 3 m.

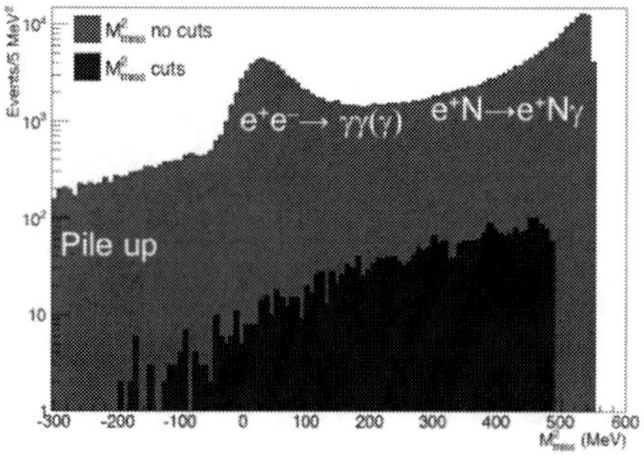

Figure 2: Montecarlo simulated squared missing mass for background events. In red with no selection cuts, in blue after all cuts are applied (see text for more details).

response time of the calorimeters. This can be achieved by reducing the beam intensity, but at the cost of increasing the running time, or by streaching the bunch. In practice, to perform the PADME experiment at the DAΦNE LINAC, the beam intensity and width have to be tuned exploiting the flexibility of the BTF beam transport system. During 2016, a new LINAC pulsing system has been installed to increase pulse duration. A dedicated machine development run has been performed and pulses of electrons of 250 ns have been produced. Further adjustments are foreseen to reduce the intensity variations and to keep the beam energy spread below 2%.

2.2 The PADME magnet

The positron beam will interact with a thin ($\sim$ 50-100 μm) carbon target passing through almost undisturbed. To drive away from the calorimeters acceptance these primary positrons, a magnetic field is necessary. Taking into account the PADME layout, this turned out in a dipole with a gap of $\pm$10 cm over a length of 1 m, providing a field of about 0.5 T.

The magnet, besides sweeping the primary beam, will play an important role in the background suppression. Scintillator strips equipping the sides of the dipole will detect and will measure the momentum of the positrons that have lost a significant fraction of their energy by Bremsstrahlung in the target.

A magnet with the requested characteristics has been found at CERN within the spare dipoles of the SPS transport line (MBP-S), and it has been shipped to LNF at the end of 2015. During 2016, the dipole vertical gap has been increased to 230 mm, to better suite the experiment needs, and the field map has been measured in different conditions (see fig. 3).

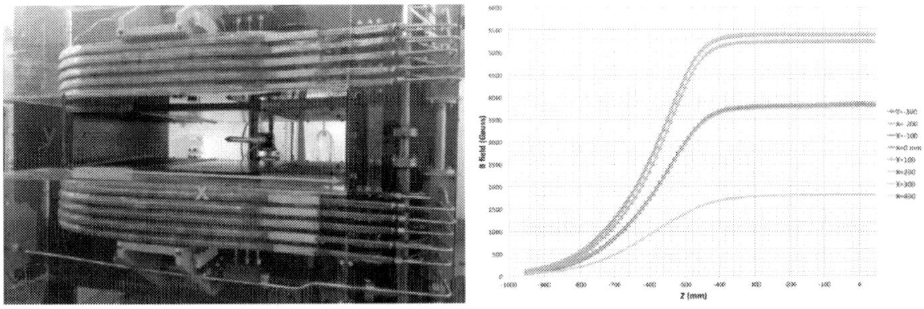

Figure 3: **left** the PADME magnet. **right** Longitudinal scan of magnetic field at different transverse positions.

2.3 The diamond target

For dark photon production and beam monitoring, a low-Z active thin target is necessary. The low-Z requirement, necessary to keep Bremsstrahlung at an acceptable level, combined with the need to monitor both the beam spot and intensity, has oriented the choice toward polycrystalline diamond sensors having on both sides orthogonal graphite strips for the readout. The advantage of this technique, developed by INFN Lecce [6], with respect to traditional metallic electrodes, lies in the fact that the target is entirely made of a homogeneous light material. The target thickness will be between 50 and 100 μm, to minimize the photon reinteraction, and the area will be about 2×2 cm^2, to match the size of the beam spot.

A first prototype of this target, 50 μm thick and with a strip pitch of 1 mm, has been built and tested at the BTF using positrons and electrons of 450 MeV (see fig. 4). The results of the test beam show a very good spatial resolution (0.2 mm RMS) for a beam intensity of 10^3 particles/bunch.

To further improve the beam monitor capability of the experiment, it is under development a system based on two planes of silicon pixel detectors placed up and down stream the diamond target. Each plane will consist of two MIMOSA 28 Ultimate chips, developed for the upgrade of the STAR vertex detector [7]. These devices integrate a Monolithic Active Pixel Sensor (MAPS) with a fast binary readout. Each sensor consists of a matrix of 928×960 pixels of 20.7 μm with a thickness of 50 μm. For the STAR experiment the chips, that dissipate ~150 mW/cm^2, operate in air without cooling. For PADME, the detectors should be placed in vacuum and a modified PCB is under development at the LNF electronics workshop. First prototypes of this new monitoring system will be tested during 2017.

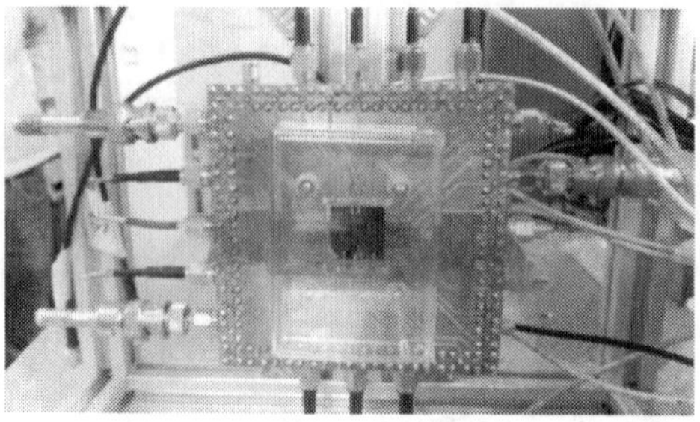

Figure 4: The first prototype of the PADME active target.

2.4 The electromagnetic calorimeters

The key point of the PADME experiment is to detect with high efficiency the energy, direction and timing of the single photons emitted from the target.

Therefore, e.m. calorimeters are necessary. The requirements are:

- high acceptance for the SM photons;
- energy resolution: 1-2% / $\sqrt{E}$ GeV;
- time resolution ~ 0.5 ns;
- angular resolution ~ 1÷2 mrad;

To avoid pile-up due to the high number of primary positrons, the main calorimeter (ECAL) will have a central hole to cut out most of the Bremsstrahlung photons produced in the interaction with the target. In the shadow of this hole, a second, small size and faster calorimeter (SAC), will be placed.

In the present design, the ECAL is a tight matrix of 616 crystals (21×21×230 mm^3) arranged in a cylinder of external radius 30 cm, with a central squared hole of 10 cm (see fig.5). The acceptance of the ECAL, for the recoil photons is 66%, for a 550 MeV positron beam.

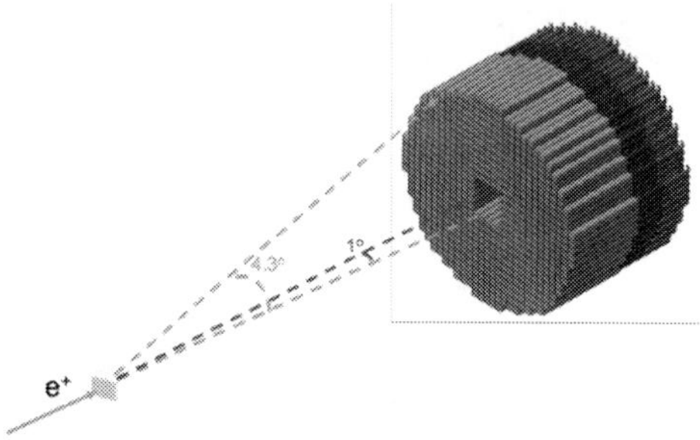

Figure 5: Layout of the PADME main e.m. calorimeter (ECAL). It consists of a tight matrix of 616 BGO crystals.

The choice of the scintillating material for the ECAL has been driven by different aspects: high density, small radiation length and Moliére radius (due to the calorimeter small dimensions), high light output (due to the low energy range), cost. Excluding the cost requirement, both LYSO and BGO would have been suitable, with a slight preference for the LYSO since it has a shorter decay time that would have mitigated the effect of event pileup caused by the beam structure. Nevertheless, the BGO crystals that were used by the L3 experiment at LEP [8], that are property of INFN, were available at the only cost of cutting them at the proper size. Therefore, after having checked that these crystals could be reconditioned and re-shaped, they have been adopted.

For the light collection, early tests, aimed at evaluating the best readout technology, showed that avalanche photodiodes (APDs), even with a relatively large area of 10×10 mm^2, have a gain that is insufficient to perform a high resolution energy measurement in the PADME energy rage (from a few to a few hundred MeV). The readout system will therefore be based on 19 mm diameter photomultipliers. On July 2016, XP1912 HZC Photonics tubes [9] have been tested at the BTF, on a calorimeter prototype made of 25 BGO crystals obtained by machining the L3 recovered ones. The 25 channels of the prototype were fed into a CAEN V1742 high-speed digitizer [10], based on the DRS4 chip, set to a sampling speed of 1 GS/s (1 ns/sample).

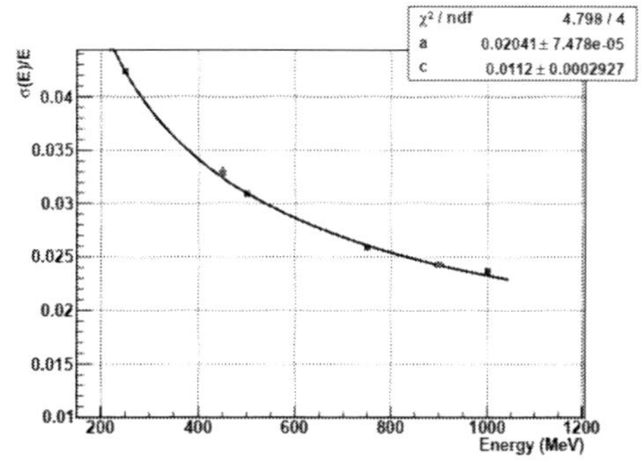

Figure 6: Energy resolution measured with a prototype of the PADME ECAL at the LNF BTF. Red and blue points refer to two different beam energies of 450 and 250 MeV, respectively. The line is the best fit curve whose parameters are quoted in the inset.

The test beam has demonstrated that the technical solutions chosen for this detector meet all the requirements. Fig. 6 shows the energy resolution obtained as a function of the photon energy. A complete overview of the test beam results can be found in [11].

Placed 50 cm downstream the ECAL, the SAC will be an array of fast detectors able to veto photons at small angles. BaF_2 or lead glass are the crystal material candidates. For the readout, fast photomultipliers will be used. A test beam to select the final components will be performed at the LNF BTF in summer 2017.

2.5 Charged particle vetoes

The detection of the positrons that have lost energy in the interaction with the target, would significantly reduce the Bremsstrahlung background. Depending on the energy of the photon radiated, the corresponding positron can be inside the magnet acceptance or close to the initial beam trajectory. This is why it is planned to place on the side walls of the dipole, and close to the positron beam exit, 3 groups of scintillator strips with SIPM readout (see fig.1). The total number of strips (10×10×200 mm^3) will be 250 (100 for the low energy positrons, 100 for the low energy electrons, 50 for the high energy positrons) and from the hit position a % accuracy on the particle momentum is expected, provided a time resolution of about 300 ps. Tests with the BTF beams of the veto detectors prototype are foreseen in 2017.

3 Experiment time line

After the approval at the end of 2015, an international collaboration has formed to perform the PADME experiment. At present it consists of about 30 people from 6 different institutions: INFN Lecce, Salento University, INFN LNF, University and INFN Roma, University of Sofia, MTA Atomki from Hungary. Contacts are ongoing with Cornell in the framework of the MAECI project PGR-226 and there are plans of cooperation to exchange expertise and knowhow. Cornell aims at creating an extracted positron beam line of 5.3 GeV on which, if realized, an upgraded version of the PADME experiment could be installed.

The design of the detector elements has been optimized and frozen in almost all aspects, and the procurement of the main components has already started in 2016.

All the crystals, necessary for the ECAL, are now ready to be machined and coupled to the corresponding photomultipliers.

A new 100 μm thick diamond target has been produced in Lecce and is currently under test with ^{241}Am radioactive source. The design of the vacuum vessel section and the actuators for the target and beam diagnostics upstream of the magnet are now being finalized.

The design of the main vacuum vessel is presently being validated by an extensive simulation campaign, and the executive design will be frozen by the beginning of the 2017, so that the construction can start.

The 15 tons magnet is ready to be moved to the BTF hall as soon as the experiment setup will be ready for the final installation.

The materials for the positron and electron scintillating veto detectors have been procured, and the construction will start in early 2017.

The requirements of the acquisition (waveform digitization) of all the sub-detectors have been defined, and the procurement of the electronics is ongoing. The acquisition and computing infrastructure preparation, including the required software, is well advanced.

In order to optimize the beam for the PADME experiment, in terms of pulse length, positron yield, spot and divergence, dedicated LINAC beam-time studies will be performed during 2017.

4 Conclusions

Searches for dark photons are well motivated by recently observed phenomena. The PADME experiment will be able to perform such a search exploiting the present LINAC of the DAΦNE facility in just one-two year of running and could be the first to directly constrain the region of invisible decays with a sensitivity on the A' mass up to 23.7 MeV$/c^2$ in a complete model independent way. With the present schedule of the detector construction the data taking could start in 2018.

An upgrade of the BTF positron beam energy will also allow to extend the sensitivity of PADME to higher mass values.

References

[1] B. Holdom, Phys. Let. **B166**, 196 (1986).
[2] M. Pospelov, Phys. Rev. **D 80**, 095002 (2009).
[3] J. Alexander *et al.*, arXiv:1608.08632[hep-ph].
[4] M. Raggi and V. Kozhuharov, Adv. High En. Phys. **2014** 959802 (2014).
[5] G. Mazzitelli *et al.*, Nucl. Instrum. Meth. **A 515**, 524 (2003).
[6] M. De Feudis *et al.*, Nuovo Cimento **39 C**, 254 (2016).

[7] I. Valin *et al.*, JINST **7**, C01102 (2012).
[8] B. Adeva *et al.*, Nucl. Instrum. Meth. **A 289**, 35 (1990).
[9] http://hzcphotonics.com.
[10] CAEN Mod. 1742, Technical Information Manual, rev6 06 February 2016.
[11] M. Raggi *et al.*, arXiv:1611.05649[physics.ins-det], submitted to Nucl. Instrum. Meth. **A**.

Kaon decays at KLOE-2

Giancarlo D'Ambrosio[1,⋆]

[1] *INFN-Sezione di Napoli, Via Cintia, 80126 Napoli, Italia*

Abstract. I review kaon decays. I introduce the flavor problem and possible solutions. Very rare kaon decays like $K \to \pi\nu\bar{\nu}$ are very important to this purpose: we study also $K \to \pi l^+ l^-$, $K \to \pi\pi ee$ where chiral dynamics is important to disentangle short distance effects. We have also studied lepton flavor (universality) violation in rare kaon decays and the Bardeen Buras Gerard approach to describe the $K^\pm \to \pi^\pm l^+ l^-$ form factor.

1 Introduction

Kaon decays are an important place to study non-perturbative aspects of QCD and test the Standard Model. Indeed some channels are completely dominated by long-distance dynamics, such as the CP-conserving amplitude for $K \to \pi\pi$ and others, like $K \to \pi\nu\bar{\nu}$ [1], which are described in terms of pure short-distance physics. In this review we will be mostly concerned with kaon decays involving electromagnetic interactions and thus long-distance phenomena are not negligible. However, as we shall see, it is still possible in these channels to extract the short-distance component with an accurate analysis. Indeed there are plenty of motivations to look for new physics (NP) in these kaon decays [2].

The channels which will be considered here are $K \to \pi\pi\gamma$, $K \to \pi\pi ee$, $K_{S,L} \to l^+l^-$, $K_{S,L} \to l^+l^-l^+l^-$ and $K_S \to \pi^+\pi^-l^+l^-$. Recently, Buras and collaborators, motivated by their previous work where weak hadronic matrix elements are evaluated in their model (BBG model) and also by recent lattice results, are investigating New Physics to explain the experimental value of ϵ' [2–5], Motivated by this interest we study also the $K^\pm \to \pi^\pm l^+l^-$ form factor in the BBG model [6].

2 THE ULTRA-RARE DECAY $K^+ \to \pi^+\nu\bar{\nu}$

Rare kaon decays furnish challenging MFV probes and will severely constrain additional flavor physics motivated by NP. SM predicts the $V - A \otimes V - A$ effective hamiltonian

$$\mathcal{H} = \frac{G_F}{\sqrt{2}}\frac{\alpha}{2\pi\sin^2\theta_W}(\underbrace{V_{cs}^* V_{cd}\, X_{NL}}_{\lambda x_c} + \underbrace{V_{ts}^* V_{td} X(x_t)}_{A^2\lambda^5\,(1-\rho-i\eta)x_t})\,\bar{s}_L\gamma_\mu d_L\,\bar{\nu}_L\gamma^\mu\nu_L, \tag{1}$$

$x_q = m_q^2/M_W^2$, θ_W the Weak angle and X's are the Inami-Lin functions with Wilson coefficients known at two-loop electroweak corrections [7]. $SU(2)$ isospin symmetry relates hadronic matrix elements

⋆e-mail: gdambros@na.infn.it

for $K \to \pi\nu\overline{\nu}$ to $K \to \pi l\overline{\nu}$ to a very good precision [8] while long distance contributions and QCD corrections are under control [7] and the main uncertainties is due to the strong corrections in the charm loop contribution. The structure in (1) leads to a pure CP violating contribution to $K_L \to \pi^0\nu\overline{\nu}$, induced only from the top loop contribution and thus proportional to $\Im m(\lambda_t)$ ($\lambda_t = V_{ts}^* V_{td}$) and free of hadronic uncertainties. This leads to the prediction [7]

$$\begin{aligned} \mathcal{B}(K_L \to \pi^0\nu\overline{\nu})_{\rm SM} = & \quad (2.9 \pm 0.2 \pm 0.0) \times 10^{-11}, \\ \mathcal{B}(K^+ \to \pi^+\nu\overline{\nu})_{\rm SM} = & \quad (8.3 \pm 0.3 \pm 0.3) \times 10^{-11}. \end{aligned} \tag{2}$$

where the first is the parametric uncertainty due to the error on $|V_{cb}|$, ρ and η, f_K, and the second error summarizes the theoretical uncertainties on non-perturbative physics and QCD higher order terms.

$K^{\pm} \to \pi^{\pm}\nu\overline{\nu}$ receives CP conserving contributions proportional to $\Re e(\lambda_c)$, and to $\Re e(\lambda_t)$ and a CP violating one proportional to $\Im m(\lambda_t)$. E949 Collaboration [9, 10] and E391a Collaboration [11] have then measured

$$\mathcal{B}(K^{\pm}) = \left(1.73^{+1.15}_{-1.05}\right) \times 10^{-10} \qquad \text{E949} \tag{3}$$

$$\mathcal{B}(K_L) < 2.6 \times 10^{-8} \text{ at } 90\% \text{ C.L.} \quad \text{E391aCollaboration} \tag{4}$$

The direct upper bound for the neutral decay can be improved with a theoretical analysis: the isospin structure of any $\overline{s}d$ operator (bilinear in the quark fields) leads to the model independent relation among $A(K_L \to \pi^0\nu\overline{\nu})$ and $A(K^{\pm} \to \pi^{\pm}\nu\overline{\nu})$ [12]; this leads to

$$\mathcal{B}(K_L \to \pi^0\nu\overline{\nu}) < 4\,\mathcal{B}(K^{\pm} \to \pi^{\pm}\nu\overline{\nu}) \tag{5}$$

The upcoming KOTO experiment [8, 13] for $K_L \to \pi^0\nu\overline{\nu}$ and NA62 at CERN [14] encourage theoretical investigations of extensions of the SM: these experiments probe deeply to the MFV scale [15]. More aggressive NP models can furnish substantial enhancements and be either discovered or ruled out [7, 16]! Further motivation to rare kaon decays program has come from the possible departure from the SM prediction of the experimental value of ϵ' [2–5, 9] : we have investigated a very predictive Susy scenario for $K \to \pi\nu\overline{\nu}$ [17, 18]

3 $K \to \pi\pi\gamma$ and $K \to \pi\pi ee$-decays

3.1 $K \to \pi\pi\gamma$

Let's discuss $K(p) \to \pi(p_1)\pi(p_2)\gamma(q)$ decays: according to gauge and Lorentz invariance we decompose $K(p) \to \pi(p_1)\pi(p_2)\gamma(q)$ decays, in electric (E) and magnetic (M) amplitudes [19]. Particularly interesting are the recent interesting NA48/2 data regarding $K^+ \to \pi^+\pi^0\gamma$ decays [20].

Due to the $\Delta I = 3/2$ suppression of the bremsstrahlung, interference between E_B and the electric dipole (E_{DE}) and magnetic transitions (M_{DE}) can be measured. Defining $z_i = p_i \cdot q/m_K^2$ $z_3 = p_K \cdot q/m_K^2$ and $z_3 z_+ = \frac{m_{\pi^+}^2}{m_K^2} W^2$, (W^2 is defined in this equation) we can study the deviation from bremsstrahlung

$$\frac{\partial^2\Gamma}{\partial T_c^* \partial W^2} = \frac{\partial^2\Gamma_{IB}}{\partial T_c^* \partial W^2}\left[1 + \frac{m_{\pi^+}^2}{m_K} 2Re\left(\frac{E_{DE}}{eA}\right) W^2 + \frac{m_{\pi^+}^4}{m_K^2}\left(\left|\frac{E_{DE}}{eA}\right|^2 + \left|\frac{M_{DE}}{eA}\right|^2\right) W^4\right], \tag{6}$$

where $A = A(K^+ \to \pi^+\pi^0)$. Study of the Dalitz plot relative to Eq. (6) has lead to measure separately the linear and the quadratic term in W^2: indeed integrating over the Dalitz plot and dividing for the IB

Branching ratio the right-hand side of Eq. (6) leads these two terms (linear and quadratic) respectively to the NA48/2 measurement of the two fractions, $Frac(INT)$ and $Frac(DE)$, (see Ref. [20])

NA48/2 $K^+ \to \pi^+\pi^0\gamma$ results

NA48/2		$T_c^* \in [0, 80]$ MeV
$Frac(DE)$	=	$(3.32\pm0.15 \pm 0.14)\times10^{-2}$
$Frac(INT)$	=	$(-2.35\pm0.35 \pm 0.39)\times10^{-2}$

3.2 $K \to \pi\pi ee$-decays

Historically kaon four body semileptonic decays, K_{e4} have been studied as a tool to tackle final state rescattering effects in $K \to \pi\pi$-decays: crucial to this goal has been finding an appropriate set of kinematical variables which would allow i) to treat the system as two body decay in dipion mass $M_{\pi\pi}$ and dilepton mass $M_{l^+l^-}$ and ii) to identify appropriate kinematical asymmetries to extract observables crucially dependent on final state interaction [21].

The traditional kinematical variables for the four body kaon semileptonic decay allow to write the four body phase space Φ in terms of the two two-body phase spaces: Φ_π and Φ_ℓ from [21, 22]. We extend the K_{e4} kinematical and dynamical description of the amplitude to describe the $K \to \pi\pi e^+e^-$ decays; in particular we write the long distance contribution as

$$\mathcal{M}_{LD} = \frac{e}{q^2}\left[\overline{u}(k_-)\gamma^\mu v(k_+)\right] H_\mu(p_1, p_2, q), \tag{7}$$

where H_μ is the electroweak hadronic vector, which can be written in terms of three form factors $F_{1,2,3}$:

$$H^\mu(p_1, p_2, q) = F_1 p_1^\mu + F_2 p_2^\mu + F_3 \varepsilon^{\mu\nu\alpha\beta} p_{1\nu} p_{2\alpha} q_\beta. \tag{8}$$

Then the kinematical distributions can be written in terms of the three form factors.

Short distance physics can be studied by the diplane angular asymmetry. However this observable is proportional to electric (bremsstrahlung) and magnetic interference, both contributions known already from $K_L \to \pi^+\pi^-\gamma$. In fact it was known that these contributions were large and they may obscure smaller but more interesting short distance physics effects.

We have performed a similar analysis for the decay $K^+ \to \pi^+\pi^0 e^+e^-$ trying to focus on i) short distance physics and ii) all possible Dalitz plot analyses to disentangle all possible interesting long and short distance effects [22]. The interesting physics is hidden by bremsstrahlung [22–24]

$$\begin{aligned} \mathcal{B}(K^+ \to \pi^+\pi^0 e^+e^-)_B &\sim (330 \pm 15) \cdot 10^{-8} \\ \mathcal{B}(K^+ \to \pi^+\pi^0 e^+e^-)_M &\sim (6.14 \pm 1.30) \cdot 10^{-8}, \end{aligned} \tag{9}$$

and so Dalitz plot analysis is necessary in order to capture the more interesting direct emission contributions. The $K^+ \to \pi^+\pi^0 e^+e^-$-amplitude is then written as in Eq. 7 optimizing the theoretical and experimental knowledge of $K(p) \to \pi(p_1)\pi(p_2)\gamma(q)$ decays.

Relatively to the $K^\pm \to \pi^\pm\pi^0\gamma$ the possibility to go kinematically at large q^2 opens the possibility to beat the bremsstrahlung: at large dilepton invariant mass the bremsstrahlung can be even 100 times smaller than the magnetic contribution, at the price, however, of decreasing the statistics. Indeed compared to $K^\pm \to \pi^\pm\pi^0\gamma$ differential decay rates we have the possibility to measure also the q^2-differential rates

$$\frac{\partial^2\Gamma_{IB}}{\partial T_c^* \partial W^2 \partial q^2} \tag{10}$$

The further advantage is that at each q^2 we can measure the different Dalitz plot. Also several short distance observables can be measured by appropriate kinematical analyses [22, 23]. In Table 1 we show the ratios of Bremsstrahlung (B) to magnetic (M) and to various interferences. Our effort has been to write the $K^{\pm} \to \pi^{\pm}\pi^{0}e^{+}e^{-}$ amplitude in terms of experimental known Bremsstrahlung (B) and DE, electric (E) and magnetic (M), transitions [22]

Table 1. Branching ratios for the Bremsstrahlung and the relative weight of the rest of the contributions for different cuts in q, starting at q_{min} (first row) and ending at 180 MeV. In the last column we have also included the parity-odd magnetic-electric interference term.

q_c (MeV)	B $[10^{-8}]$	B/M	B/E	B/BE	B/BM
$2m_l$	418.27	71	4405	128	208
55	5.62	12	118	38	44
100	0.67	8	30	71	36
180	0.003	12	5	-19	44

The big news is that NA48/2 has reported a final measurement of this branching: the experiment selects 3 reconstructed tracks coming from one decay vertex, then Particle ID for $e^{\pm}/\pi^{\pm}$ separation; then two reconstructed γ clusters are compatible with π^0 mass [25]. The number of $K^{\pm}$ decays (kaon flux) is measured by using the reference channel $K^{\pm} \to \pi^{\pm}\pi^{0}(\gamma)$. 5076 events have been selected with 289 background; the error is dominated by external error on $\mathcal{B}(\pi^0 \to e^+e^-\gamma)$

$$(4.06 \pm 0.06_{stat.} \pm 0.04_{syst.} \pm 0.13_{ext.}) \times 10^{-6}$$

to be compared to our prediction dominated from Bremsstrahlung [22, 23]

$$\mathcal{B}(IB) = (4.19) \times 10^{-6} \quad \text{no Isospin breaking} \quad \mathcal{B}(TOT) = (4.29) \times 10^{-6}$$

$$\mathcal{B}(IB) = (4.10) \times 10^{-6} \quad \text{with Isospin breaking} \quad \mathcal{B}(TOT) = (4.19) \times 10^{-6}$$

4 Bardeen Buras Gerard approach and $K^{\pm} \to \pi^{\pm}l^{+}l^{-}$ form factor

Attempting to match long distance to short distance Buras and collaborators [3, 4] compute loop calculations employing a cut-off regulator instead of the dimensional regularization method. Consequently, their results exhibit a quadratic dependence on the physical cut-off M which is lost in the usual chiral perturbative calculations. This quadratic dependence is a crucial ingredient in the matching of the meson and quark pictures. They argue that one can obtain a parametrization of non-perturbative QCD effects by matching a low-energy Lagrangian valid up to the scale M, to the logarithmic behavior of relevant Wilson coefficients at high-energy. In this work we refer to this computational method as the *Bardeen-Buras-Gérard framework* (BBG). Recentely Buras has advocated this determination to claim New Physics to explain the experimental value of ϵ' [5]. BBG evaluate $K \to \pi\pi$ chiral loop with a dimensional cut-off (M) and match the quadratic divergences of the $Q_i(M)$ weak matrix elements with the log of the short distance Wilson coefficient as described by the equation

$$H_{\text{eff}} = \sum_i c_i(\mu)\, Q_i(\mu) \tag{11}$$

We have studied in the BBG contest the $K^{\pm} \to \pi^{\pm}l^{+}l^{-}$ form factor described in CHPT by a loop function, $W_{+}^{\pi\pi}(z)$ and a polynomial in z with experimentally determined coefficients a_+ and b_+ [26].

	PDG	Prospects
$K_S \to \mu\mu$	$< 9 \times 10^{-9}$ at 90% CL	(LD)$(5.0 \pm 1.5) \cdot 10^{-12}$ NP $< 10^{-11}$
$K_L \to \mu\mu$	$(6.84 \pm 0.11) \times 10^{-9}$	difficult : SD << LD
$K_S \to \mu\mu\mu\mu$	–	SM LD $\sim 2 \times 10^{-14}$
$K_S \to ee\mu\mu$	–	$\sim 10^{-11}$
$K_S \to eeee$	–	$\sim 10^{-10}$
$K_S \to \pi^+\pi^-\mu^+\mu^-$	–	SM LD $\sim 10^{-14}$

Table 2. Interesting channels: PDG values vs theoretical estimates

These parameterize the intermediate region between low and high energy regimes. Our goal is to predict the values of these two coefficients using BBG framework [6].

5 Lepton flavor (universality) violation in rare kaon decays

Recent anomalies in the decays of B-mesons and the Higgs boson provide hints towards lepton flavor (universality) violating physics beyond the Standard Model. We observe that 4-fermion operators which can explain the B-physics anomalies have corresponding analogs in the kaon sector, and we analyze their effect on $K \to \pi\ell\ell'$ and $K \to \ell\ell'$ decays ($\ell = \mu, e$). For these processes, we point out the corresponding physics opportunities at the NA62 and KOTO experiments. In particular, assuming minimal flavor violation, we comment on the required improvements in sensitivity necessary to test the B-physics anomalies in the kaon sector [27].

6 $K_{S,L} \to l^+l^-$, $K_{S,L} \to l^+l^-l^+l^-$ and $K_S \to \pi^+\pi^-l^+l^-$

The recent LHCB limit on $K_S \to \mu\bar{\mu}$ [9] in Table 6 is close to test interesting New Physics (NP) models [28]. A high precision measurement can test the short distance (SD) SM but it requires to improve the long distance (LD) prediction [28, 29] with auxiliary channels [30]. $K_L \to \mu\mu$: the small ratio SD/LD $\sim \frac{1}{30}$ may obscure an experimental improvement on the rate [28]. The situation would be a bit ameliorated if the still unknown sign of $A(K_L \to \gamma\gamma)$ would be either theoretically or experimentally determined. Help to this ambiguity could come from the experimental study of $K_{S,L} \to l^+l^-l^+l^-$ [30] As shown in Table 6 these channels are at reach in a high intensity machine and they may also give LD distance info needed for a better control of $K_L \to \mu\mu$. These four body decays have also a peculiar feature, similarly to $K_{S,L} \to \pi^+\pi^-e^+e^-$, the two different helicity amplitudes interfere; then one can measure the sign $K_L \to\to \gamma^*\gamma^* \to l^+l^-l^+l^-$ by studying the time interference K_S K_L which it has a decay length $2\Gamma_S$ [30].

Actually only $K^+ \to \pi^+\pi^0e^+e^-$ has been studied so far [22] while $K_S \to \pi^+\pi^-\mu^+\mu^-$, more feasible experimentally, is in progress; however generic features can be already extracted from Table I in Ref. [22]: we have less than 10 MeV phase space which can be extracted from the last lines of the Table I in Ref. [22] telling us that i) the Branching is expected $O(10^{-14})$ and the novel purely electric and magnetic contribution are relatively enhanced with respect to the less interesting bremsstrahlung.

7 Conclusions

In these proceedings we have studied theoretical issues in radiative rare kaon decays. An interesting channel, $K^+ \to \pi^+\pi^0e^+e^-$, is studied also in connection with the recent experimental NA48 results.

Motivated by LHCB results on $K_S \to \mu^+\mu^-$ we discuss other channels like $K_{S,L} \to l^+l^-l^+l^-$. Motivated by recent theoretical work by Buras and collaborators we study also the $K^\pm \to \pi^\pm l^+l^-$ form factor; in this channel may be interesting to study forward-backward asymmetry [26].

Acknowledgements

I thank the organizers of the workshop, in particular Antonio Di Domenico. Supported in part by MIUR under project 2015P5SBHT and by the INFN research initiative ENP.

References

[1] V. Cirigliano, G. Ecker, H. Neufeld, A. Pich and J. Portoles, Rev. Mod. Phys. **84**, 399 (2012) doi:10.1103/RevModPhys.84.399 [arXiv:1107.6001 [hep-ph]].

[2] A. J. Buras, arXiv:1609.05711 [hep-ph].

[3] W. A. Bardeen, A. J. Buras and J. M. Gerard, Nucl. Phys. B **293** (1987) 787. doi:10.1016/0550-3213(87)90091-5; A. J. Buras, J. M. Gérard and W. A. Bardeen, "Large N Approach to Kaon Decays and Mixing 28 Years Later: $\Delta I = 1/2$ Rule, $\hat{B}_K$ and ΔM_K", Eur. Phys. J. C **74**, 2871 (2014)

[4] J. M. Gerard, Acta Phys. Polon. B **21** (1990) 257.

[5] A. J. Buras, M. Gorbahn, S. Jager and M. Jamin, JHEP **1511**, 202 (2015) doi:10.1007/JHEP11(2015)202 [arXiv:1507.06345 [hep-ph]].

[6] E. Coluccio Leskow, G. D'Ambrosio, D. Greynat and A. Nath, Phys. Rev. D **93**, no. 9, 094031 (2016) doi:10.1103/PhysRevD.93.094031 [arXiv:1603.09721 [hep-ph]].

[7] M. Gorbahn, M. Patel and S. Robertson, arXiv:1104.0826 [hep-ph]: J. Brod, M. Gorbahn and E. Stamou, Phys. Rev. **D83**, 034030 (2011); M. Antonelli, D. M. Asner, D. A. Bauer, T. G. Becher, M. Beneke, A. J. Bevan, M. Blanke and C. Bloise *et al.*, Phys. Rept. **494**, 197 (2010) [arXiv:0907.5386 [hep-ph]].

[8] T. K. Komatsubara, Prog. Part. Nucl. Phys. **67**, 995 (2012) [arXiv:1203.6437 [hep-ex]].

[9] C. Patrignani *et al.* [Particle Data Group], Chin. Phys. C **40**, no. 10, 100001 (2016). doi:10.1088/1674-1137/40/10/100001

[10] A. V. Artamonov *et al.* [E949 Collaboration], Phys. Rev. Lett. **101**, 191802 (2008) [arXiv:0808.2459 [hep-ex]].

[11] J. K. Ahn *et al.* [E391a Collaboration], Phys. Rev. D **81**, 072004 (2010) [arXiv:0911.4789 [hep-ex]].

[12] Y. Grossman and Y. Nir, Phys. Lett. B **398**, 163 (1997) [hep-ph/9701313].

[13] `http://koto.kek.jp/`

[14] F. Newson *et al.*, arXiv:1411.0109 [hep-ex]; A. Romano, arXiv:1411.6546 [hep-ex].

[15] G. D'Ambrosio, G. F. Giudice, G. Isidori and A. Strumia, *Nucl.Phys.* **B 645** (2002) 155, [hep-ph/0207036].

[16] D. M. Straub, arXiv:1012.3893 [hep-ph]; G. Isidori, F. Mescia, P. Paradisi, C. Smith and S. Trine, JHEP **0608**, 064 (2006) [hep-ph/0604074].

[17] A. Crivellin, G. D'Ambrosio, T. Kitahara and U. Nierste, arXiv:1703.05786 [hep-ph].

[18] M. Endo, T. Kitahara, S. Mishima and K. Yamamoto, arXiv:1612.08839 [hep-ph].

[19] G. D'Ambrosio and D. N. Gao, JHEP **0010**, 043 (2000) doi:10.1088/1126-6708/2000/10/043 [hep-ph/0010122]; L. Cappiello and G. D'Ambrosio, Phys. Rev. D **75**, 094014 (2007) [arXiv:hep-ph/0702292].

[20] J. R. Batley *et al.* [NA48/2 Collaboration], Eur. Phys. J. C **68**, 75 (2010) [arXiv:1004.0494 [hep-ex]].

[21] J. Bijnens, G. Colangelo, G. Ecker and J. Gasser, arXiv:hep-ph/9411311 and references there

[22] L. Cappiello, O. Cata, G. D'Ambrosio and D. -N. Gao, Eur. Phys. J. C **72**, 1872 (2012) [arXiv:1112.5184 [hep-ph]].

[23] S. R. Gevorkyan and M. H. Misheva, Eur. Phys. J. C **74**, no. 5, 2860 (2014) [arXiv:1403.1053 [hep-ph]].

[24] H. Pichl, Eur. Phys. J. C **20**, 371 (2001) [hep-ph/0010284].

[25] Brigitte Bloch-Devaux On behalf of the NA48/2 collaboration "*Study of the* $K^{\pm} \to \pi^{\pm}\pi^{0}e^{+}e^{-}$ *decay with NA48/2 @ CERN* ": these proceedings.

[26] G. D'Ambrosio, G. Ecker, G. Isidori and J. Portoles, JHEP **9808**, 004 (1998) doi:10.1088/1126-6708/1998/08/004 [hep-ph/9808289].

[27] A. Crivellin, G. D'Ambrosio, M. Hoferichter and L. C. Tunstall, Phys. Rev. D **93**, no. 7, 074038 (2016) doi:10.1103/PhysRevD.93.074038 [arXiv:1601.00970 [hep-ph]].

[28] G. Isidori and R. Unterdorfer, JHEP **0401**, 009 (2004) [hep-ph/0311084].

[29] G. Ecker and A. Pich, Nucl. Phys. B **366**, 189 (1991);

[30] G. D'Ambrosio, D. Greynat and G. Vulvert, Eur. Phys. J. C **73**, 2678 (2013) [arXiv:1309.5736 [hep-ph]].

KLOE Results on Hadron Physics and Perspectives for KLOE-2

Simona Giovannella[1,*] *on behalf of the KLOE-2 Collaboration*

[1] *Laboratori Nazionali di Frascati dell'INFN, Frascati, Italy*

Abstract. Recent results obtained with KLOE data on light meson decays, together with perspectives for the current KLOE-2 run, are reviewed.

1 Introduction

The KLOE-2 experiment aims to enlarge and extend the KLOE physics programs with a larger data set and an upgraded detector. The KLOE detector [1] has collected 2.5 fb^{-1} at the e^+e^- collider DAΦNE [2], running at the peak of the ϕ resonance, $M_\phi \sim 1020$ MeV. An off-peak run provided also 250 pb^{-1} at 1 GeV. The detector consists of a large cylindrical drift chamber [3] and an electromagnetic calorimeter [4], surrounded by a magnetic field of 0.52 T. The trigger [5] uses information from both the calorimeter and the drift chamber. Collected data are analyzed by an event classification filter [6], which selects and streams various categories of events in different output files.

A new beam crossing scheme, allowing for a reduced beam size and increased luminosity, is now operating at DAΦNE [7]. The KLOE-2 detector is collecting e^+e^- collision data at center of mass energy equal to M_ϕ, with the aim of a final data set of ~5 fb^{-1}. Four tag stations [8], the High Energy Taggers (HET) and the Low Energy Taggers (LET), have been installed to detect electrons and positrons from the reaction $e^+e^- \to e^+e^-\gamma^*\gamma^* \to e^+e^-X$, to investigate $\gamma^*\gamma^* \to \pi^0/\pi\pi/\eta/\eta\pi$ physics at the ϕ resonance. An inner tracker [9] has been installed between the beam pipe and the inner wall of the DCH to increase the acceptance for low transverse momentum tracks and improve charged vertex reconstruction. Photon detection has been improved by means of a small crystal calorimeter in the forward direction [10] and of a tungsten-scintillating tile sampling device instrumenting the low-beta quadrupoles located inside the detector [11]. A detailed description of the extended experimental physics program can be found in Ref. [12].

In this paper recent results obtained with KLOE data on the decay dynamics of $\eta \to \pi^+\pi^-\pi^0$ (Sec. 2) and on $\phi \to \eta e^+e^-$, $\phi \to \pi^0 e^+e^-$ transition form factors (Sec. 3) are reported. Perspectives on light meson spectroscopy at KLOE-2 are discussed in Sec. 4, while $\gamma\gamma$ physics at KLOE and KLOE-2 is described in Sec. 5.

2 Decay dynamics of $\eta \to \pi^+\pi^-\pi^0$

The $\eta \to \pi^+\pi^-\pi^0$ decay is fully dominated by the isospin violating part of the strong interactions. A high precision measurement of the $\eta \to \pi^+\pi^-\pi^0$ decay dynamics allows to access the light quark mass

*e-mail: simona.giovannella@lnf.infn.it

 https://doi.org/10.1051/epjconf/201816600011

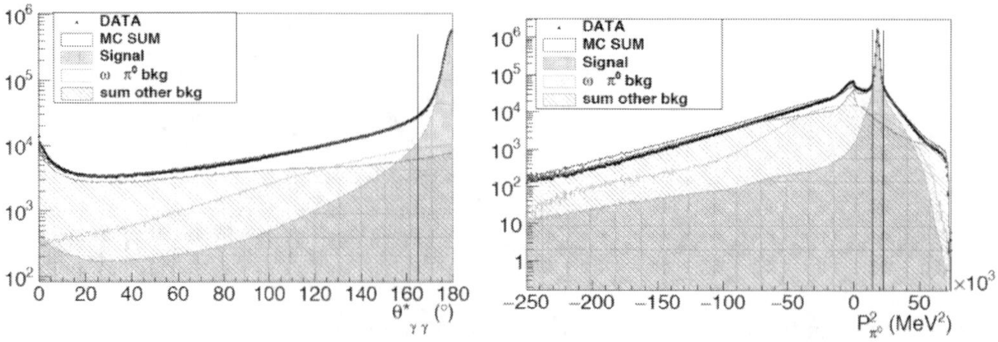

Figure 1. Dalitz plot analysis of $\eta \to \pi^+\pi^-\pi^0$: data-MC comparison for the missing mass squared of the π^0 (left) and the opening angle between photons in the π^0 rest frame (right). The vertical lines represent the selection cuts.

difference [13]. The Dalitz plot density is commonly described by a polinomial expansion in the X and Y variables

$$X = \sqrt{3}\,(T_+ - T_-)/Q_\eta\,, \tag{1}$$

$$Y = 3T_0/Q_\eta\,. \tag{2}$$

In these relations, T is the kinetic energy of the pions in the η rest frame while $Q_\eta = m_\eta - 2m_{\pi^\pm} - m_{\pi^0}$. The squared amplitude of the decay can then be extracted with a fit:

$$|A(X,Y)|^2 \simeq N(1 + aY + bY^2 + cX + dX^2 + eXY + fY^3 + gX^2Y + hXY^2 + lX^3 + ...) \tag{3}$$

In 2008, the KLOE experiment measured for the first time the Dalitz Plot parameters up to the term f (KLOE08) [14]. These results have been recently confirmed, although with less precision, by the WASA and BESIII experiments [16, 17]. A new measurement has been carried out with KLOE data (KLOE16), with an independent and ~ 4 times larger data set (1.7 fb^{-1}), a new analysis scheme and an improved Monte Carlo simulation, providing with improved accuracy the parameters of the decay matrix [15].

In KLOE, light mesons are produced via radiative decays of the ϕ and are tagged by identifying the recoil monochromatic photon, E_{recoil}. The $\eta \to \pi^+\pi^-\pi^0$ selection requires two additional prompt neutral clusters from the π^0 and two tracks with opposite curvature in the drift chamber pointing to the IP. Decay kinematics is then exploited to constrain E_{recoil} and to assign photons to π^0. Background scaling factors are obtained by fitting data with MC distribution for two variables: the missing mass squared of the π^0 and the opening angle between photons in the π^0 rest frame (Fig. 1). Cuts on these variables are used to reduce the background contamination. The resulting efficiency for signal events is 37.6%, with a background contamination less than 1%.

The resulting Dalitz plot density has been fitted, after background subtraction, with the complete third order polinomial expansion of Eq. (3), folded with smearing matrix and analysis efficency. Bin size is about three times the X, Y resolution. Fitting with the whole polinomial expansion, the c, e, h and l parameters are consistent with zero, as expected from C-invariance. Fixing them to zero and comparing with the previous KLOE measurement (Tab. 1), the statistical uncertainty is reduced by about a factor of two, while improving also the systematic uncertainties, which are in some cases reduced by a factor of 2 ÷ 3. The major improvement in the systematic uncertainties comes from the analysis of the effect of the event classification with an unbiased prescaled data sample. When

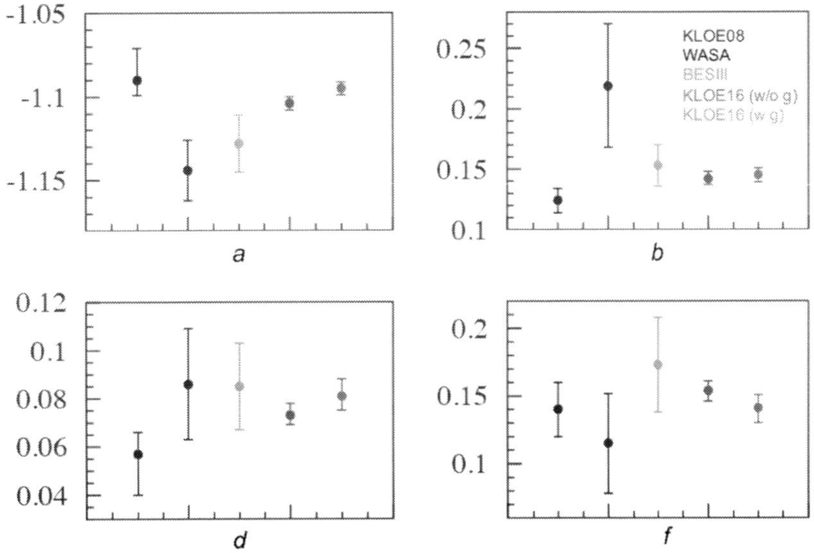

Figure 2. Comparison of a, b, d, f parameters extracted from the fit to the Dalitz plot distribution of $\eta \to \pi^+\pi^-\pi^0$ decay. KLOE results [14, 15] are compared with recent measurements from WASA [16] and BESIII [17].

the g parameter is included in the fit, its value is different from zero at 3σ level, improving the χ^2 probability from 24% to 56%. Comparison of KLOE, WASA and BESIII results is reported in Fig. 2, showing a good agreement among experimental results, with KLOE16 being the most precise one.

Table 1. Fit results for $\eta \to \pi^+\pi^-\pi^0$ Dalitz plot analysis, compared to previous KLOE measurement.

	a	b	d	f	g
KLOE08	$-1.090(5)(^{+8}_{-19})$	$0.124(6)(10)$	$0.057(6)(^{+7}_{-16})$	$0.14(1)(2)$	–
KLOE16	$-1.104(3)(2)$	$0.142(3)(^{+5}_{-4})$	$0.073(3)(^{+4}_{-3})$	$0.154(6)(^{+4}_{-5})$	–
KLOE16	$-1.095(3)(^{+3}_{-2})$	$0.145(3)(5)$	$0.081(3)(^{+6}_{-5})$	$0.141(7)(^{+7}_{-8})$	$-0.044(9)(^{+12}_{-13})$

The smearing matrix of the $\eta \to \pi^+\pi^-\pi^0$ Dalitz plot is very close to diagonal. For this reason, acceptance corrected data have been used to directly fit Eq. (3). The extracted parameters are in agreement with the values obtained using the whole smearing matrix. Dalitz plot acceptance corrected data is provided as supplementary material in [15].

The unbinned integrated left-right (A_{LR}), quadrant (A_Q) and sextant (A_S) charge asymmetries provide a more sensitive test of C parity conservation with respect to the fit to the Dalitz plot. The values extracted from the analysis of the new KLOE data set are consistent with zero at 10^{-4} level, thus improving existent evaluations [18–20]. Systematic uncertainties are of the same size of the statistical ones except for A_{LR}, where the error is dominated by the description of the Bhabha background. Experimental results are reported in Tab. 2.

Table 2. Measurements of the charge asymmetries for $\eta \to \pi^+\pi^-\pi^0$.

Experiment	$A_{LR}(10^{-2})$	$A_Q(10^{-2})$	$A_S(10^{-2})$
Gormley68	1.5 ± 0.5	—	0.5 ± 0.5
Layter72	-0.05 ± 0.22	-0.07 ± 0.22	0.10 ± 0.22
Jane74	0.28 ± 0.26	-0.30 ± 0.25	0.20 ± 0.25
KLOE08	$0.09 \pm 0.10^{+0.09}_{-0.14}$	$-0.05 \pm 0.10^{+0.03}_{-0.05}$	$0.08 \pm 0.10^{+0.08}_{-0.13}$
KLOE16	$-0.050 \pm 0.045^{+0.05}_{-0.11}$	$0.020 \pm 0.045^{+0.048}_{-0.023}$	$0.004 \pm 0.045^{+0.033}_{-0.035}$

3 Dalitz decays of the ϕ meson

Vector Meson Dominance (VMD) model fails to describe the vector to pseudoscalar transition form factor (TFF) for the process $\omega \to \pi^0\mu^+\mu^-$, as measured by the NA60 collaboration [21]. The only other existing experimental result for $VP\gamma^*$ TFF comes from the SND experiment, which has measured the M_{ee} invariant mass distribution of the $\phi \to \eta e^+e^-$ decay on the basis of 213 events [22], providing too large statistical error to confirm the NA60 evidence. New measurements of $VP\gamma^*$ transitions are therefore needed. A detailed study of the $\phi \to \eta e^+e^-$ and $\phi \to \pi^0 e^+e^-$ decays has been performed with 1.7 fb^{-1} of KLOE data [23, 24].

3.1 The $\phi \to \eta e^+e^-$ decay

The $\phi \to \eta e^+e^-$ decay has been studied using the $\eta \to \pi^0\pi^0\pi^0$ final state. Preselection cuts require: (i) two tracks of opposite sign originated from the interaction point (IP) plus six prompt photon candidates; (ii) a loose cut on the six photon invariant mass: $400 < M_{6\gamma} < 700$ MeV; (iii) a 3 σ cut on the recoil mass against the e^+e^- pair, $M_{\text{recoil}}(ee)$. A residual background contamination, due to $\phi \to \eta\gamma$ events with photon conversion on beam pipe (BP) or drift chamber walls (DCW), is rejected by extrapolating back to BP/DCW surfaces the tracks of the e^+, e^- candidates and then reconstructing their invariant mass and distance. Both quantities are small for the events coming from photon conversion. $\phi \to K\bar{K}$ and $\phi \to \pi^+\pi^-\pi^0$ events surviving analysis cuts have more than two pions in the final state. They are rejected using time-of-flight to the calorimeter. When an EMC cluster is connected to a track, the arrival time to the calorimeter is evaluated both with calorimeter (T_{cluster}) and drift chamber (T_{track}) information. Events with an e^+, e^- candidate outside a $3\,\sigma$'s window on the $DT = T_{\text{track}} - T_{\text{cluster}}$ variable are rejected. Comparison between data and Monte Carlo events at different steps of the analysis is reported in Fig. 3.

At the end of the analysis chain, 30,577 events are selected, with a residual background contamination of $\sim$ 3%. After background subraction, the measured branching fraction for the $\phi \to \eta e^+e^-$ process is:

$$BR(\phi \to \eta e^+e^-) = (1.075 \pm 0.007 \pm 0.038) \times 10^{-4}, \tag{4}$$

much more precise compared with the present PDG average of $(1.15 \pm 0.10) \times 10^{-4}$.

The slope of the transition form factor, $b_{\phi\eta}$, has been obtained from a fit to the di-lepton invariant mass using the differential cross section from Ref. [25] and the transition form factor in one-pole parametrization:

$$F(q^2) = \frac{1}{1 - q^2/\Lambda^2}, \tag{5}$$

where the slope parameter is defined as $b = [dF(q^2)/dq^2]|_{q^2=0} = \Lambda^{-2}$. This results in:

$$b_{\phi\eta} = (1.17 \pm 0.10\,^{+0.07}_{-0.11})\ \text{GeV}^{-2}, \tag{6}$$

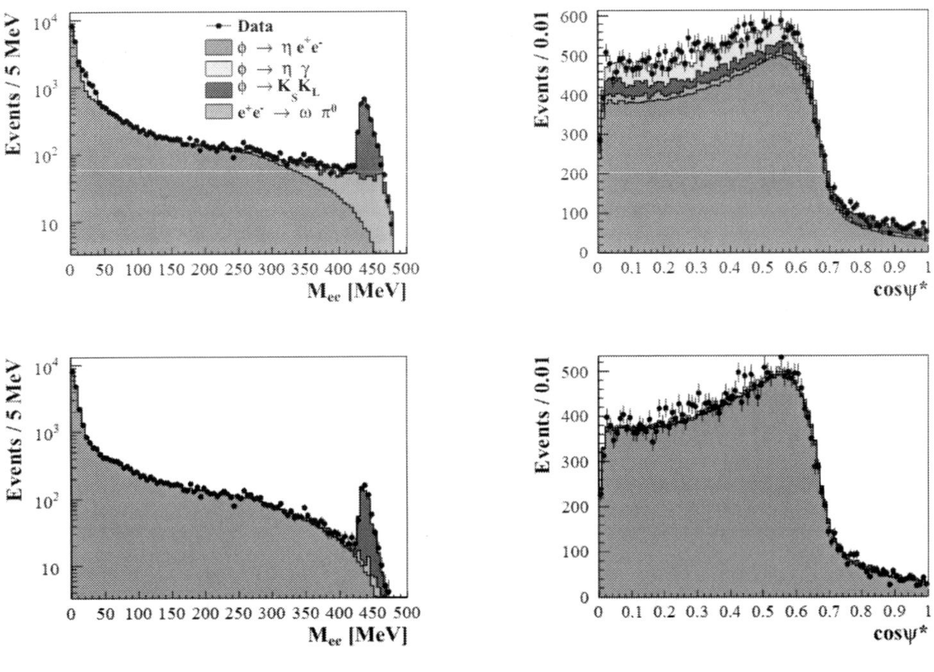

Figure 3. Analysis of the $\phi \to \eta e^+e^-$ decay channel: data-MC comparison for the di-lepton invariant mass (left) and the $\cos(\psi^*)$ variable (right), after the cut on $M_{\rm recoil}(ee)$ (top) and at the end of the analysis chain (bottom).

in agreement with VMD predictions ($b_{\phi\eta} = 1\ \mathrm{GeV}^{-2}$). Fit results are reported in Fig. 4 left.

The squared modulus of the transition form factor, $|F_{\phi\eta}(q^2)|^2$, as a function of the e^+e^- invariant mass (Fig. 4 right) has been obtained by dividing the M_{ee} spectrum bin by bin with the corresponding distribution obtained for MC events generated with a constant transition form factor. The value of $b_{\phi\eta}$ extracted from the fit is in agreement with Eq. (6).

3.2 The $\phi \to \pi^0 e^+e^-$ decay

No data are available on the transition form factor of the $\phi \to \pi^0 e^+e^-$ decay. Differently from the $\phi \to \eta e^+e^-$ channel, a large background contamination is still present after preselection cuts, requiring two tracks and two photon candidates in the final state. Dedicated analysis cuts strongly reduce the main background component of Bhabha scattering events to $\sim 20\%$, which dominates for $M_{ee} > 300$ MeV (Fig. 5 left). The other residual relevant background contribution is from ϕ radiative decays. At the end of the analysis, about 14,500 events are selected, with a total background contamination of $\sim 30\%$. Data-MC comparison is shown in Fig. 5 for different kinematical variables. The background contribution is removed bin-by-bin by subtracting the fits to each single background component from data points in the $M_{\rm recoil}(ee)$ distribution (Fig. 6 left).

The branching ratio of the available q^2 range has been obtained from the background subtracted e^+e^- mass spectrum by applying an efficiency correction evaluated bin by bin:

$$BR(\phi \to \pi^0 e^+e^-;\ \sqrt{q^2} < 700\ \mathrm{MeV}) = (1.19 \pm 0.05^{+0.05}_{-0.10}) \times 10^{-5}\,. \quad (7)$$

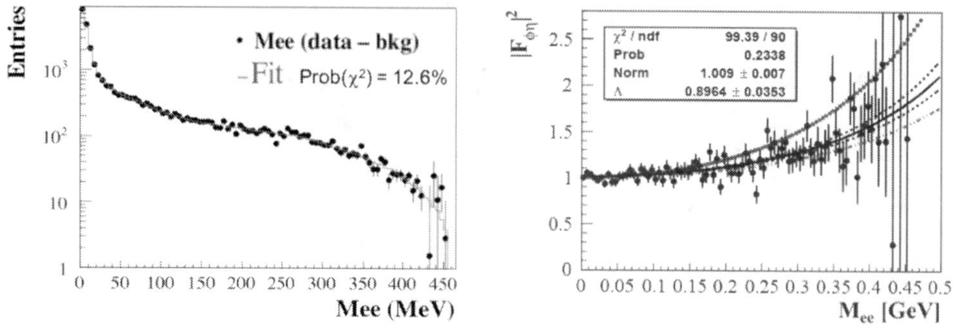

Figure 4. Left: fit to the di-lepton invariant mass for the $\phi \to \eta e^+ e^-$ decay channel. Right: $\phi\eta$ form factor as a function of the di-lepton invariant mass. The blue curve is the fit result, with its uncertainty; expectations from VMD and Ref. [26] are reported in red and pink, respectively.

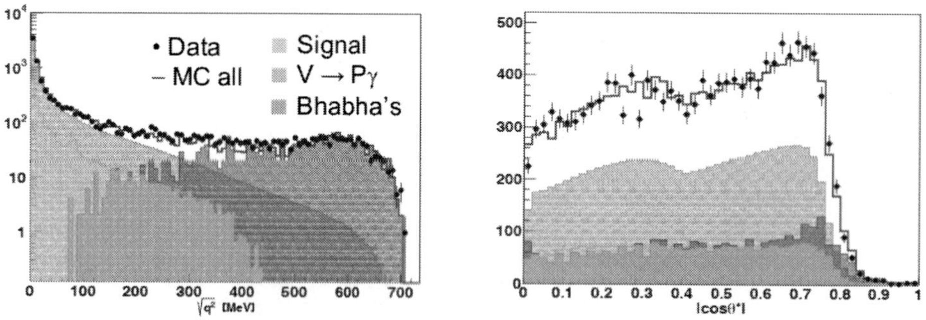

Figure 5. Analysis of the $\phi \to \pi^0 e^+ e^-$ decay channel: data-MC comparison for the di-lepton invariant mass (left) and the $\cos(\psi^*)$ variable (right) at the end of the analysis chain.

The first error includes the statistical and normalization errors, while the second one is due to systematics on analysis cuts and background subtraction. This result can be extrapolated to the full $\sqrt{q^2}$ range by using the teoretical description of the decay that best fit our transition form factor data [27]:

$$BR(\phi \to \pi^0 e^+ e^-) = (1.35 \pm 0.05^{+0.05}_{-0.10}) \times 10^{-5}\,. \tag{8}$$

The di-lepton invariant mass after background subtraction and efficiency correction is reported in Fig. 6 right and compared with theoretical predictions [27–29]. The slope of the transition form factor has been extracted by fitting this curve in the one-pole appoximation: $b_{\phi\pi^0} = (2.02 \pm 0.11)\ \text{GeV}^{-2}$.

4 Perspectives from e^+e^- interactions with KLOE-2 data

The KLOE-2 detector is taking data at the update DAΦNE e^+e^- collider, aiming to collect 5 fb^{-1} at the ϕ peak. This larger data sample will allow a deeper investigation of ligth meson properties, decay dynamics and transition form factors. About 1.5×10^{10} ϕ mesons will be produced, which will be used to complete and extend the study of the transition form factors from the Dalitz decays of vector

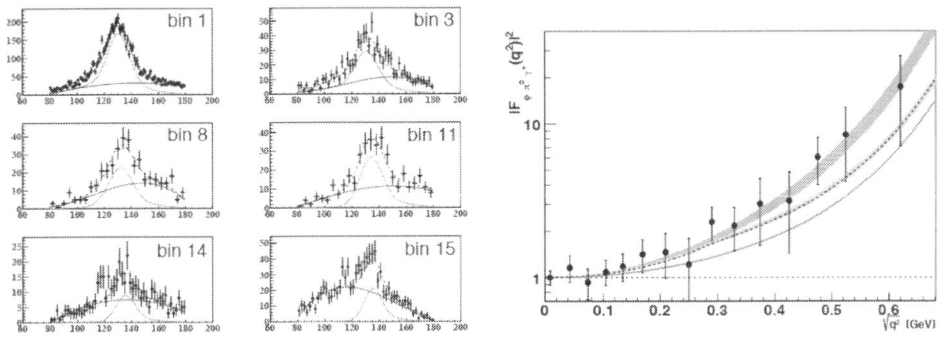

Figure 6. Analysis of the $\phi \to \pi^0 e^+ e^-$ decay channel. Left: fit to the $M_{recoil}(ee)$ distributions for different M_{ee} values, used to extract signal (blue) and background (red) components. The green lines represent the sum of the two. Right: $\phi\pi^0$ form factor as a function of the di-lepton invariant mass. Predictions from dispersive analyses [28, 29] (orange, light blue, blue dashed), chiral theory approach [27] (green) and the one-pole VMD model (red) are also reported.

mesons. As an example, the first evidence of the $\phi \to \eta\pi^+\pi^-$ and $\phi \to \eta\mu^+\mu^-$ decays are expected at KLOE-2. Currently, both decays have been searched by the CMD-2 experiments, that set an upper limit of [30, 31]: $BR(\phi \to \eta\pi^+\pi^-) < 1.8 \times 10^{-5}$ @ 90% C.L., $BR(\phi \to \eta\mu^+\mu^-) < 9.4 \times 10^{-6}$ @ 90% C.L.

A sample of $\sim 2.5 \times 10^6$ ω mesons will be also produced through the Initial State Radiation (ISR) process $e^+e^- \to \omega\gamma_{ISR}$. This will provide a large statistics data sample for the study of the $\omega \to \pi^+\pi^-\pi^0$ decay dynamics. The peak obtained with 300 pb^{-1} of KLOE data is visible in Fig. 7 left. It has been selected by requiring two track and three photons in the final state and applying a kinematic fit. No specific background rejection has been performed.

Radiative decays of the ϕ mesons will provide the largest data set of η meson: additional 2.5×10^8 will be produced in 5 fb^{-1}. A very clean η sample is tagged by means of the monocromatic recoil photon, 363 MeV, which is clearly identified as the most energetic neutral cluster in case of three or more particles in the final state (see Fig. 7 right). This data will be used to improve existing limits on η violating modes, listed in Tab. 3. Existing limits and expectations combining KLOE and KLOE-2 data are also reported.

Table 3. Current upper limit for η violating modes and extrapolations for KLOE plus KLOE-2 data set.

Decay	Test	UL [32]	UL (KLOE)	KLOE statistics	Scaled @ 5+2 fb^{-1}
$\eta \to \gamma\gamma\gamma$	C violation	1.6×10^{-5}	1.6×10^{-5}	410 pb^{-1}	3.9×10^{-6}
$\eta \to \pi^+\pi^-$	P, CP violation	1.3×10^{-5}	1.3×10^{-5}	350 pb^{-1}	2.9×10^{-6}
$\eta \to \pi^0\pi^0$	P, CP violation	3.5×10^{-4}	—		
$\eta \to \pi^0\pi^0\gamma$	C violation	5.0×10^{-4}	—		

The $\eta \to \pi^0\gamma\gamma$ decay is an important test of χ_{PT} because of its sensitivity to p^6 on both the branching ratio (BR) and the $M_{\gamma\gamma}$ spectrum. [33, 34]. The most accurate determination of the branching fraction has been obtained by the Crystal Ball experiment at MAMI [35], which measured $BR(\eta \to \pi^0\gamma\gamma) = (25.2 \pm 2.5) \times 10^{-5}$. A preliminary KLOE measurement [36], based on 70 signal events, 3 σ signal evidence, provided a 4 σ lower value: $BR(\eta \to \pi^0\gamma\gamma) = (8.4 \pm 2.7_{stat} \pm 1.4_{syst}) \times 10^{-5}$. The completely different environment of KLOE/KLOE-2 with respect to all other $\eta \to \pi^0\gamma\gamma$ measure-

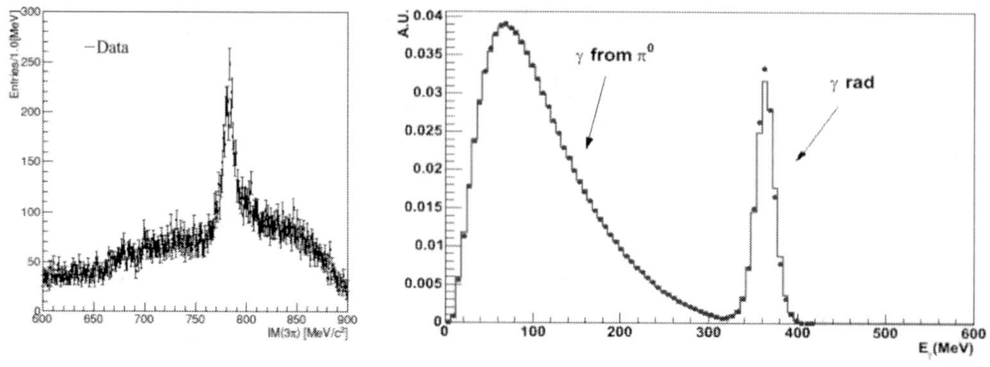

Figure 7. Left: peak of the ω meson produced through $e^+e^- \to \omega\gamma_{ISR}$ interactions. Right: photon energy for the $\phi\eta\gamma$, $\eta \to \pi^0\pi^0\pi^0$ decay. Blue dots are data while the red solid line represent MC expectations. Photons from π^0's are clearly separated from the primary photon.

ments, performed at hadron machines, asks for an upated measurement with larger statistics. using KLOE BR, a signal events comparable with Crystal Ball ($\sim$ 1200 signal events) is expected merging KLOE and KLOE-2 statistics. This measurement will benefit of the new CCALT calorimeter, which is design to detect photons from the iteraction region produced at low polar angle. It will substantially improve the rejection of the most relevant background, $\eta \to \pi^0\pi^0\pi^0$ with merged/lost photons. MC simulation shows that 83% of the residual $\eta \to 3\pi^0$ events has at least one photon lost. Half of them has photons in CCALT acceptance. KLOE/KLOE-2 statistics can provide the di-photon invariant mass to test theoretical models. As an example, a 3 σ separation is expected in the VMD+a_0 model in case of destructive and constructive interference.

KLOE-2 can improve the knowledge of η decays in four charged particles, listed in Tab. 4. These decays probe the structure of the η meson [25] and test the theoretical predictions of the branching ratio [39–42]. Moreover, the $\pi^+\pi^-e^+e^-$ final state could reveal CP violation beyond the prediction of the Standard Model by measuring the angular asymmetry A_ϕ between pions and electrons decay planes [43]. KLOE has already measured the first two processes of Tab. 4, obtaining the most precise measurement of the BRs and the first measurement of A_ϕ with a statistical sensitivity of 2.5×10^2 [37, 38]. The larger KLOE-2 data set, together with the increased track acceptance of the Inner Tracker, will improve all these measurements, allowing also to get the first evidence for the decay $\eta \to \mu^+\mu^-e^+e^-$.

Table 4. Measured and theoretical branching fractions for eta decays in four charged particles.

Decay	BR_{exp}	BR_{theory}
$\eta \to \pi^+\pi^-e^+e^-$	$(26.8 \pm 1.1) \times 10^{-5}$	$(25.7 \div 30) \times 10^{-5}$
$\eta \to e^+e^-e^+e^-$	$(2.4 \pm 0.2) \times 10^{-5}$	$(2.5 \div 2.6) \times 10^{-5}$
$\eta \to \mu^+\mu^-e^+e^-$	$BR < 1.6 \times 10^{-4}$ @ 90% C.L.	$(1.6 \div 2.2) \times 10^{-5}$
$\eta \to \pi^+\pi^-\mu^+\mu^-$	$BR < 3.6 \times 10^{-4}$ @ 90% C.L.	8×10^{-7}

In the $\phi \to K\bar{K}\gamma$ process, never been observed, the $K\bar{K}$ system has scalar quantum numbers, so that it is expected to proceeds through the $\phi \to [a_0(980) + f_0(980)]\gamma \to K\bar{K}\gamma$ decay. KLOE searched for this channel using the $\phi \to K_SK_S\gamma \to \pi^+\pi^-\pi^+\pi^-\gamma$ decay chain. The main backgrounds are the

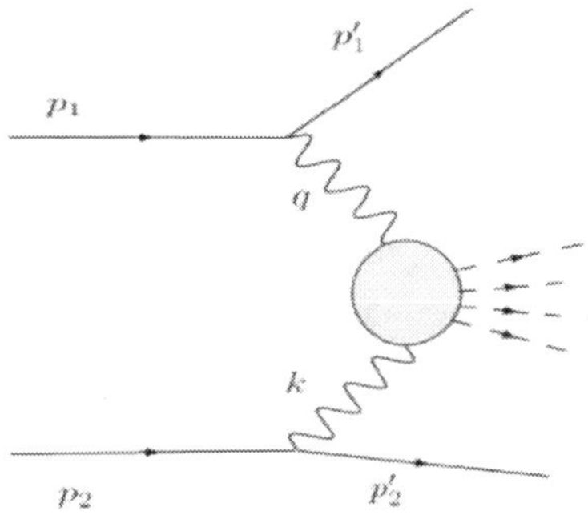

Figure 8. Diagram to produce X final state through $\gamma\gamma$ interaction.

resonant $e^+e^- \to \phi\gamma \to K_S K_L \gamma$ with an ISR photon and the continuum $e^+e^- \to \pi^+\pi^-\pi^+\pi^-\gamma$ processes. MC signal has been simulated according to the phase space and the radiative decay dynamics. Selection cuts have been optimized by MC. From an integrated luminosity of 2.2 fb^{-1}, 5 candidate events are found in data, while 3.2 ± 0.7 background events are expected from MC. This leads to BR($\phi \to K\bar{K}\gamma$) $< 1.9\times10^{-8}$ at the 90% CL [44]. This upper limit is consistent with the BR($\phi \to K\bar{K}\gamma$) prediction computed with the $f_0(980)$ and $a_0(980)$ couplings measured by KLOE [45–47]. Scaling these numbers to the KLOE+KLOE-2 statistics a sensitivity of $BR(\phi \to K^0\bar{K}^0\gamma) < 1.1 \times 10^{-8}$ is expected. The inner tracker will provide better vertex resolution, which is beneficial for the rejection capability of the background. This could lead to the first observation of such process.

5 $\gamma\gamma$ interactions at KLOE and KLOE-2

The $\gamma\gamma$ couplings and partial widths of mesons provide information about their structure and can be measured in the $e^+e^- \to e^+e^-\gamma^*\gamma^* \to e^+e^-X$ processes, Fig. 8, where X is a generic $J^{PC} = 0^{\pm+}, 2^{\pm+}$ final state. In case of a single particle R produced in the final state, the cross section of the decays is proportional to the transition form factor:

$$\sigma_{\gamma\gamma\to R}(q_1, q_2) \propto \Gamma_{R\to\gamma\gamma} \frac{8\pi^2}{M_R}\, \delta\left((q_1 + q_2)^2 - M_R^2\right) |F(q_1^2, q_2^2)|^2 \,. \tag{9}$$

A precise knowledge of the $\gamma\gamma$ decay width provides a measurement of the transition form factor, which is very important for the determination of the hadronic light-by-light contribution to the anomalous magnetic moment of the muon.

5.1 Measurement of $\Gamma(\eta \to \gamma\gamma)$ at KLOE

At KLOE, where there is no tagging of the outgoing e^+e^-, $\gamma\gamma$ interactions have been studied using off-peak data (240 pb^{-1} collected at $\sqrt{s} = 1$ GeV), to avoid backgrounds from ϕ decays. The η partial width, $\Gamma(\eta \to \gamma\gamma)$, is extracted from the measurement of the $e^+e^- \to e^+e^-\eta$ cross section, using both neutral and charged $\eta \to \pi\pi\pi$ decay channels [48]. The main background is due to the $e^+e^- \to \eta\gamma$ reaction, with an undetected recoil photon. After reducing the background components with specific kinematical cuts, signal events are extracted by fitting with the expected Monte Carlo components the two-dimentional plot $M^2_{\rm miss}$–$p_{L/T}$ (Figs. 9), where $M^2_{\rm miss}$ is the squared missing mass and $p_{L/T}$ is the η

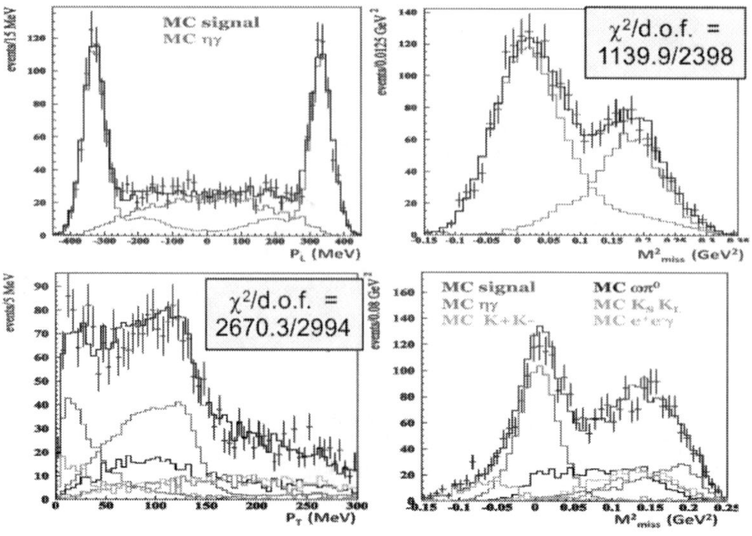

Figure 9. Top: η longitudinal momentum (left) and $M^2_{\rm miss}$ distribution (right) for $\gamma\gamma \to \eta \to \pi^0\pi^0\pi^0$ events. Bottom: η tranverse momentum (left) and $M^2_{\rm miss}$ distribution (right) for $\gamma\gamma \to \eta \to \pi^+\pi^-\pi^0$ events. Points with error bars are data, black solid histograms are fit result. Different fit components are reported in colors.

longitudinal/transverse momentum in the $\pi^0\pi^0\pi^0/\pi^+\pi^-\pi^0$ decay. Combining the two measurements, the extracted value for the production cross section results to be:

$$\sigma(e^+e^- \to e^+e^-\eta) = (32.7 \pm 1.3_{\rm stat} \pm 0.7_{\rm syst})\ {\rm pb} \tag{10}$$

This value is used to extract the most precise measurement to date of the $\eta \to \gamma\gamma$ partial width:

$$\Gamma(\eta \to \gamma\gamma) = (520 \pm 20_{\rm stat} \pm 13_{\rm syst})\ {\rm eV}\,. \tag{11}$$

5.2 The π^0 Transition Form Factor at KLOE-2

The upgrade of the KLOE-2 detector, with four detectors installed to tag electrons and positrons from the reaction $e^+e^- \to e^+e^-\gamma^*\gamma^* \to e^+e^-X$, will give the opportunity to investigate $\gamma\gamma$ physics also at the ϕ resonance for the reactions $\gamma\gamma \to \pi^0/\pi\pi/\eta/\eta\pi$ [12]. Single pseudoscalar production will improve the determination of the two-photon decay widths of these mesons, $\Gamma_{\gamma\gamma}$. For the π^0, the most precise measurement is obtained exploiting the Primakoff effect, reaching an accuracy of 2.8% [49]. At KLOE-2, the coincidence between the KLOE central detector and the HET taggers will provide a very clean sample of $\sim$ 1900 $\gamma\gamma \to \pi^0$ events per fb^{-1}, with background from radiative Bhabha scattering events being rejected by using the coincidence between the central detector and the HET stations. An accuracy of 1% on $\Gamma_{\gamma\gamma}(\pi^0)$ is reachable with 5-6 fb^{-1}, matching the current theory precision (Fig. 10 left). With the same amount of data, the measurement of the $\pi^0 \to \gamma\gamma^*$ transition form factor in the space-like region at low momentum transfer for the virtual photon will be possible with 5-6% accuracy. The KLOE-2 measurement will cover an unexploited region of the momentum transfer, as shown in Fig. 10 right. For the form factor measurement, the coincidence between the central detector and one of the HET stations will be used. These two proposed measurements are relevant for the theoretical evaluation of the hadronic light-by-light contribution to the muon magnetic anomaly, that is limited by the knowledge of the pseudoscalar transition form factor [50].

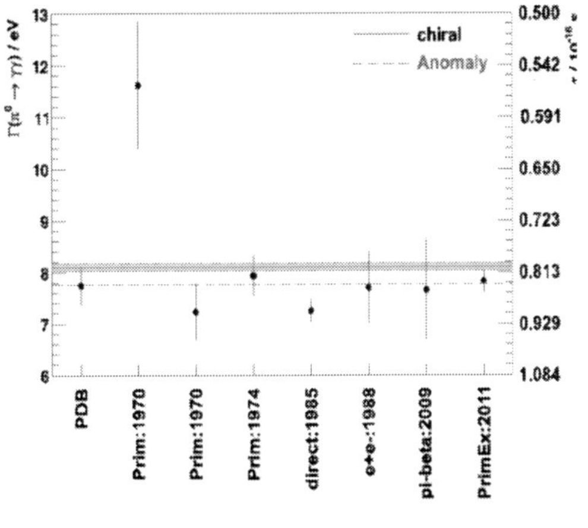

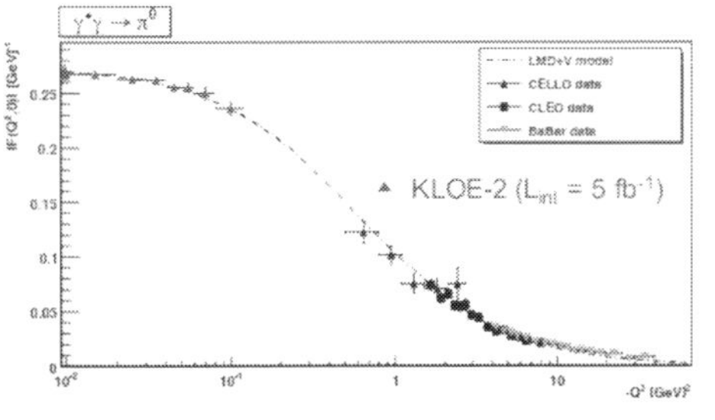

Figure 10. Left: experimental measurements of $\Gamma(\pi^0 \to \gamma\gamma)$ (black dots) and predictions from χPT with (black line) and without (blue dotted line) chiral anomaly. The theoretical error band is displayed in grey. Right: KLOE-2 expectations for $\pi^0\gamma\gamma^*$ TFF (red triangles), compared with existing measurments from CELLO [51] (black triangles), CLEO [52] (blue squares) and Babar [53] (green triangles).

6 Conclusions

The large data sample of light mesons collected with the KLOE detector provided several state-of-the-art measurements on the properties of light scalar, pseudoscalar and vector mesons. In the last year, the most precise measurements of the $\eta \to \pi^+\pi^-\pi^0$ decay dynamics and of the $\phi \to \eta e^+e^-$, $\phi \to \pi^0 e^+e^-$ transition form factors have been published.

Currently, the KLOE-2 run is in progress aiming to collect at least 5 fb^{-1} within the first months of year 2018. This data sample will allow to extend the high precision investigation of light meson properties. KLOE-2 will be an ideal tool to study in details the π^0 and η transition form factor at momentum transfer below 1 GeV^2.

References

[1] KLOE Collaboration, LNF-92/019 (IR) (1992) and LNF-93/002 (IR) (1993)
[2] A. Gallo *et al.*, Conf. Proc. C060626 (2006) 604, SLAC-PUB-12093
[3] M. Adinolfi *et al.*, Nucl. Instr. Meth. **A 488** 51 (2002)
[4] M. Adinolfi *et al.*, Nucl. Instr. Meth. **A 482** 364 (2002)
[5] M. Adinolfi *et al.*, Nucl. Inst. Meth. **A 492** 134 (2002)
[6] F. Ambrosino *et al.*, Nucl. Inst. Meth. **A 534** 403 (2004)
[7] C. Milardi *et al.*, JINST **7** T03002 (2012)
[8] D. Babusci *et al.*, Acta Phys. Polon. **B 46** 81 (2015)
[9] A. Di Cicco *et al.*, Acta Phys. Polon. **B 46** 73 (2015)
[10] M. Cordelli *et al.*, Nucl. Instr. Meth. **A 718** 81-82 (2013)
[11] F. Happacher *et al.*, Acta Phys. Polon. **B 46** 87 (2015)
[12] G. Amelino-Camelia *et al.*, Eur. Phys. J. **C 68** 619 (2010)
[13] H. Leutwyler, Mod. Ph. Lett. **A 28** 1360014 (2013) 1360014
[14] F. Ambrosino *et al.*, JHEP **05** 006 (2008)

[15] A. Anastasi *et al.*, JHEP **05** 019 (2016)
[16] P. Adlarson *et al.*, Phys. Rev. **C 90** 045207 (2014)
[17] M. Ablikim *et al.*, Phys. Rev. **D 92** 012014 (2015)
[18] M. Gormley *et al.*, Phys. Rev. Lett. **21** 402 (1968)
[19] J. G. Layter *et al.*, Phys. Rev. Lett. **29** 316 (1972)
[20] M. R. Jane *et al.*, Phys. Lett. **B 48** 260 (1974)
[21] S. Damjanovic *et al.*, Phys. Lett. B 677 (2009) 260.
[22] M. N. Achasov *et al.*, Phys. Lett. B 504 (2001) 275.
[23] D. Babusci *et al.*, Phys. Lett. **B 742** 1 (2015)
[24] A. Anastasi *et al.*, Phys. Lett. **B 757** 362 (2016)
[25] L.G. Landsberg, Phys. Rep. **128** 301 (1985)
[26] C. Terschlusen and S. Leupold, Phys. Lett. **B 691** 191 (2010)
[27] S. Ivashyn, Prob. Atomic. Sci. Technol. **2012N1** 179 (2012)
[28] S. P. Schneider, B. Kubis, F. Niecknig, Phys. Rev. **D 86** 054013 (2012)
[29] I. Danilkin *et al.*, Phys. Rev. **D 91** 094029 (2015)
[30] R. R. Akhmetshin *et al.*, Phys. Lett. **B 491** 81 (2000)
[31] R. R. Akhmetshin *et al.*, Phys. Lett. **B 591** 191 (2001)
[32] C. Patrignani *et al.* (Particle Data Group), Chin. Phys. **C 40** 100001 (2016)
[33] J. Bijnens, Phys. Scripta **T 99** 34 (2002)
[34] E. Oset, J. R. Pelaez, L. Roca, Phys. Rev. **D 67** 073013 (2003)
[35] B. M. K. Nefkens *et al.*, Phys. Rev. **C 90** 025206 (2014)
[36] B. Di Micco *et al.*, Acta Phys. Slov. **56** 403 (2006)
[37] F. Ambrosino *et al.*, Phys. Lett. **B 675** 283 (2009)
[38] F. Ambrosino *et al.*, Phys. Lett. **B 702** 324 (2011)
[39] C. Jarlskog and H. Pilkuhn, Nucl. Phys. **B 1** 264 (1967)
[40] A. Faessler, C. Fuchs and M. I. Krivoruchenko, Phys. Rev. **C 61** 035206 (2000)
[41] C. Picciotto and S. Richardson, Phys. Rev. **D 48** 3395 (1993)
[42] B. Borasoy and R. Nissler, Eur. Phys. J. **A 33** 95 (2007)
[43] D. N. Gao, Mod. Phys. Lett. **A 17** 1583 (2002)
[44] F. Ambrosino *et al.*, Phys. Lett. **B 679** 10 (2009)
[45] F. Ambrosino *et al.*, Phys. Lett. **B 634** 148 (2006)
[46] F. Ambrosino *et al.*, Eur. Phys. J. **C 49** 473 (2007)
[47] F. Ambrosino *et al.*, Phys. Lett. **B 681** 5 (2009)
[48] D. Babusci *et al.*, JHEP **01** 199 (2013)
[49] I. Larin *et al.*, Phys. Rev. Lett. **106** 162303 (2011)
[50] D. Babusci *et al.*, Eur. Phys. J. **C 72** 1917 (2012)
[51] H. J. Behrend *et al.*, Z. Phys. **C 49** 401 (1991)
[52] J. Gronberg *et al.*, Phys. Rev. **D 57** 33 (1998)
[53] (B. Aubert *et al.*, Phys. Rev. **D 80** 052002 (2009)

Towards a dispersive determination of the η and η' transition form factors

Bastian Kubis[1,⋆]

[1] *Helmholtz-Institut für Strahlen- und Kernphysik (Theorie) and Bethe Center for Theoretical Physics, Universität Bonn, 53115 Bonn, Germany*

Abstract. We discuss status and prospects of a dispersive analysis of the η and η' transition form factors. Particular focus is put on the various pieces of experimental information that serve as input to such a calculation. These can help improve on the precision of an evaluation of the η and η' pole contributions to hadronic light-by-light scattering in the anomalous magnetic moment of the muon.

1 Hadronic light-by-light scattering and the anomalous magnetic moment of the muon

There is a by now long-standing discrepancy between the experimental determination of the anomalous magnetic moment of the muon as measured by the BNL E821 experiment [1], and its calculation within the Standard Model [2], with both experiment and theory affected by a comparable uncertainty. With two different proposals to further improve on the accuracy of the measurement well under way [3, 4], it is of utmost importance to also improve the accuracy of the theoretical Standard-Model prediction before a persistent discrepancy of potentially increased significance can be interpreted in terms of physics beyond the Standard Model.

The Standard-Model uncertainty is entirely dominated by hadronic contributions. While the dominant hadronic vacuum polarization can be expressed, via a dispersion relation, in terms of measurable $e^+e^- \to$ hadrons total cross sections, which therefore can be further improved upon by a strong experimental effort to determine many of the exclusive channels contributing therein with yet improved precision [5], the situation for the α_{QED}-suppressed hadronic light-by-light scattering is less straightforward, relying until recently to a much larger extent on modeling [6], with badly controlled errors.

This presentation is part of an effort to analyze also the hadronic light-by-light scattering tensor using dispersion theory [7, 8]. Dispersion theory makes maximal use of *analyticity* and *unitarity*, two fundamental consequences of relativistic quantum field theories that mathematically incorporate the principles of *causality* and *probability conservation*. The various (in principle infinitely many different) contributions to hadronic light-by-light scattering are organized in terms of their analytic structure, according to the cuts and poles in the different energy variables that are dictated by the above principles. This has the advantage that all contributions will be given in terms of on-shell form factors and scattering amplitudes, hence observables that can to a large extent again be related to experimentally accessible quantities [9]. An ordering principle will be to consider intermediate

⋆e-mail: kubis@hiskp.uni-bonn.de

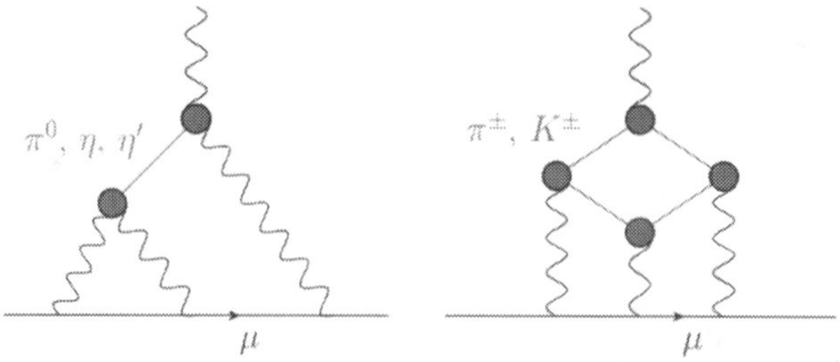

Figure 1. Dominant contributions to hadronic light-by-light scattering: pseudoscalar pole terms (left) and pion/kaon loop graphs (right).

states in terms of increasing masses, hence the lightest pseudoscalar pole terms (π^0, η, η') as well as pion–pion intermediate states are expected to give the largest individual contributions; see Fig. 1. In addition, all such form factors and scattering amplitudes can in turn also be analyzed using dispersion theory: they are reconstructed from their discontinuities or imaginary parts, whose most important contributions are determined by the leading hadronic intermediate states. One therefore interrelates a larger set of possible data, and often can avoid the direct use of amplitudes with leptons in the final state that are strongly suppressed by powers of the fine structure constant α_{QED}; see the discussion in Sect. 2.3. We will here concentrate on the transition form factors of the η and the η' that determine the corresponding pole terms; parallel efforts similar in style exist for the π^0 transition form factor [10–12] as well as for the pion loop including rescattering effects [13, 14].

2 Dispersion relation for the η and η' singly-virtual transition form factors

2.1 Definition, intermediate states

The $\eta^{(\prime)}$ transition form factor that describes the process $\eta^{(\prime)} \to \gamma^*\gamma^*$ is defined according to

$$\int \mathrm{d}^4 x e^{iq_1 x} i\langle 0|T\{j_\mu(x) j_\nu(0)\}|\eta^{(\prime)}(q_1+q_2)\rangle = -\epsilon_{\mu\nu\alpha\beta} q_1^\alpha q_2^\beta F_{\eta^{(\prime)}\gamma^*\gamma^*}(q_1^2, q_2^2), \tag{1}$$

where

$$j_\mu = e \sum_f Q_f \bar{q}_f \gamma_\mu q_f \tag{2}$$

denotes the electromagnetic vector current for quarks of all flavors f, weighted with the respective electric quark charges Q_f. In order to write down dispersion relations in the two virtualities $q_{1/2}^2$, we first decompose the form factors in terms of isospin: while the neutral pion always needs to decay into one isovector and one isoscalar photon (vs), combined in both possible ways, the $\eta^{(\prime)}$ is given by the sum of two distinct contributions, an isovector–isovector (vv) and an isoscalar–isoscalar (ss) one,

$$F_{\eta^{(\prime)}\gamma^*\gamma^*}(q_1^2, q_2^2) = F_{vv}(q_1^2, q_2^2) + F_{ss}(q_1^2, q_2^2), \qquad F_{\pi^0\gamma^*\gamma^*}(q_1^2, q_2^2) = F_{vs}(q_1^2, q_2^2) + F_{vs}(q_2^2, q_1^2). \tag{3}$$

We need to identify the dominant intermediate states at low energies, where high precision is of paramount importance. These are depicted in Fig. 2. For the isovector photons, the most important intermediate state consists of a pair of charged pions, while for the isoscalar ones, three pions are the lightest option. The isovector dispersion relation will then express the transition form factor in

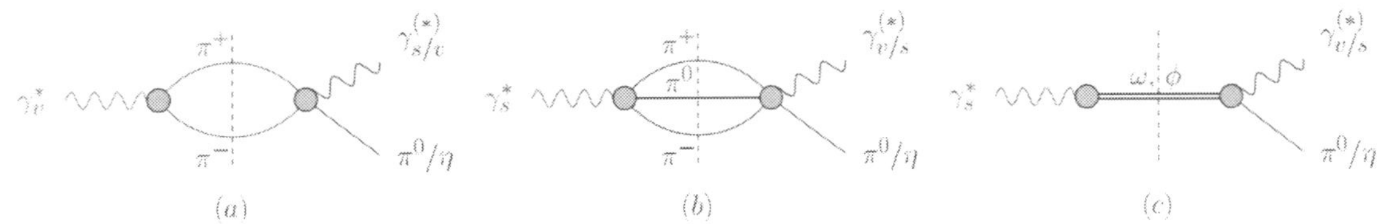

Figure 2. Leading intermediate hadronic states in π^0 and η transition form factors: (a) due to two pions for an isovector photon, (b) due to three pions for an isoscalar photon, (c) the three-pion intermediate state approximated by the lightest isoscalar vector resonances ω and ϕ.

terms of the product of the pion vector form factor $F_\pi^V(s)$ and the decay amplitude $\eta^{(\prime)} \to \pi^+\pi^-\gamma^{(*)}$ (the amplitude $\gamma^{(*)}\pi^0 \to \pi^+\pi^-$ in the case of the π^0)—this is what we will discuss in more detail below. The most general description of the three-pion intermediate state for the isoscalar photons, in contrast, would be a lot more complicated, and certain approximations at least for the transitions $3\pi \to \gamma^{(*)}\pi^0/\eta^{(\prime)}$ are almost unavoidable (see e.g. the corresponding discussion in Ref. [11]). Fortunately, the vector–isoscalar spectral function at low energies is strongly dominated by the narrow ω and ϕ resonances, see Fig. 2(c), such that a vector-meson-dominance (VMD) approximation is justified here to a large extent. In the context of the π^0, the corresponding $\omega \to \pi^0\gamma^*$ and $\phi \to \pi^0\gamma^*$ transition form factors have hence been treated dispersively [15–17] (see Ref. [18] for an extension even to the $J/\psi \to \pi^0\gamma^*$ transition), while a pure VMD description for $F_{ss}(q_1^2, q_2^2)$ was so far deemed sufficient for the $\eta^{(\prime)}$ transition form factors.

In the following, we will discuss how the different hadronic sub-amplitudes that are required as input to the unitarity relations can in turn be reconstructed based on dispersive methods.

2.2 Universality of final-state interactions, $\eta \to \pi^+\pi^-\gamma$

Final-state interactions between two strongly interacting particles as asymptotic states can be described in terms of form factors, which in turn can be linked to the properties of scattering amplitudes using analyticity and unitarity. As illustrated in Fig. 3, the unitarity relation for a pion form factor $F_J^I(s)$ of isospin I and angular momentum J reads

$$\operatorname{disc} F_J^I(s) = 2i \operatorname{Im} F_J^I(s) = 2i\, F_J^I(s) \times \theta(s - 4M_\pi^2) \times \sin\delta_J^I(s) e^{-i\delta_J^I(s)} , \tag{4}$$

from which one immediately deduces Watson's final-state theorem [19]: the form factor shares the phase $\delta_J^I(s)$ of the (elastic) scattering amplitude. The solution to Eq. (4) is obtained in terms of the

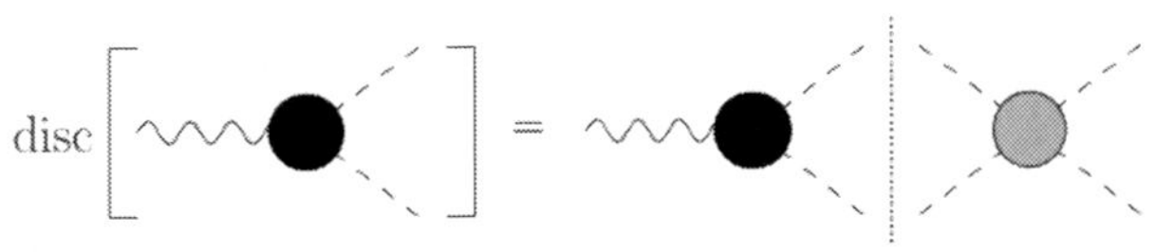

Figure 3. Graphical representation of the discontinuity relation for pion form factors. The black disc denotes the form factor, while the gray disc denotes the pion–pion scattering T-matrix, projected onto the appropriate partial wave.

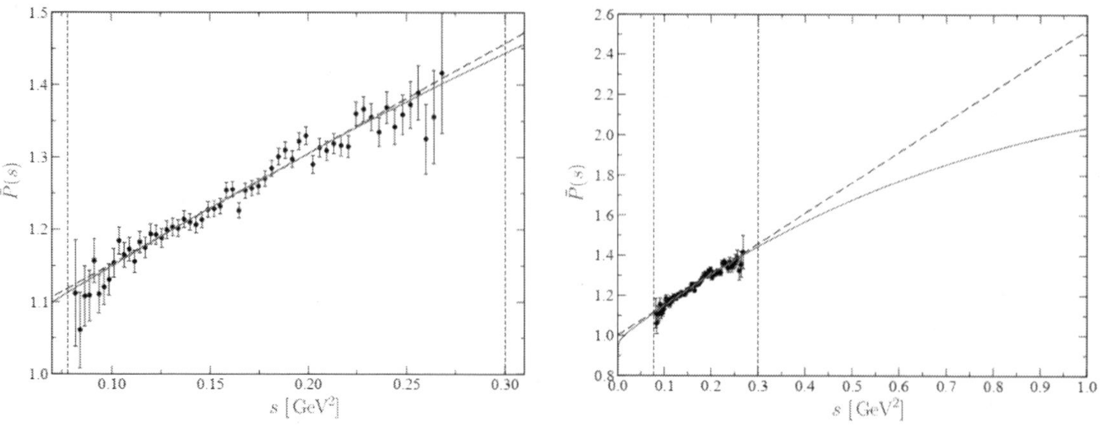

Figure 4. Left panel: representation of the decay distribution $\eta \to \pi^+\pi^-\gamma$ to allow for comparison to the polynomial $\bar{P}(s) = 1 + \alpha s$ (blue dashed curve). The full red curve includes the effects of a_2-exchange in addition; compare the discussion in Sect. 2.4. The vertical dashed lines denote the boundaries of phase space at $s = 4M_\pi^2$ and $s = M_\eta^2$. Right panel: the same distribution, extended up to $s = 1\,\text{GeV}^2$.

Omnès function $\Omega_J^I(s)$ [20],

$$F_J^I(s) = P_J^I(s)\Omega_J^I(s)\,, \quad \Omega_J^I(s) = \exp\left\{\frac{s}{\pi}\int_{4M_\pi^2}^{\infty} \mathrm{d}x \frac{\delta_J^I(x)}{x(x-s)}\right\}, \tag{5}$$

where $P_J^I(s)$ is a polynomial whose coefficients need to be determined by other methods, e.g. by matching to chiral perturbation theory near $s = 0$. The Omnès function is entirely given in terms of the appropriate pion–pion phase shift, which is particularly useful as we today have excellent information on pion–pion scattering at our disposal [21–24]. The pion vector form factor $F_\pi^V(s)$ as extracted from $\tau^- \to \pi^-\pi^0\nu_\tau$ decays, e.g., can be described very accurately by a representation (5) up to $\sqrt{s} = 1\,\text{GeV}$, employing a *linear* polynomial $R(s) \equiv P_1^1(s) = 1 + \alpha_V s$—at higher energies, the nonlinear effects of higher, inelastic (ρ', ρ'') resonances become important [25]. The slope parameter α_V therein is relatively small, $\alpha_V \sim 0.1\,\text{GeV}^{-2}$. For the pion vector form factor as measured in $e^+e^- \to \pi^+\pi^-$, the isospin-violating mixing effect with the ω-meson needs to be taken into account, see e.g. the extensive recent discussion in Ref. [26].

The power of the universality of final-state interactions lies in the fact that an Omnès representation similar to the one for the vector form factor will apply everywhere where two pions are produced from a point source in a relative *P*-wave; the process-dependence can be reduced to the coefficients of the multiplicative polynomial. It was pointed out that such a representation can in particular be used for the decays $\eta^{(\prime)} \to \pi^+\pi^-\gamma$ [27]: they are driven by the chiral anomaly and require the pion pair to be in an odd partial wave, hence the assumption of dominance by the *P*-wave $f_1(s)$ is entirely justified. For the decay of the η, an ansatz with a linear polynomial,

$$f_1(s) = A(1 + \alpha s)\Omega_1^1(s)\,, \tag{6}$$

was shown to be sufficient to describe the data in the physical decay region [28, 29], however, with one important difference to the vector form factor: α turns out to be large, $\alpha \sim 1.5\,\text{GeV}^{-2}$ [29, 30]; see Fig. 4.

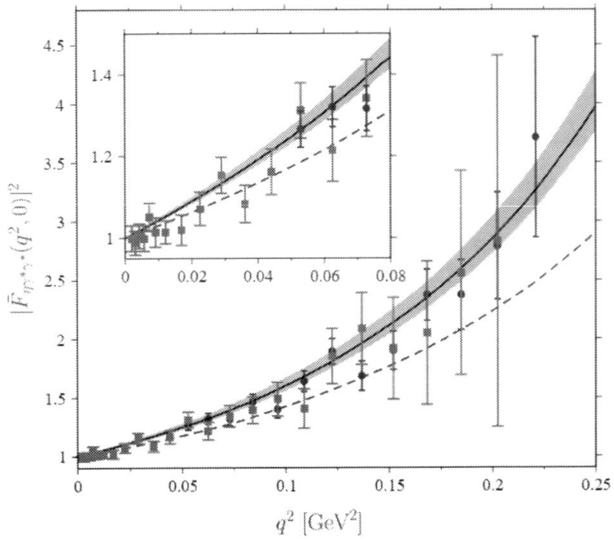

Figure 5. Comparison of the dispersive prediction for $|\bar{F}_{\eta\gamma^*\gamma^*}(q^2,0)|^2$ to data from the A2 [31] (green) and NA60 [34] (blue) collaborations. Figure courtesy of C. Hanhart.

2.3 Dispersive prediction for $\eta \to \ell^+\ell^-\gamma$

With the input to the isovector dispersion relation fixed, we can therefore calculate the singly-virtual η transition form factor: employing a once-subtracted dispersion relation for the normalized form factor $\bar{F}_{\eta\gamma^*\gamma^*}(q^2,0) = F_{\eta\gamma^*\gamma^*}(q^2,0)/F_{\eta\gamma^*\gamma^*}(0,0)$, one finds [25]

$$\bar{F}_{\eta\gamma^*\gamma^*}(q^2,0) = 1 + \frac{e\,q^2}{96\pi^2 A^{\eta}_{\gamma\gamma}} \int_{4M_\pi^2}^{\infty} \frac{\mathrm{d}x}{x-q^2}\left(1-\frac{4M_\pi^2}{x}\right)^{3/2} F_\pi^{V*}(x) f_1(x) + \Delta F^{I=0}_{\eta\gamma^*\gamma^*}(q^2,0)\,, \qquad (7)$$

where $A^{\eta}_{\gamma\gamma}$ is related to the $\eta \to 2\gamma$ real-photon decay width, and the last term denotes the (small) isoscalar contribution; see Refs. [25, 30] for details. The resulting prediction includes the propagated uncertainties both from the experimental input employed and due to the high-energy continuation of the dispersion integral; see Ref. [25] for a detailed discussion. It can be compared to the experimental data on the decays $\eta \to \ell^+\ell^-\gamma$, which have been obtained both for the electron–positron [31, 32] and the dimuon [33, 34] final states. In Fig. 5, we observe one of the main strengths of the dispersive approach: by fixing the input to the singly-radiative decay $\eta \to \pi^+\pi^-\gamma$ (with the decay rate scaling according to $\propto \alpha_{\mathrm{QED}}$), we gain a huge statistical advantage over the direct measurements of $\eta \to \ell^+\ell^-\gamma$ (rate $\propto \alpha^3_{\mathrm{QED}}$). A direct measurement of the transition form factor of comparable precision to the theoretical calculation based on dispersion relations will be enormously difficult.

2.4 Left-hand cuts

We have remarked above that the interpretation of the polynomial parameters multiplying the universal Omnès function has to be obtained case by case, and observed that the slope parameters α_V for the pion vector form factor and α in the $\eta \to \pi^+\pi^-\gamma$ decay are very different. While α_V can consistently be described as the low-energy tail of higher resonances in the region $1\,\mathrm{GeV} \lesssim \sqrt{s} \lesssim 2\,\mathrm{GeV}$, such an interpretation seems implausible for the $\eta \to \pi^+\pi^-\gamma$ decay. Furthermore, it is known that a linear

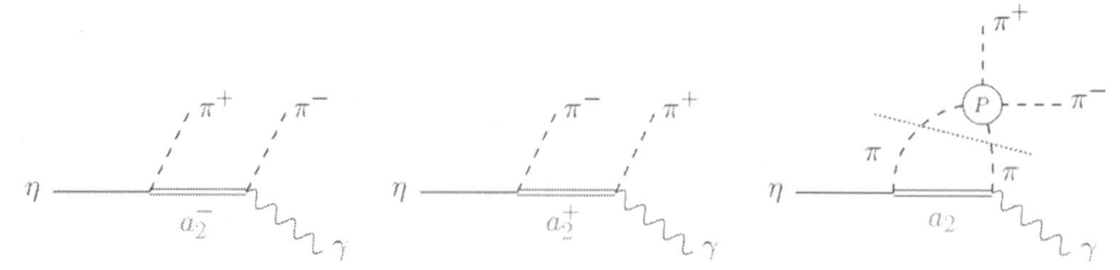

Figure 6. a_2 contributions to $\eta \to \pi^+\pi^-\gamma$, both at tree level and including pion–pion P-wave rescattering. The combination of both is required to preserve the final-state theorem.

polynomial $P(s)$ can only be a low-energy approximation: with the pion–pion P-wave Omnès function dropping like $1/s$ for large energies, the polynomial multiplying it should rather approach a constant asymptotically. Given that the dispersion integral to calculate the transition form factor formally extends to infinity, deviations from the linear rise in $P(s)$ may become important. It is therefore crucial to search for opportunities to investigate the amplitude $f_1(s)$ outside the relatively narrow η decay region.

One remarkable property of the amplitude (6) is the fact that it displays a zero for negative values of s at $s = -1/\alpha \approx -0.66\,\text{GeV}^2$, which therefore occurs at rather low energies. This kinematic configuration can be tested in the crossed process $\gamma\pi^- \to \pi^-\eta$, which can be investigated experimentally in a Primakoff reaction, i.e. the scattering of highly energetic charged pions in the strong Coulomb field of a heavy nucleus, e.g. at COMPASS; see Ref. [35] for an overview of the COMPASS Primakoff program and Ref. [36] for a first attempt to access the $\gamma\pi^- \to \pi^-\eta$ reaction. The zero would then occur in the angular distribution at fixed energies in the crossed center-of-mass system. In order to judge how reliable such a prediction is, we need to consider the possible effects of nontrivial $\pi\eta$ dynamics—something that has been entirely neglected in the discussion of the decay $\eta \to \pi^+\pi^-\gamma$ so far. A $\pi\eta$ S-wave is forbidden in this anomalous process. The P-wave is of exotic quantum numbers, $J^{PC} = 1^{-+}$, i.e. there are no resonances possible in a quark model of mesons, and no resonant states have been identified unambiguously so far [37, 38]; hence the P-wave is expected to be very weak at low energies. The first resonance to occur in $\gamma\pi^- \to \pi^-\eta$ is therefore the D-wave $a_2(1320)$. At the same time, it provides the dominant left-hand-cut contribution to the decay $\eta \to \pi^+\pi^-\gamma$ that has not been considered so far: a dispersion relation including the effects of a_2-exchange, see Fig. 6, is more complicated than the form factor relation (4), and requires a more complicated solution; for details, see Ref. [30]. There also the predictions for the Primakoff reaction $\gamma\pi^- \to \pi^-\eta$ are discussed: the zero of the simple amplitude representation (6) partially survives in the form of a strong amplitude suppression in backward direction in the energy region between threshold and the a_2 resonance, that can be explained as a P-D-wave interference effect. Here we only discuss the compatibility of this more refined model with the decay data: if the very precise data on $\eta \to \pi^+\pi^-\gamma$ by the KLOE collaboration [29] were already accurately described by the simpler amplitude of the form $P(s) \times \Omega_1^1(s)$, is this accuracy preserved when including a_2 effects? The answer is shown in Fig. 4: the left-hand cut induces a curvature in $P(s)$ that has only very small effects in the decay region, resulting in a data fit of equal quality; however, plotted over a larger energy region, it becomes clear that the effect on the dispersion integrals for the η transition form factor is nonnegligible. If one evaluates a sum rule for the slope of the form factor at $q^2 = 0$ that is easily obtained from Eq. (7), one finds that the a_2 effects reduce this slope by 7–8%.

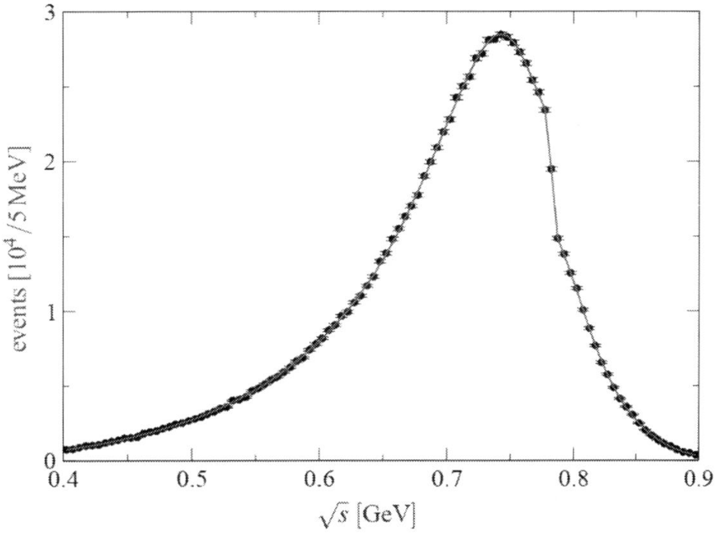

Figure 7. Differential decay rate $d\Gamma(\eta' \to \pi^+\pi^-\gamma)/d\sqrt{s}$. Pseudo-data generated according to preliminary BESIII results [40] is compared to the best fit according to Eq. (8). Figure taken from Ref. [26].

2.5 $\eta' \to \pi^+\pi^-\gamma$ and η' transition form factor

All the steps discussed in the previous subsections can in principle be carried over to an analysis of the η' transition form factor in a straightforward manner. While old data on $\eta' \to \pi^+\pi^-\gamma$ by the Crystal Barrel Collaboration [39] were not precise enough for an advanced analysis, a new, forthcoming measurement by BESIII [40] is in certain ways even more conclusive than the η decay data, as the larger phase space of the η' decay allows one to see deviations from the assumption of a linear polynomial much more clearly. It has been demonstrated in Ref. [30] that, despite the occurrence of a left-hand cut, in fact the ratio $f_1(s)/\Omega_1^1(s)$ in the model including the a_2 can be approximated very accurately by a *quadratic* polynomial within the physical decay region, i.e. for $4M_\pi^2 \leq s \leq M_{\eta'}^2$. The preliminary BESIII data [40] demonstrate the need of such a quadratic term to very high significance; moreover, they are so precise that even the isospin-breaking ρ–ω-mixing effect is clearly discernible (see Ref. [26] for details on how to extract the mixing strength and subsequently the partial width $\Gamma(\omega \to \pi^+\pi^-)$ from this decay). The full representation of the $\eta' \to \pi^+\pi^-\gamma$ *P*-wave amplitude is therefore of the form [26]

$$f_1(s) = \left[A(1 + \alpha s + \beta s^2) + \frac{\kappa_2}{m_\omega^2 - s - im_\omega\Gamma_\omega}\right] \times \Omega_1^1(s)\,. \tag{8}$$

The fit to pseudo-data generated according to preliminary results is shown in Fig. 7. The leading left-hand-cut contribution provided by a_2-exchange gives an estimate of the parameter $\beta = (-1.0 \pm 0.1)\,\text{GeV}^{-4}$ [30], which yields the correct sign and order of magnitude, but is somewhat larger than what the new data suggest [26].

The combination of the very precise representation of the $\eta' \to \pi^+\pi^-\gamma$ decay amplitude as well as the pion vector form factor then once more yields a dispersive prediction of the isovector part of the η' (singly-virtual) transition form factor according to the analogous formula to Eq. (7) [41]. In comparison to the η form factor, the isoscalar contribution is larger; however, this can be modeled well using the vector-meson-dominance assumption, with the necessary coupling constants extracted from

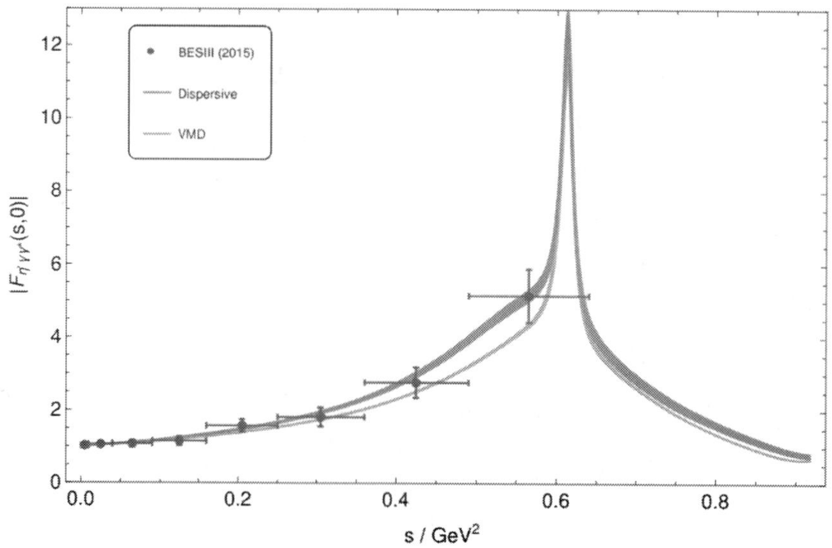

Figure 8. Dispersion-theoretical prediction for the singly-virtual η' transition form factor (blue band), compared to a vector-meson-dominance model (orange), as well as the data points taken by the BESIII collaboration [42]. Figure courtesy of S. Holz.

the partial decay widths for $\eta' \to \omega\gamma$ and $\omega \to \ell^+\ell^-$ as well as $\phi \to \eta'\gamma$ and $\phi \to \ell^+\ell^-$. The combined prediction for the singly-virtual η' transition form factor is shown in Fig. 8 as the blue band [41]. Both the broad resonance shape due to the ρ enhancement and the narrow ω peak are clearly visible. The error band comprises the propagated uncertainties due to different pion form factor data sets, different pion–pion scattering phase shifts, and different assumptions on the high-energy continuation of the input to the dispersion integral. However, despite these uncertainties we still note significant deviations of the dispersive prediction from a pure VMD model (with a simple finite-width ρ^0 pole as the isovector contribution), which is due to the much more sophisticated treatment of the two-pion cut contribution. The available data on $\eta' \to e^+e^-\gamma$ from BESIII [42] are not yet precise enough to differentiate between the two theoretical curves; new data also from other laboratories are eagerly awaited.

3 Towards a description of the doubly-virtual $\eta^{(\prime)}$ transition form factors

In the previous section, we have discussed in detail that dispersion theory provides a very accurate description of the singly-virtual $\eta^{(\prime)}$ transition form factors, based on high-precision experimental data that allows us to fix a small number of free polynomial parameters in the various dispersive amplitude representations. We have confined our discussion to experimental tests of the timelike form factors that can be measured in the decays $\eta^{(\prime)} \to \ell^+\ell^-\gamma$; we wish to emphasize that as the form factors by construction have the correct analytic structure, they can be analytically continued into the spacelike region without difficulties, i.e. Eq. (7) is easily evaluated for $q^2 < 0$. We expect this description to be reliable in a similar range of $|q^2|$ as the timelike momenta that enter the description of the spectral function, $-q^2 \lesssim 1\,\text{GeV}^2$; compare the related discussion of the π^0 transition form factor for spacelike arguments in Ref. [11].

The light-by-light loop integral that determines the pseudoscalar pole contributions to the anomalous magnetic moment of the muon, however, also requires the doubly-virtual transition form factors, see Fig. 1, hence the next challenge is to extend the formalism presented above to this more complicated case.

3.1 Testing factorization in $e^+e^- \to \eta\pi^+\pi^-$

One possible reaction in which the doubly-virtual (timelike) η transition form factor is directly accessible experimentally is obviously $e^+e^- \to \eta e^+e^-$. However, this process is strongly suppressed in the fine structure constant, and in the spirit of the preceding, we point to the much more promising possibility to investigate $e^+e^- \to \eta\pi^+\pi^-$ instead, thus beating the suppression by α^2_{QED}, and subsequently using a dispersion relation to reconstruct the dependence on the second (outgoing) photon's virtuality by means of the dominant $\pi^+\pi^-$ intermediate state.

One assumption frequently employed in the description of doubly-virtual transition form factors at low-to-moderate energies is the one of factorization: the dependence on the two virtualities is simplified into the product of two functions of a single variable each. For the example of the isovector–isovector contribution to the η transition form factor, the assumption for the normalized function $\bar{F}_{vv}(q_1^2, q_2^2) = F_{vv}(q_1^2, q_2^2)/F_{vv}(0,0)$ reads

$$\bar{F}_{vv}(q_1^2, q_2^2) \stackrel{?!}{=} \bar{F}_{vv}(q_1^2, 0) \times \bar{F}_{vv}(0, q_2^2) = \bar{F}_{vv}(q_1^2, 0) \times \bar{F}_{vv}(q_2^2, 0)\,. \tag{9}$$

We can test this factorization hypothesis also in $e^+e^- \to \eta\pi^+\pi^-$ [43]. Denoting the $\pi^+\pi^-$ squared invariant mass by $s_{\pi\pi}$ and the dilepton/the $\eta\pi^+\pi^-$ invariant mass squared by Q_2^2, it translates into

$$F_{\eta\pi\pi\gamma^*}(s_{\pi\pi}, Q_2^2) \stackrel{?!}{=} F_{\eta\pi\pi\gamma}(s_{\pi\pi}) \times F_{\eta\gamma\gamma^*}(Q_2^2)\,. \tag{10}$$

In other words: for fixed Q_2^2, factorization predicts the same dependence on $s_{\pi\pi}$ as observed in the real-photon case, i.e. in the decay $\eta \to \pi^+\pi^-\gamma$. Our strategy is therefore as follows: we allow for a similar functional form for $F_{\eta\pi\pi\gamma}(s_{\pi\pi})$ as in Sects. 2.2 and 2.4, i.e. an Omnès function multiplied with a linear polynomial in the one case, and including the effects of left-hand cuts due to a_2-exchange in the other. The subtraction constants can then be fitted to data and subsequently compared to the real-photon decay amplitude. As Q_2^2 is necessarily large—the reaction threshold is around $\sqrt{Q_2^2} = 0.83$ GeV, and the cross section is dominated by the ρ' or $\rho(1450)$ resonance—we for the moment simply parametrize $F_{\eta\gamma\gamma^*}(Q_2^2)$ by a sum of two Breit–Wigner functions [for the $\rho(770)$ and the $\rho(1450)$].

Unfortunately, the BaBar data [44] that we have analyzed for this purpose are not truly doubly differential in $s_{\pi\pi}$ and Q_2^2: only the total cross section $\sigma_{\mathrm{tot}}(Q_2^2)$ as well as the $\pi\pi$ spectrum $d\Gamma/d\sqrt{s_{\pi\pi}}$, integrated over $1\,\mathrm{GeV} \le \sqrt{Q_2^2} \le 4.5\,\mathrm{GeV}$, are available. The best fit to the data is shown in Fig. 9. We find that fitting a linear polynomial in $F_{\eta\pi\pi\gamma}(s_{\pi\pi})$ leads to polynomial parameters that are incompatible with the ones found in $\eta \to \pi^+\pi^-\gamma$, hence seemingly pointing towards a large violation of the factorization assumption. In contrast, preliminary results including effects of a_2-exchange indicate that such a violation is much attenuated in this case [43].

3.2 Double spectral function based on $\eta' \to \pi^+\pi^-\pi^+\pi^-$

We have emphasized repeatedly how it is advantageous to use hadronic amplitudes, which can be measured much more precisely due to the statistical advantage of lack of suppression in α_{QED}, as input to dispersion relations instead of measuring radiative amplitudes directly; therefore, we have based

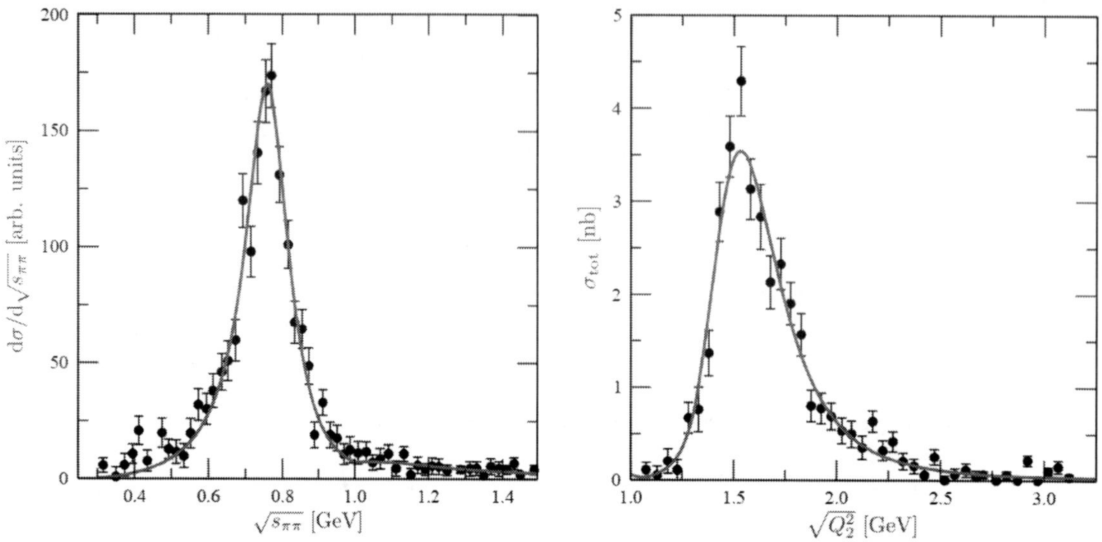

Figure 9. Result of the best fit to both the $\pi\pi$ spectrum (left panel) and the total cross section (right panel) for the reaction $e^+e^- \to \eta\pi^+\pi^-$. The data are from Ref. [44]. Figure taken from Ref. [43].

predictions for $\eta^{(\prime)} \to \ell^+\ell^-\gamma$ on precision data for $\eta^{(\prime)} \to \pi^+\pi^-\gamma$. In order to calculate the doubly-virtual transition form factor (or, more precisely, the isovector–isovector contribution therein), we may therefore consider going even one step further and base the construction of a double spectral function on a model for the purely hadronic decay $\eta' \to \pi^+\pi^-\pi^+\pi^-$. This decay has been analyzed in Ref. [45], combining the chiral expansion (the leading decay amplitude only occurs at $\mathcal{O}(p^6)$) with vector-meson resonance saturation of the low-energy constants involved. It was found that the (virtual) decay chain $\eta' \to 2\rho^0 \to 2(\pi^+\pi^-)$ entirely dominates the reaction. The prediction for the branching fraction, $\mathcal{B}(\eta' \to \pi^+\pi^-\pi^+\pi^-) = (10 \pm 3) \times 10^{-5}$ [45], has subsequently been confirmed by BESIII [46] with a smaller uncertainty,

$$\mathcal{B}(\eta' \to \pi^+\pi^-\pi^+\pi^-) = (8.7 \pm 0.7 \pm 0.6) \times 10^{-5}, \tag{11}$$

hence a theoretical refinement of the amplitude analysis is in order.

However, even based on the preliminary model of Ref. [45], we can attempt to assess the size of nonfactorizing contributions to the doubly-virtual η' transition form factor, see Fig. 10, by combining $\pi^+\pi^-$ pairs into virtual photons in the sense of a dispersion relation: it naturally yields a (dominant) factorizing term that closely resembles a VMD model, see the left diagram in Fig. 10, and a nonfactorizing one, where the charged pions stemming from different intermediate ρ^0 resonances are recombined [Fig. 10 (right)]. First steps towards such an investigation indicate that these nonfactorizing terms are small [47].

4 Summary, and a wish list for improved experimental input

We have shown that dispersion relations allow for high-precision predictions of the singly-virtual $\eta^{(\prime)}$ transition form factors, based on high-precision data for $\eta \to \pi^+\pi^-\gamma$ by KLOE [29] and for $\eta' \to \pi^+\pi^-\gamma$ by BESIII [40], as well as data on the pion vector form factor that is known to excellent precision, largely due to the strong experimental campaign to reduce the error on hadronic vacuum polarization in the muon's anomalous magnetic moment [48–53].

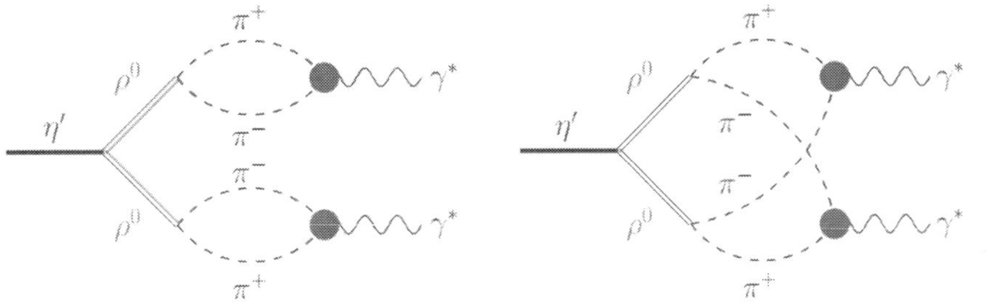

Figure 10. Factorizing (left) and nonfactorizing (right) contributions to the doubly-virtual η' transition form factor, based on a vector-meson-dominance model for the amplitude $\eta' \to \pi^+\pi^-\pi^+\pi^-$.

The following additional pieces of experimental information will help improve these model-independent, data-driven determinations, largely with the aim to extend the existing analysis to the doubly-virtual case.

1. A measurement of the Primakoff-type process $\gamma\pi^- \to \pi^-\eta$ allows for better constrained high-energy input to the existing analysis.

2. More differential data on $e^+e^- \to \eta\pi^+\pi^-$, i.e., measurements of the dipion distribution for smaller bins in the total e^+e^- energy, or even $\eta\pi^+\pi^-$ Dalitz plot distributions, would be extremely helpful to understand deviations from the factorization hypothesis for the doubly-virtual form factor, which is known to be invalid at high energies.

3. More detailed, differential decay data on $\eta' \to \pi^+\pi^-\pi^+\pi^-$ may allow us to construct a double spectral function for the doubly-virtual η' transition form factor; the precise form in which such differential data for a four-body decay ought to be analyzed should be developed in close collaboration between experimentalists and theorists.

4. Finally, given the fantastic precision of the upcoming BESIII data on $\eta' \to \pi^+\pi^-\gamma$, a direct measurement of $\eta' \to \pi^+\pi^-e^+e^-$ with good precision may be feasible.

The combination of further theoretical development along the lines sketched above, and improved experimental input for such a data-driven approach will therefore pave the way to a determination of the $\eta^{(\prime)}$ pole contributions to hadronic light-by-light scattering with controlled uncertainties.

Acknowledgements I would like to thank the organizers of the *KLOE-2 Workshop on e^+e^- collider physics at 1 GeV* for the invitation to this exciting workshop, and Christoph Hanhart for useful comments on this writeup. This research is supported in part by the DFG and the NSFC through funds provided to the Sino–German CRC 110 "Symmetries and the Emergence of Structure in QCD" (NSFC Grant No. 11621131001, DFG Grant No. TRR110).

References

[1] G. W. Bennett *et al.* [Muon $g-2$ Collaboration], Phys. Rev. D **73**, 072003 (2006)
[2] F. Jegerlehner and A. Nyffeler, Phys. Rept. **477**, 1 (2009)
[3] J. Grange *et al.* [Muon $g-2$ Collaboration], arXiv:1501.06858 [physics.ins-det]
[4] N. Saito [J-PARC $g-2$/EDM Collaboration], AIP Conf. Proc. **1467**, 45 (2012)

[5] T. Blum, A. Denig, I. Logashenko, E. de Rafael, B. Lee Roberts, T. Teubner and G. Venanzoni, arXiv:1311.2198 [hep-ph]
[6] J. Prades, E. de Rafael and A. Vainshtein, Adv. Ser. Direct. High Energy Phys. **20**, 303 (2009)
[7] G. Colangelo, M. Hoferichter, M. Procura and P. Stoffer, JHEP **1409**, 091 (2014)
[8] G. Colangelo, M. Hoferichter, M. Procura and P. Stoffer, JHEP **1509**, 074 (2015)
[9] G. Colangelo, M. Hoferichter, B. Kubis, M. Procura and P. Stoffer, Phys. Lett. B **738**, 6 (2014)
[10] M. Hoferichter, B. Kubis and D. Sakkas, Phys. Rev. D **86**, 116009 (2012)
[11] M. Hoferichter, B. Kubis, S. Leupold, F. Niecknig and S. P. Schneider, Eur. Phys. J. C **74**, 3180 (2014)
[12] S. Leupold *et al.*, these proceedings
[13] G. Colangelo, M. Hoferichter, M. Procura and P. Stoffer, arXiv:1701.06554 [hep-ph]
[14] M. Procura *et al.*, these proceedings
[15] F. Niecknig, B. Kubis and S. P. Schneider, Eur. Phys. J. C **72**, 2014 (2012)
[16] S. P. Schneider, B. Kubis and F. Niecknig, Phys. Rev. D **86**, 054013 (2012)
[17] I. V. Danilkin, C. Fernández-Ramírez, P. Guo, V. Mathieu, D. Schott, M. Shi and A. P. Szczepaniak, Phys. Rev. D **91**, 094029 (2015)
[18] B. Kubis and F. Niecknig, Phys. Rev. D **91**, 036004 (2015)
[19] K. M. Watson, Phys. Rev. **95**, 228 (1954)
[20] R. Omnès, Nuovo Cim. **8**, 316 (1958)
[21] B. Ananthanarayan, G. Colangelo, J. Gasser and H. Leutwyler, Phys. Rept. **353**, 207 (2001)
[22] G. Colangelo, J. Gasser and H. Leutwyler, Nucl. Phys. B **603**, 125 (2001)
[23] R. García-Martín, R. Kamiński, J. R. Peláez, J. Ruiz de Elvira and F. J. Ynduráin, Phys. Rev. D **83**, 074004 (2011)
[24] I. Caprini, G. Colangelo, and H. Leutwyler, Eur. Phys. J. C **72**, 1860 (2012)
[25] C. Hanhart, A. Kupść, U.-G. Meißner, F. Stollenwerk and A. Wirzba, Eur. Phys. J. C **73**, 2668 (2013) [Erratum: Eur. Phys. J. C **75**, 242 (2015)]
[26] C. Hanhart, S. Holz, B. Kubis, A. Kupść, A. Wirzba and C. W. Xiao, Eur. Phys. J. C **77**, 98 (2017)
[27] F. Stollenwerk, C. Hanhart, A. Kupść, U.-G. Meißner and A. Wirzba, Phys. Lett. B **707**, 184 (2012)
[28] P. Adlarson *et al.* [WASA-at-COSY Collaboration], Phys. Lett. B **707**, 243 (2012)
[29] D. Babusci *et al.* [KLOE Collaboration], Phys. Lett. B **718**, 910 (2013)
[30] B. Kubis and J. Plenter, Eur. Phys. J. C **75**, 283 (2015)
[31] P. Aguar-Bartolome *et al.* [A2 Collaboration], Phys. Rev. C **89**, 044608 (2014)
[32] P. Adlarson *et al.*, arXiv:1609.04503 [hep-ex]
[33] R. Arnaldi *et al.* [NA60 Collaboration], Phys. Lett. B **677**, 260 (2009)
[34] R. Arnaldi *et al.* [NA60 Collaboration], Phys. Lett. B **757**, 437 (2016)
[35] N. Kaiser and J. M. Friedrich, Eur. Phys. J. A **36**, 181 (2008)
[36] E. Altenbach, *Diploma thesis* (University of Bonn, 2016)
[37] C. Adolph *et al.* [COMPASS Collaboration], Phys. Lett. B **740**, 303 (2015)
[38] D. Schott [CLAS Collaboration], PoS **Confinement X**, 106 (2012)
[39] A. Abele *et al.* [Crystal Barrel Collaboration], Phys. Lett. B **402**, 195 (1997)
[40] S.-s. Fang [BESIII Collaboration], PoS **CD 15**, 032 (2016)
[41] S. Holz, *Bachelor thesis* (University of Bonn, 2016)

[42] M. Ablikim *et al.* [BESIII Collaboration], Phys. Rev. D **92**, 012001 (2015)
[43] C. W. Xiao, T. Dato, C. Hanhart, B. Kubis, U.-G. Meißner and A. Wirzba, arXiv:1509.02194 [hep-ph]
[44] B. Aubert *et al.* [BaBar Collaboration], Phys. Rev. D **76**, 092005 (2007) [Erratum: Phys. Rev. D **77**, 119902 (2008)]
[45] F.-K. Guo, B. Kubis and A. Wirzba, Phys. Rev. D **85**, 014014 (2012)
[46] M. Ablikim *et al.* [BESIII Collaboration], Phys. Rev. Lett. **112**, 251801 (2014) [Addendum: Phys. Rev. Lett. **113**, 039903 (2014)]
[47] T. Dato, *Master thesis* (University of Bonn, 2016)
[48] M. N. Achasov *et al.*, J. Exp. Theor. Phys. **103**, 380 (2006) [Zh. Eksp. Teor. Fiz. **130**, 437 (2006)]
[49] R. R. Akhmetshin *et al.* [CMD-2 Collaboration], Phys. Lett. B **648**, 28 (2007)
[50] B. Aubert *et al.* [BaBar Collaboration], Phys. Rev. Lett. **103**, 231801 (2009)
[51] F. Ambrosino *et al.* [KLOE Collaboration], Phys. Lett. B **700**, 102 (2011)
[52] D. Babusci *et al.* [KLOE Collaboration], Phys. Lett. B **720**, 336 (2013)
[53] M. Ablikim *et al.* [BESIII Collaboration], Phys. Lett. B **753**, 629 (2016)

Towards a dispersive determination of the pion transition form factor

Stefan Leupold[1,⋆], *Martin* Hoferichter[2], *Bastian* Kubis[3], *Franz* Niecknig[3], and *Sebastian P.* Schneider[3]

[1] *Institutionen för fysik och astronomi, Uppsala Universitet, Box 516, 75120 Uppsala, Sweden*
[2] *Institute for Nuclear Theory, University of Washington, Seattle, WA 98195-1550, USA*
[3] *Helmholtz-Institut für Strahlen- und Kernphysik (Theorie) and Bethe Center for Theoretical Physics, Universität Bonn, 53115 Bonn, Germany*

Abstract. We start with a brief motivation why the pion transition form factor is interesting and, in particular, how it is related to the high-precision standard-model calculation of the gyromagnetic ratio of the muon. Then we report on the current status of our ongoing project to calculate the pion transition form factor using dispersion theory. Finally we present and discuss a wish list of experimental data that would help to improve the input for our calculations and/or to cross-check our results.

1 Motivation

This presentation is closely related to the talks by B. Kubis and M. Procura; see also their contributions to these proceedings. The first parts of this presentation are based on [1] to which we refer for further details.

The pion transition form factor is defined by

$$\int d^4x\, e^{iq_1\cdot x}\, i\,\langle 0|T\, j_\mu(x)\, j_\nu(0)|\pi^0(q_1+q_2)\rangle = -\varepsilon_{\mu\nu\alpha\beta}\, q_1^\alpha\, q_2^\beta\, F_{\pi^0\gamma^*\gamma^*}(q_1^2, q_2^2) \tag{1}$$

where

$$j_\mu = e\sum_f Q_f\, \bar{q}_f \gamma_\mu q_f \tag{2}$$

denotes the electromagnetic current carried by the quarks and Q_f the electric charge of the quark of flavor f (in units of the proton charge e).

Experimentally the pion transition form factor $F_{\pi^0\gamma^*\gamma^*}(q_1^2, q_2^2)$ can be probed for various virtualities q_1^2 and q_2^2 of the electromagnetic currents; see also [2] and references therein. For the decay $\pi^0 \to 2\gamma$ the quantity $F_{\pi^0\gamma^*\gamma^*}(0,0)$ enters. The Dalitz decay $\pi^0 \to \gamma\, e^+e^-$ probes the *singly* virtual pion transition form factor $F_{\pi^0\gamma^*\gamma^*}(q^2, 0)$ for small timelike di-electron virtualities q^2 satisfying $4m_e^2 \leq q^2 \leq m_{\pi^0}^2$. Larger virtualities of this singly virtual pion transition form factor are probed, e.g., by the production process $e^+e^- \to \pi^0\gamma$; here $s = q^2 \geq m_{\pi^0}^2$. The corresponding processes to explore the *doubly* virtual

⋆e-mail: stefan.leupold@physics.uu.se

case in the timelike region are $\pi^0 \to 2e^+ 2e^-$ and $e^+e^- \to \pi^0\ell^+\ell^-$ with a lepton ℓ. Finally, the doubly virtual pion transition form factor in the spacelike region $q_1^2, q_2^2 < 0$ is probed, e.g., by reactions $2e^- \to \pi^0 2e^-$ and $e^+e^- \to \pi^0 e^+e^-$ where the pion can be produced via the fusion of two virtual photons.

Like for any quantum-field theoretical quantity there are kinematical regions of the doubly virtual pion transition form factor that are not or not easily experimentally accessible, for instance if q_1^2 is spacelike and q_2^2 timelike. For the pion transition form factor there is an additional challenge: It is of order e^2, i.e. already small at the amplitude level. Observables where the pion transition form factor enters quadratically are therefore very much suppressed. As we will see below, dispersion theory offers a way to circumvent the direct determination of such small quantities.

In general, form factors describe the deviation from a pointlike behavior, i.e. for a pointlike pion the normalized transition form factor $F_{\pi^0\gamma^*\gamma^*}(q_1^2, q_2^2)/F_{\pi^0\gamma^*\gamma^*}(0,0)$ would be unity, independent of the virtualities q_1^2 and q_2^2 of the electromagnetic currents. In other words, form factors encode the information about the composite structure of an object. Thus, from the point of view of hadron physics, the pion transition form factor tells us something about the intrinsic (quark-gluon) structure of the neutral pion. Yet there is a broader interest in this particular form factor beyond the motivation to explore the composite structure of strongly interacting matter. One promising way to look for physics beyond the standard model is to compare experimental results and corresponding standard-model calculations for selected observables where one can achieve high precision both on the experimental side and on the theoretical (standard model) side. A clear (statistically significant) deviation between theory (standard model) and experiment would point towards an additional impact on the respective observable from physical effects beyond the standard model. For this endeavor the gyromagnetic ratio of the muon provides a very promising observable [3, 4]; see also the presentations by B. Kubis, M. Procura, M. Knecht, and F. Jegerlehner in these workshop proceedings. The pion transition form factor is a crucial ingredient to improve the standard-model prediction for this quantity. We note in passing that also the rare decay $\pi^0 \to e^+e^-$ is sensitive to the pion transition form factor and might have some potential to reveal effects from physics beyond the standard model; see [2, 5] and references therein.

At present, the largest uncertainty in the standard-model prediction for the gyromagnetic ratio of the muon resides in the so-called hadronic vacuum-polarization contribution depicted on the left hand side of figure 1. It turned out that a significant part of this contribution comes from hadronic physics

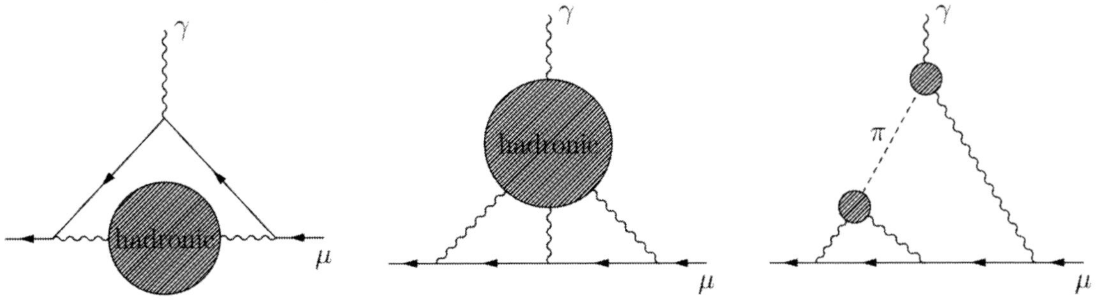

Figure 1. *Left:* Hadronic vacuum-polarization contribution to the gyromagnetic ratio of the muon. *Middle:* Hadronic light-by-light scattering contribution. *Right:* Pion-pole part. Note that in the dispersive formulation the pion is onshell [6].

below 2 GeV. Thus perturbative QCD cannot be used to calculate this contribution. Also the non-perturbative regime of the strong interaction is probed here. On the other hand, a phenomenological hadronic model might produce a number for the hadronic vacuum-polarization contribution but not a serious theory uncertainty estimate. Given that one wants to test the standard model and not the quality of a phenomenological hadronic model, one has to resort to methods firmly based on QCD or quantum field theory in general. Fortunately, the hadronic vacuum-polarization contribution can be directly related via dispersion theory to the cross section $e^+e^- \to$ hadrons [3]. Thus, instead of modeling this contribution with not determinable accuracy one can use data with experimentally known accuracy.

This interrelation between the hadronic vacuum-polarization contribution and hadronic data has triggered significant experimental activities as can be deduced from numerous presentations in these proceedings. As a consequence of these on-going activities it can be expected that the uncertainty in the standard-model prediction originating in the hadronic vacuum-polarization contribution will be significantly reduced in the near future. This development moves the focus to the next hadronic contribution to the gyromagnetic ratio of the muon: the hadronic light-by-light scattering contribution depicted in the middle of figure 1. So far, the size of this contribution has been determined by hybrids of hadronic and quark models and attaching a rather conservative and therefore big model uncertainty to the results [3, 4].

Given the ongoing experimental activities to improve on the direct determination of the gyromagnetic ratio of the muon and on the input for the hadronic vacuum-polarization contribution, it is high time to improve also on the standard-model prediction for the hadronic light-by-light scattering contribution including a reliable uncertainty estimate. In [6–9] a dispersive framework has been proposed to address the hadronic light-by-light scattering contribution. This framework is based on data and fundamental principles of quantum field theory, QED and QCD. It is expected that the numerically most important part of the hadronic light-by-light scattering contribution comes from processes where the hadronic blob in the middle diagram of figure 1 is replaced by the lowest-mass hadronic state, the pion. Such a process, the "pion-pole term", is sketched by the right diagram of figure 1. Other important parts of the hadronic light-by-light scattering contribution come from processes where the hadronic blob in the middle diagram of figure 1 is replaced by an η- or η'-meson or by a pair of pions. These processes are discussed in the contributions of B. Kubis and M. Procura. The central quantity that needs to be determined for the calculation of the pion-pole term is just the doubly virtual pion transition form factor. Note that in the dispersive framework the pion in figure 1 is onshell while the photons in the loops can have arbitrary virtuality; see the corresponding discussion in [6]. This matches with the definition (1) of the pion transition form factor where the pion is an external, i.e. physical state.

2 Dispersion theory

The ambition is to determine via dispersion theory and experimental data the pion-pole term, which constitutes an important part of the hadronic light-by-light scattering contribution to the gyromagnetic ratio of the muon. The aim is a numerical value for this contribution *and* an associated reliable theory uncertainty estimate. Using dispersion theory provides the advantage that this framework and therefore also the uncertainty estimate is solely based on QED, QCD and fundamental principles of quantum field theory. In turn quantum field theory constitutes the basis of the standard model. Thus an uncertainty estimate based on this data-driven dispersive approach can truly be regarded as a standard-model uncertainty estimate.

Two central aspects of relativistic local quantum field theory are (a) the probabilistic interpretation of scattering amplitudes, implied by quantum theory, and (b) the locality, which ensures the demand from the theory of relativity that information cannot propagate faster than the speed of light. Unitarity and analyticity of scattering amplitudes are directly related to these fundamental aspects (a) and (b), respectively.

Transitions from any initial to any final state can be encoded in the S-matrix. Demanding that the total probability to obtain an arbitrary final state must be 1 (probabilistic interpretation of quantum mechanics) leads to the unitarity condition

$$S\,S^\dagger = 1\,. \tag{3}$$

The S-matrix contains the possibility that no interaction has taken place. This trivial part can be split off by introducing the T-matrix via $S =: 1 + iT$. The unitarity condition (3) provides a profound relation for T, the optical theorem:

$$2\,\mathrm{Im}T = T\,T^\dagger\,. \tag{4}$$

Note that this is a matrix equation. For an initial state A and a final state B the relation (4) reads

$$2\,\mathrm{Im}T_{A\to B} = \sum_X T_{A\to X}\,T^\dagger_{X\to B}\,. \tag{5}$$

The locality requirement of relativistic quantum field theory demands that scattering amplitudes consist of polynomials (from the *local* interaction vertices), propagators and integrals thereof. The poles of the propagators are related to physical states. Consequently scattering amplitudes are analytic except for poles and cuts, which have a one-to-one correspondence to physical intermediate states. Schematically this can be expressed via the *dispersion relation*[1]

$$T(s) = \frac{1}{\pi}\int\limits_{-\infty}^{\infty} ds'\,\frac{\mathrm{Im}T(s')}{s'-s-i\epsilon}\,. \tag{6}$$

Thus, based on analyticity, one can obtain the whole amplitude from its imaginary part. On the other hand, based on unitarity, one can obtain the imaginary part of the amplitude from other amplitudes on account of (5).[2]

One situation where such a framework is of particular importance is the case where $T_{A\to B}$ is very small. For instance, the pion transition form factor is doubly suppressed by the electromagnetic coupling constant e. Here one might identify A with γ^* and B with $\gamma^*\pi^0$. Because of the overwhelming strength of the strong interaction the sum of intermediate states X in (5) is dominated by hadrons. Thus, while the amplitude $T_{A\to B}$ is of order e^2, the amplitudes $T_{A\to X}$ and $T_{X\to B}$ are both of order e, i.e. less suppressed and therefore more easily accessible by experiment.

Of course, there are practical limitations for such a dispersive reconstruction of amplitudes. First of all, one formally needs to know $\mathrm{Im}T$ in (6) up to infinitely large energies. This can never be achieved by experiment. Second, it seems that it is necessary to know/measure the amplitudes $T_{A\to X}$ and $T_{X\to B}$ for *all* intermediate states X. Both problems can be tamed by subtracted dispersion relations and by restricting the attention to not too large values of $|s|$ in (6). In principle, a systematic framework to determine hadronic scattering amplitudes at very low energies is chiral perturbation theory. However, it turned out that the energy regime covered by chiral perturbation theory is not large enough to provide a reliable calculational framework for the pion-pole term (right diagram in figure 1). The dispersive framework is in practice also restricted to low energies, but the results remain reliable in a regime that is larger than the one covered by chiral perturbation theory.

As already spelled out, one can utilize subtracted dispersion relations to suppress the influence of the largely unknown high-energy region and to restrict the sum over all intermediate states X to a

[1] In reality a scattering amplitude depends on several kinematical variables. For instance a partial-wave decomposition might boil it down to a one-variable function.

[2] Note that this is a schematic, oversimplified discussion of dispersion theory. For the full-fledged formalism we refer to [1] and references therein.

manageable amount. For simplicity we specify the scattering amplitude to some extent by studying a transition form factor F_{BC} related to $\langle C|j|B\rangle$ where j denotes a quark current. To make contact with the previous formulae one should identify schematically A from (5) with j and B from (5) with BC. Furthermore we assume that the lowest-mass intermediate states X are given by 2-pion states. Schematically a twice-subtracted dispersion relation is given by

$$F_{BC}(s) = F_{BC}(0) + F'_{BC}(0)\, s + \frac{s^2}{\pi} \int_{4m_\pi^2}^{\infty} ds' \, \frac{T^*_{BC\pi\pi}(s')\, F_{\pi\pi}(s')}{(s')^2\,(s' - s - i\epsilon)} + \ldots . \tag{7}$$

The general idea is to obtain the subtraction constants $F_{BC}(0)$, $F'_{BC}(0)$ from data and/or chiral perturbation theory and the amplitudes $T_{BC\pi\pi}$ and $F_{\pi\pi}$ from data or other dispersion relations. Note that the s' integral starts at the threshold of the 2-pion states. The dots in (7) refer to other intermediate states different from 2-pion states. These additional states produce corresponding s'-integrals. It is crucial to understand that these additional states are heavier and therefore contribute only for large values of s'. We want to know the transition form factor F_{BC} for small values of $|s|$. Then all contributions from large values of s' are highly suppressed by $\sim (1/s')^3$. This should be compared to the unsubtracted dispersion relation (6) where there is only a $1/s'$ suppression of the high-energy part. Subtractions help to focus on the low-energy information. The price to pay are the subtraction constants that need to be obtained from other sources. In addition, the high-energy behavior of the amplitude is modified. Typically general constraints from scattering theory and asymptotic QCD are at odds with a polynomial growth of the scattering amplitudes. Thus especially for the uncertainty estimates it must be carefully monitored to which extent an incorrect high-energy behavior influences the results for observables.

In (7) all the quantities that we do not know so well — the contributions from other intermediate states and the amplitudes $T_{BC\pi\pi}(s')$ and $F_{\pi\pi}(s')$ for large values of s' — are highly suppressed. In practice one can introduce a high-energy cutoff below which we trust our amplitudes $T_{BC\pi\pi}(s')$ and $F_{\pi\pi}(s')$. Here "trust" means that we can quantify the uncertainties of these amplitudes. Varying this cutoff in a reasonable range and exploring different high-energy extrapolations of the amplitudes allows one to quantify the overall uncertainty of the quantity of interest, namely $F_{BC}(s)$ for low values of $|s|$. In practice, $\sqrt{|s|}$ should be smaller than about 1 GeV, where, for instance, the 4-pion continuum starts to become noticeable.

3 First results

For the pion transition form factor one might schematically identify B with γ^* and C with π in (7). Thus one needs the amplitude $T_{\gamma^* 3\pi}$ and the pion vector form factor $F_{\pi\pi}$. For both quantities one can set up again a dispersive framework. At low energies the relevant intermediate states are again 2-pion states. Thus the pion vector form factor can be determined from the 2-pion (p-wave) phase shift — and possibly a low-order polynomial accounting for not explicitly considered other intermediate states. These states cannot create a strongly varying function at low energies. Thus a low-order polynomial should be sufficient. The pion phase shifts have been determined from (dispersively improved) data by two different groups [10, 11]. The differences between the two analyses can be used as an uncertainty estimate for the input pion phase shift. The pion vector form factor is also experimentally very well known. It agrees fairly well with the dispersive reconstruction from the pion p-wave phase shift. The data can be used to pin down the low-order polynomial; see [12] for details.

The amplitude $T_{\gamma^* 3\pi}$ depends on several variables. The 3-pion invariant mass agrees with the virtuality of the photon. The variation of the amplitude in this variable is sensitive to the 3-pion correlations. In contrast, the 2-pion invariant masses (Mandelstam variables) depend on the 2-pion correlations, i.e. on the pion phase shifts. As already discussed, the pion phase shifts are under control [10, 11]. With this input dispersion theory can be used to determine the dependence of $T_{\gamma^* 3\pi}$ on the 2-pion invariant masses for any photon virtuality [1, 13]. Physically this accounts for pion rescattering and cross-channel rescattering. What remains to be determined is the dependence of the amplitude $T_{\gamma^* 3\pi}$ on the photon virtuality, i.e. on the genuine 3-pion correlations. Here "genuine" means the correlations not caused by a succession of cross-channel 2-pion correlations.

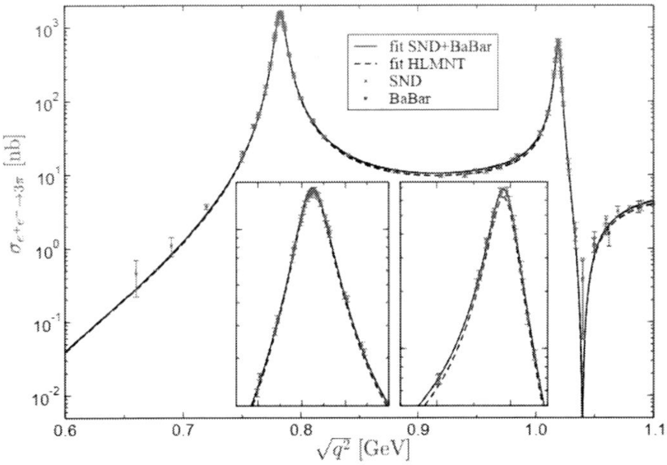

Figure 2. Fit to the cross section $e^+e^- \to 3\pi$. Figure taken from [1]. There is a tension between different data sets. This causes a significant theory uncertainty in our calculations.

Starting from an amplitude that depends on several variables we have reduced the problem to the determination of a one-parameter function that accounts for the genuine 3-pion correlations. This can be achieved by fitting to a one-variable observable, namely to the energy dependence of the total cross section for the reaction $e^+e^- \to 3\pi$. Physically it is known that 3 pions with the quantum numbers of a photon are strongly correlated to the ω and ϕ mesons. Thus we parametrize our one-parameter fit function accordingly by dispersively improved Breit-Wigner functions and a polynomial. The quality of our fit can be inspected in figure 2.

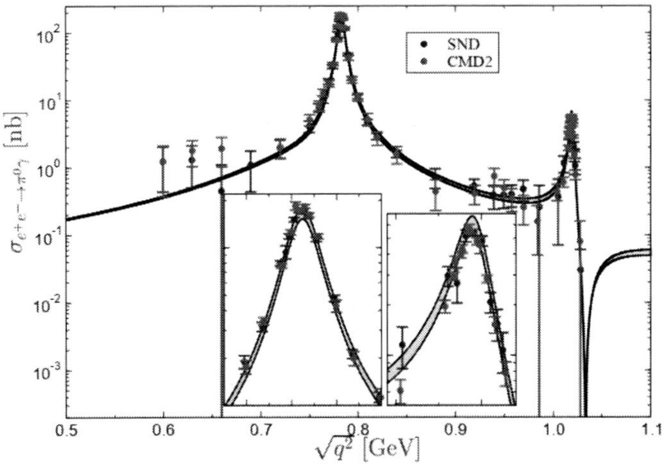

Figure 3. Postdiction for the cross section $e^+e^- \to \pi^0\gamma$ (probes the timelike pion transition form factor). Figure taken from [1]. Theory uncertainties from different data sets for $e^+e^- \to 3\pi$ (main uncertainty), different pion phase shifts, other intermediate states than 2π neglected.

The present status of our dispersive calculation of the pion transition form factor is that we have determined the singly virtual transition form factor in the timelike and spacelike region for momenta below 1 GeV. The dispersive reconstruction of the doubly virtual pion transition form factor is presently investigated. In the timelike region the singly virtual transition form factor is directly related to the cross section for the reaction $e^+e^- \to \pi^0\gamma$. Figure 3 shows that we have obtained an excellent description of this cross section. Our prediction for the singly virtual pion transition form factor in the spacelike region is shown in figure 4. We are looking forward to experimental cross-checks of our prediction.

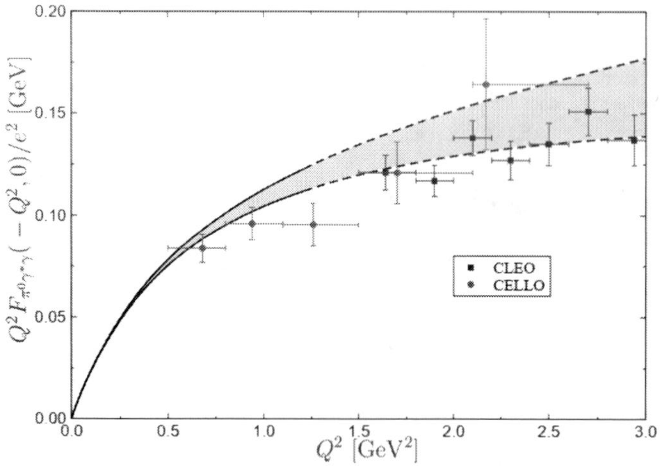

Figure 4. Prediction for the space-like pion transition form factor. Figure taken from [1]. Theory uncertainties from different data sets for $e^+e^- \to 3\pi$ (main uncertainty), different pion phase shifts, other intermediate states than 2π neglected.

The final aim of our endeavor is to determine with a reliable (and hopefully small) uncertainty estimate the pion-pole term as one important part of the hadronic light-by-light scattering contribution to the standard-model prediction for the gyromagnetic ratio of the muon. Clearly the quality of our results hinges on the quality of the experimental input that we use. We will spend the rest of this presentation to spell out a wish list with interesting observables that could help to improve the input for the dispersive calculations and/or to further scrutinize our framework.

4 Interesting observables

It would improve the input for our dispersive framework [7], if (1) remaining ambiguities in the data for the cross section $e^+e^- \to 3\pi$ could be resolved, (2) a Dalitz plot for the decay $\omega \to 3\pi$ could be provided with the same accuracy as already achieved for $\phi \to 3\pi$, (3) the cross section for the reaction $\gamma\pi \to 2\pi$ could be provided.

Cross-checks of our approach can be obtained by improved determinations of the transition form factors $\omega \to \pi^0 e^+e^-$ and $\phi \to \pi^0 e^+e^-$ in the respective full kinematically accessible Dalitz decay regions.

4.1 Resolve ambiguities in the cross section $e^+e^- \to 3\pi$

As a matter of fact the various data sets for the cross section of the reaction $e^+e^- \to 3\pi$ are not fully consistent with each other. Fits to different data sets lead to significant uncertainties for our determination of the singly virtual pion transition form factor in the timelike and spacelike region. Other uncertainties are for instance caused by the high-energy extrapolations of the scattering amplitudes. A first impression of the differences induced by the ambiguities in the cross section $e^+e^- \to 3\pi$ can be deduced from figure 2 where the full line shows a fit to one set of data while the dashed line shows the fit to the data base that essentially contains all data. For instance, the inlay of the ϕ region (right one of the two inlays) shows visible differences at the ϕ peak. For a more detailed discussion of all the uncertainties we refer to [1].

Some uncertainties of the $e^+e^- \to 3\pi$ cross section data and maybe discrepancies between the data sets might be caused by the extrapolation of the actual 3-pion measurements to the full 4π angular coverage. In that context we note that the dispersive framework of [13] predicts the angular distribution of the three pions for a given collision energy of the e^+e^- system, see also figure 5 below. Such information might help for cross-checking the experimental acceptance corrections and for the extrapolation to the full 4π angle.

4.2 Dalitz plot $\omega \to 3\pi$

The amplitude $T_{\gamma^* 3\pi}$ discussed in section 3 is also probed in the decays of vector mesons V to three pions. Here the virtuality of the photon (better to say: of the electromagnetic current) is fixed to the respective mass of the vector meson. If one utilizes a once subtracted dispersion relation, then one can predict the Dalitz decay distribution of the decay $V \to 3\pi$; see the left hand side of figure 5. The subtraction constant is fixed by the corresponding decay width. For the decay $\phi \to 3\pi$ it turns out that the data have such an excellent quality — see the top right panel of figure 5 — that a better description of the data can be achieved by a twice subtracted dispersion relation [13]. The two subtraction constants are determined by a combined fit to the decay width and the distribution itself. As we will argue now, we would like to know the values of these subtraction constant pair also for the decay $\omega \to 3\pi$ and for the reaction $\gamma\pi \to 2\pi$.

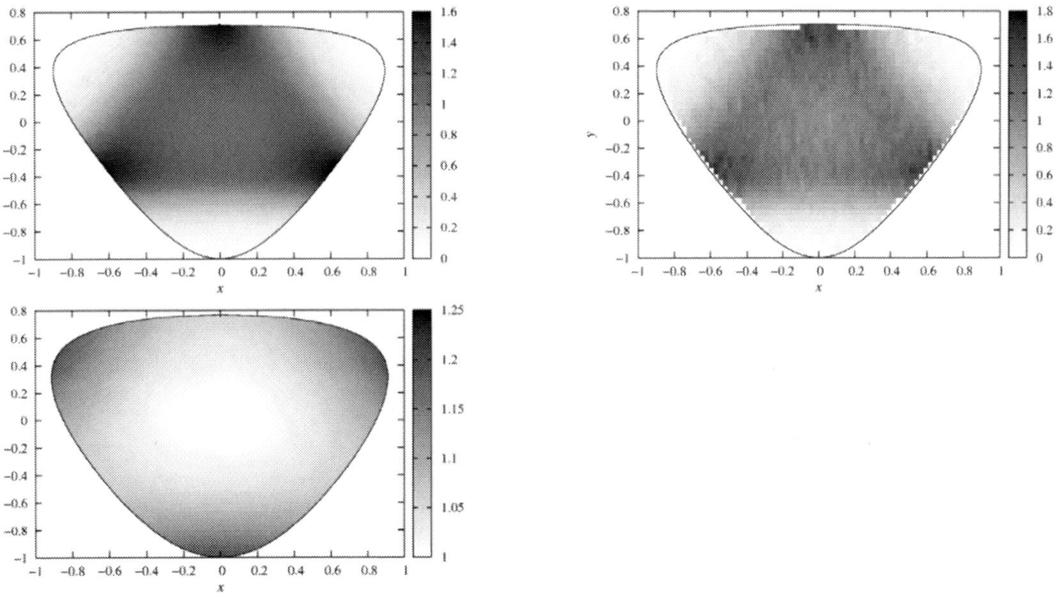

Figure 5. *Top, left:* Dispersive calculation of the Dalitz decay distribution for $\phi \to 3\pi$. Figure taken from [13]. *Top, right:* KLOE data for $\phi \to 3\pi$ with comparable quality. Figure taken from [14]. *Bottom, left:* Dispersive calculation of the Dalitz decay distribution for $\omega \to 3\pi$. Figure taken from [13]. *Bottom, right:* Missing, i.e. no data with comparable quality for $\omega \to 3\pi$.

In [1] we have determined the amplitude $T_{\gamma^* 3\pi}$ for arbitrary photon virtualities from a once subtracted dispersion relation. Since the considered intermediate states, cf. (5), are 2-pion states, the variable in the dispersion integral is one of the 2-pion invariant masses, not the 3-pion invariant mass. However, the subtraction constant depends on the 3-pion invariant mass, i.e. on the photon virtuality. As described in the previous section we have related this dependence of the subtraction constant to the cross section of the reaction $e^+e^- \to 3\pi$.

If one could determine the amplitude $T_{\gamma^* 3\pi}$ from a twice instead of a once subtracted dispersion relation, then the unknown high-energy parts would be further suppressed, cf. the discussion around (7). In turn this would reduce our theory uncertainty. But then we would need to get an idea about the dependence of the second subtraction constant on the photon virtuality. So far, we only know the value at one point, where the virtuality coincides with the ϕ mass. As discussed above, this information comes from the high-quality Dalitz decay distribution $\phi \to 3\pi$. If we could deduce the corresponding second subtraction constant also from a comparable high-quality Dalitz decay distribution $\omega \to 3\pi$ and from the cross section of the reaction $\gamma\pi \to 2\pi$, then one could interpolate the dependence of this second subtraction constant on the photon virtuality from zero up to the ϕ mass.

Unfortunately there exists no $\omega \to 3\pi$ Dalitz plot with an accuracy comparable to the achievements for $\phi \to 3\pi$. Actually until recently there was no experimental indication at all that the pions in the decay $\omega \to 3\pi$ show any final-state interaction. Experimental data could not distinguish between pure phase space and pion rescattering. In [15] first steps beyond pure phase space are documented. Following [13] a parameter α has been introduced to parametrize for the Dalitz plot of $\omega \to 3\pi$ the main deviation from pure phase space. A vanishing α would signal the absence/smallness of pion rescattering. This possibility has been ruled out by [15]; see figure 6. The experimental value of α is definitely different from zero and lies in the ballpark of recent theory predictions. The dispersive prediction is labeled by "Bonn" in figure 6. The future experimental challenge is to produce an α value with a better accuracy than the one obtained by dispersion theory. Only with such a quality the parameters of a twice subtracted dispersion relation could be pinned down.

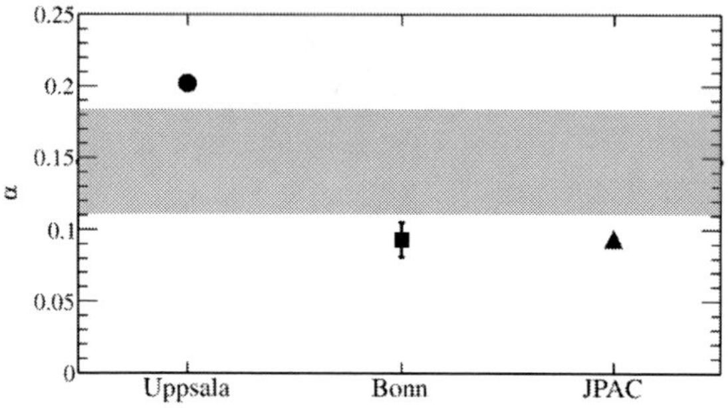

Figure 6. $\omega \to 3\pi$ — first steps beyond pure phase space. Experimental verification (yellow band) of a Dalitz plot shape beyond pure phase space, i.e. $\alpha \neq 0$. Also shown is a comparison to recent theory predictions for α. "Uppsala" refers to [16], "Bonn" to [13], and "JPAC" to [17]. Figure taken from [15].

4.3 Cross section $\gamma + \pi \to 2\pi$

As already noted a remaining piece of information should come from the cross section of the reaction $\gamma + \pi \to 2\pi$. This could be measured with a pion beam by a Primakoff reaction, for instance by the COMPASS experiment at CERN. Actually such a measurement is interesting for several reasons. For low-momentum pions the interaction strength of $\gamma 3\pi$ is dictated by the chiral anomaly. Needless to say that it is very important to experimentally check clean predictions from QCD. Interestingly one can avoid the experimental challenge to extrapolate data to low pion momenta, i.e. to the reaction threshold [18]. Using dispersion theory and pion phase-shift data, the pion-pion interactions in the reaction $\gamma + \pi \to 2\pi$ are under control for all energies where additional intermediate states like 4π can be safely neglected. The only unknown quantity is a subtraction constant which can be identified with the chiral anomaly, up to chiral corrections and uncertainties from the neglected states. As shown in figure 7 one can use the whole energy range up to the ρ-meson mass to pin down the value of the anomaly.

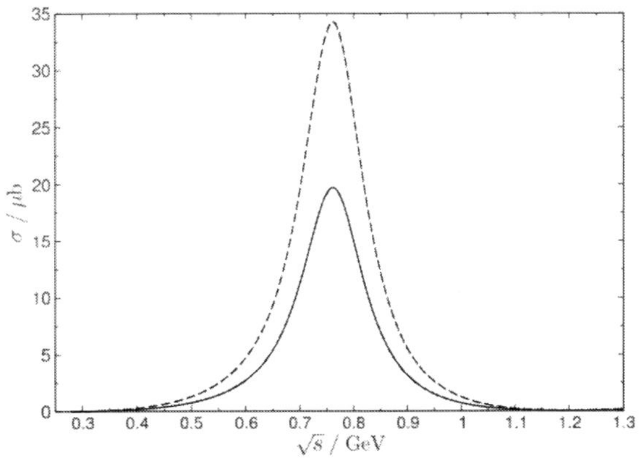

Figure 7. Prediction of the cross section $\gamma + \pi \to 2\pi$ based on dispersion theory. *Solid line:* prediction from anomaly; *dashed line:* size of anomaly scaled up by about 30% . Figure taken from [18].

Concerning our amplitude $T_{\gamma^* 3\pi}$ we use the chiral anomaly to pin down a subtraction constant. A better procedure would be to use the corresponding physical quantity that includes chiral corrections caused by the non-vanishing light quark masses.[3] With data from the reaction $\gamma + \pi \to 2\pi$ or maybe from lattice QCD [19] one could improve on this aspect. With higher-quality data at hand one could even use a twice subtracted dispersion relation and obtain additional knowledge and lower theory uncertainties as pointed out in the previous subsection.

4.4 Transition form factor $\omega \to \pi^0 \ell^+ \ell^-$

The decay of a vector meson to $\pi^0 \ell^+ \ell^-$, where ℓ denotes a lepton, probes a transition form factor that has the same quantum numbers as the pion transition form factor. Consequently the same dispersive framework can be used to determine the form factors of these electromagnetic transitions from a vector meson to a pion [12]. Experimentally one has to determine the differential decay width as a function of the dilepton mass.

For the decay $\omega \to \pi^0 \mu^+ \mu^-$ the dispersive results deviate significantly from the data obtained by the NA60 collaboration [20, 21]. On the other hand, recent data for the decay $\omega \to \pi^0 e^+ e^-$ obtained by the A2 collaboration [22] lie right on top of the dispersive prediction; see figure 8.

If one just looks at the various data sets and their error bars in figure 8 — without the theory lines that might misguide the eye — then one might still get the impression that the data are not in complete contradiction to each other, though the trends differ. Yet it is remarkable that no hadronic theory has been able to fully explain the NA60 data and according to [23] part of these data disagree with general information obtained from QCD and quantum field theory. Thus it would be very illuminating to have additional data that could help to resolve the tension between NA60 on the one side and the dispersive calculation and A2 on the other side.

From a dispersive point of view the input for the transition form factor probed by $\omega \to \pi^0 \ell^+ \ell^-$ comes from the amplitude $\omega\, 3\pi$; the reader may just replace $B \to \omega$ and $C \to \pi$ in (7). The amplitude $\omega\, 3\pi$ can also be determined dispersively [13], but it would be good to have an experimental cross-check on this input for the transition form factor. Thus we repeat our call from subsection 4.2 for a high-quality Dalitz plot for the decay $\omega \to 3\pi$. The situation is somewhat different if one replaces ω by ϕ, as we will discuss next.

[3] The chiral-anomaly prediction is based on the chiral limit.

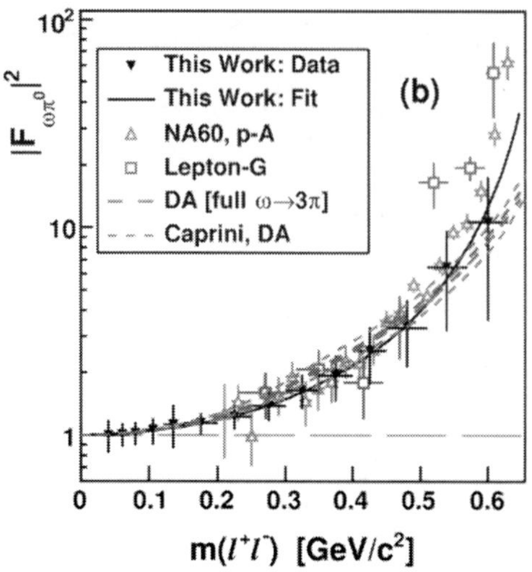

Figure 8. The electromagnetic form factor of the transition ω to π^0 from various experiments and theory approaches. The figure is adopted from the recent work [22] of the A2 collaboration and "This work" refers to it. The magenta lines show the dispersive calculation from [12].

We note in passing that in principle the cross section for the reaction $e^+e^- \to \pi^0\omega$ probes the very same transition form factor, albeit in a kinematically different regime. For the energies relevant for this reaction one cannot expect that the relevant intermediate states, cf. (5), are just the 2-pion states. Thus this reaction cannot be used to scrutinize the present dispersive framework, which applies to reaction energies below 1 GeV, not above.

4.5 Transition form factor $\phi \to \pi^0 e^+e^-$ in ρ-peak region

Another closely related transition form factor is probed by the decay $\phi \to \pi^0\ell^+\ell^-$. A dispersive prediction has been provided in [12], see figure 9. The input for the dispersive calculation is provided by the amplitude $\phi\,3\pi$, which can also be determined dispersively. The latter has been cross-checked successfully against high-quality data on the Dalitz plot $\phi \to 3\pi$ [13]. Thus a very firm prediction for $\phi \to \pi^0\ell^+\ell^-$ has been obtained. Of course, it would be interesting to see if one finds for the ϕ a deviation between data and theory that resembles the deviation between the NA60 data and theory for the ω case.

In contrast to the decay $\omega \to \pi^0\ell^+\ell^-$ there is a resonance in the region that is kinematically allowed for the ϕ decay — the ρ meson. Of course, most discriminative power between various theory approaches is achieved around the resonance peak, where the differential decay rate turns large. Recently first data on the differential decay width for $\phi \to \pi^0 e^+e^-$ have been published by the KLOE-2 collaboration [24] and the corresponding transition form factor has been extracted; see the right hand side of figure 9. So far, the data are restricted to the region below the ρ-meson peak and do not point to a significant disagreement between theory and experiment. But given the puzzle for the ω transition form factor it would be highly desirable to explore the full kinematically accessible range of the $\phi \to \pi^0 e^+e^-$ decay.

All the data called for in the present section would help to cross-check and/or improve the precision of our dispersive framework. Dispersion theory combined with high-quality data has many applications but one concrete ongoing project is to obtain a numerical value with a reliable uncertainty estimate for the pion-pole part of the hadronic light-by-light scattering contribution to the gyromagnetic ratio of the muon.

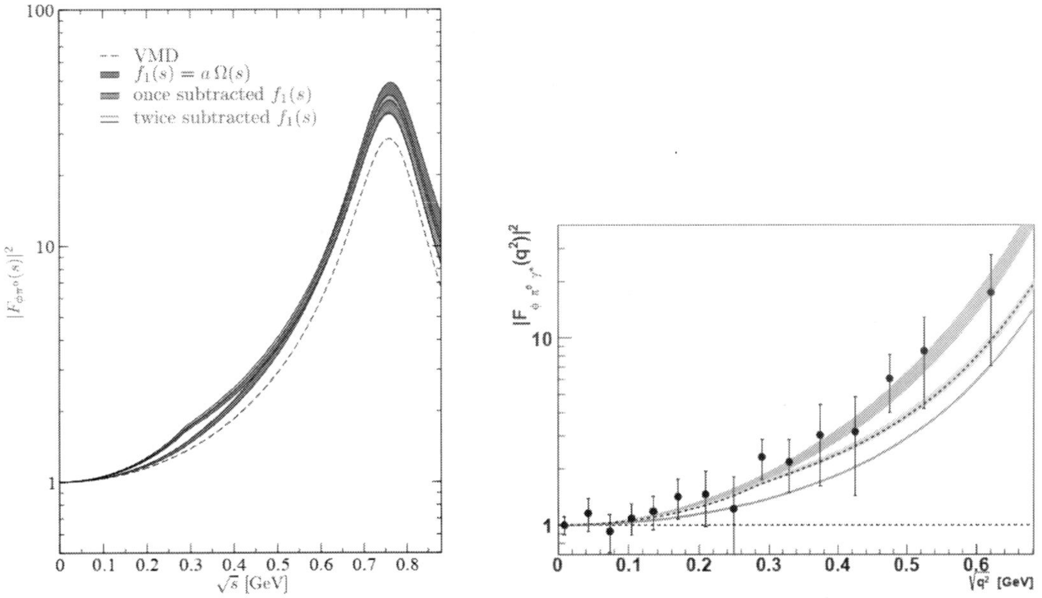

Figure 9. *Left:* Prediction for the electromagnetic form factor of the transition ϕ to π^0 based on dispersion theory [12]. The full calculation is shown by the yellow band. The complete kinematically allowed region for the decay $\phi \to \pi^0 e^+ e^-$ is shown. *Right:* Extraction of the same form factor from KLOE-2 data in the low-energy region. The yellow band from the plot on the left hand side translates here to the middle curve. Figure taken from [24].

Acknowledgments: The speaker (SL) thanks the organizers for an inspiring workshop in a very enjoyable atmosphere. He also thanks P. Adlarson, S. Damjanovic, L. Heijkenskjöld, A. Kupść, V. Metag, P. Salabura, and H. Specht for many illuminating discussions. This work has been supported by the project "Study of Strongly Interacting Matter" (HadronPhysics3, Grant Agreement No. 283286) under the 7th Framework Program of the EU.

References

[1] M. Hoferichter, B. Kubis, S. Leupold, F. Niecknig and S. P. Schneider, Eur. Phys. J. C **74**, 3180 (2014)

[2] E. Czerwiński, S. Eidelman, C. Hanhart, B. Kubis, A. Kupść, S. Leupold, P. Moskal and S. Schadmand, "MesonNet Workshop on Meson Transition Form Factors," arXiv:1207.6556 [hep-ph].

[3] F. Jegerlehner and A. Nyffeler, Phys. Rept. **477**, 1 (2009)

[4] J. Bijnens and J. Prades, Mod. Phys. Lett. A **22**, 767 (2007)

[5] T. Husek and S. Leupold, Eur. Phys. J. C **75**, 586 (2015)

[6] G. Colangelo, M. Hoferichter, M. Procura and P. Stoffer, JHEP **1409**, 091 (2014)

[7] G. Colangelo, M. Hoferichter, B. Kubis, M. Procura and P. Stoffer, Phys. Lett. B **738**, 6 (2014)

[8] G. Colangelo, M. Hoferichter, M. Procura and P. Stoffer, JHEP **1509**, 074 (2015)

[9] G. Colangelo, M. Hoferichter, M. Procura and P. Stoffer, arXiv:1701.06554 [hep-ph].

[10] B. Ananthanarayan, G. Colangelo, J. Gasser and H. Leutwyler, Phys. Rept. **353**, 207 (2001)

[11] R. García-Martín, R. Kamiński, J. R. Peláez, J. Ruiz de Elvira and F. J. Ynduráin, Phys. Rev. D **83**, 074004 (2011)

[12] S. P. Schneider, B. Kubis and F. Niecknig, Phys. Rev. D **86**, 054013 (2012)
[13] F. Niecknig, B. Kubis and S. P. Schneider, Eur. Phys. J. C **72**, 2014 (2012)
[14] A. Aloisio *et al.* [KLOE Collaboration], Phys. Lett. B **561**, 55 (2003) Erratum: [Phys. Lett. B **609**, 449 (2005)]
[15] P. Adlarson *et al.* [WASA-at-COSY Collaboration], B. Kubis, S. Leupold, arXiv:1610.02187 [nucl-ex].
[16] C. Terschlüsen, B. Strandberg, S. Leupold and F. Eichstädt, Eur. Phys. J. A **49**, 116 (2013)
[17] I. V. Danilkin, C. Fernández-Ramírez, P. Guo, V. Mathieu, D. Schott, M. Shi and A. P. Szczepaniak, Phys. Rev. D **91**, 094029 (2015)
[18] M. Hoferichter, B. Kubis and D. Sakkas, Phys. Rev. D **86**, 116009 (2012)
[19] R. A. Briceño, J. J. Dudek, R. G. Edwards, C. J. Shultz, C. E. Thomas and D. J. Wilson, Phys. Rev. D **93**, 114508 (2016)
[20] R. Arnaldi *et al.* [NA60 Collaboration], Phys. Lett. B **677**, 260 (2009)
[21] R. Arnaldi *et al.* [NA60 Collaboration], Phys. Lett. B **757**, 437 (2016)
[22] P. Adlarson *et al.*, arXiv:1609.04503 [hep-ex].
[23] B. Ananthanarayan, I. Caprini and B. Kubis, Eur. Phys. J. C **74**, 3209 (2014)
[24] A. Anastasi *et al.* [KLOE-2 Collaboration], Phys. Lett. B **757**, 362 (2016)

Dispersion relations for hadronic light-by-light scattering and the muon $g-2$

Massimiliano Procura[1,⋆], Gilberto Colangelo[2,⋆⋆], Martin Hoferichter[3,⋆⋆⋆], and Peter Stoffer[4,5,⋆⋆⋆⋆]

[1] *Theoretical Physics Department, CERN, Geneva, Switzerland*
[2] *Albert Einstein Center for Fundamental Physics, Institute for Theoretical Physics, University of Bern, Sidlerstrasse 5, 3012 Bern, Switzerland*
[3] *Institute for Nuclear Theory, University of Washington, Seattle, WA 98195-1550, USA*
[4] *Helmholtz-Institut für Strahlen- und Kernphysik (Theory) and Bethe Center for Theoretical Physics, University of Bonn, 53115 Bonn, Germany*
[5] *Department of Physics, University of California at San Diego, La Jolla, CA 92093, USA*

Abstract. The largest uncertainties in the Standard Model calculation of the anomalous magnetic moment of the muon $(g-2)_\mu$ come from hadronic effects, and in a few years the subleading hadronic light-by-light (HLbL) contribution might dominate the theory error. We present a dispersive description of the HLbL tensor, which is based on unitarity, analyticity, crossing symmetry, and gauge invariance. This opens up the possibility of a data-driven determination of the HLbL contribution to $(g-2)_\mu$ with the aim of reducing model dependence and achieving a reliable error estimate.

Our dispersive approach defines unambiguously the pion-pole and the pion-box contribution to the HLbL tensor. Using Mandelstam double-spectral representation, we have proven that the pion-box contribution coincides exactly with the one-loop scalar-QED amplitude, multiplied by the appropriate pion vector form factors. Using dispersive fits to high-statistics data for the pion vector form factor, we obtain $a_\mu^{\pi\text{-box}} = -15.9(2)\times 10^{-11}$. A first model-independent calculation of effects of $\pi\pi$ intermediate states that go beyond the scalar-QED pion loop is also presented. We combine our dispersive description of the HLbL tensor with a partial-wave expansion and demonstrate that the known scalar-QED result is recovered after partial-wave resummation. After constructing suitable input for the $\gamma^*\gamma^* \to \pi\pi$ helicity partial waves based on a pion-pole left-hand cut (LHC), we find that for the dominant charged-pion contribution this representation is consistent with the two-loop chiral prediction and the COMPASS measurement for the pion polarizability. This allows us to reliably estimate S-wave rescattering effects to the full pion box and leads to $a_\mu^{\pi\text{-box}} + a_{\mu,J=0}^{\pi\pi,\pi\text{-pole LHC}} = -24(1)\times 10^{-11}$.

⋆e-mail: massimiliano.procura@cern.ch
⋆⋆e-mail: gilberto@itp.unibe.ch
⋆⋆⋆e-mail: mhofer@uw.edu
⋆⋆⋆⋆e-mail: stoffer@hiskp.uni-bonn.de

1 Introduction

The anomalous magnetic moment of the muon $(g-2)_\mu$ has been measured [1] and computed to very high precision of about 0.5 ppm (see e.g. [2]). For more than a decade, a discrepancy has persisted between the experiment and the Standard Model prediction, now of about 3σ. Forthcoming measurements at FNAL [3] and J-PARC [4] will update the experimental value. The aim is to increase the precision by a factor of 4 and check for systematic errors.

The main uncertainty of the theory prediction is due to strong interaction effects. At present, the largest error arises from hadronic vacuum polarisation, which, however, forthcoming data from e^+e^- experiments [2] may help reduce. Thus in a few years, the subleading[1] hadronic light-by-light contribution might dominate the theory error. In present calculations of the HLbL contribution, systematic errors are difficult, if not impossible, to quantify, due to model dependence. A new strategy is required to provide a solid estimate of the theory uncertainties and reduce them. Lattice QCD is making remarkable progress in this direction, and may play a leading role in this field in the near future [7–13]. In [14, 15], we have presented the first dispersive description of the HLbL tensor.[2] By making use of the fundamental principles of unitarity, analyticity, crossing symmetry, and gauge invariance, we provide an approach that reduces model dependence and allows for a more data-driven determination of the HLbL contribution to $(g-2)_\mu$.

Here, we report on a several improvements of our dispersive framework [17–19]. We have constructed a generating set of Lorentz structures for the HLbL tensor that is free of kinematic singularities and zeros. This simplifies significantly the calculation of the HLbL contribution to $(g-2)_\mu$. Within our dispersive formalism, the definitions of both the pion-pole and pion-box topologies are unambiguous. By constructing a Mandelstam representation for the scalar functions, we prove that the box topologies coincide with the scalar-QED (sQED) contribution multiplied by pion vector form factors. Here we present a numerical evaluation of the pion box using a form factor fit to high-statistics data, in turn using a dispersive representation to analytically continue the time-like data into the space-like region required for the $(g-2)_\mu$ integral and show that this contribution can be calculated with negligible uncertainties. We then present a first numerical evaluation of S-wave $\pi\pi$-rescattering effects, which unitarize the pion-pole contribution to $\gamma^*\gamma^* \to \pi\pi$. This constitutes the first step towards a full treatment of the $\gamma^*\gamma^* \to \pi\pi$ partial waves [20–22]. In particular, our calculation settles the role of the pion polarizability, which enters at next-to-leading order in the chiral expansion of the HLbL amplitude [23–25] and has been suspected to produce sizable corrections in [24].

2 Lorentz structure of the HLbL tensor

In order to study the HLbL contribution to $(g-2)_\mu$, we need a description of the HLbL tensor, namely the hadronic Green's function of four light-quark electromagnetic currents, evaluated in pure QCD:

$$\Pi^{\mu\nu\lambda\sigma}(q_1,q_2,q_3) = -i\int d^4x\, d^4y\, d^4z\, e^{-i(q_1\cdot x+q_2\cdot y+q_3\cdot z)}\langle 0|T\{j^\mu_{\mathrm{em}}(x)j^\nu_{\mathrm{em}}(y)j^\lambda_{\mathrm{em}}(z)j^\sigma_{\mathrm{em}}(0)\}|0\rangle. \quad (1)$$

Gauge invariance requires the HLbL tensor to satisfy the Ward-Takahashi (WT) identities

$$\{q_1^\mu, q_2^\nu, q_3^\lambda, q_4^\sigma\}\Pi_{\mu\nu\lambda\sigma}(q_1,q_2,q_3) = 0, \quad (2)$$

where $q_4 = q_1 + q_2 + q_3$. The HLbL tensor can be written a priori in terms of 138 basic Lorentz structures built out of the metric tensor and the four-momenta [26]. Our first task is to write the

[1] Even higher-order hadronic contributions have been considered in [5, 6].

[2] A different approach, which aims at a dispersive description of the muon vertex function instead of the HLbL tensor, has been presented in [16].

HLbL tensor in terms of Lorentz structures that satisfy the WT identities, while at the same time the scalar functions that multiply these structures must be free of kinematic singularities and zeros. A recipe for the construction of these structures has been given by Bardeen, Tung [27], and Tarrach [28] (BTT) for generic photon amplitudes. Gauge invariance imposes 95 linear relations between the 138 initial scalar functions. A generating set[3] consisting of 43 elements can be constructed following Bardeen and Tung [27]. However, as it was shown by Tarrach [28], such a set is not free of kinematic singularities and has to be supplemented by additional structures. We find a redundant BTT generating set of dimension 54:

$$\Pi^{\mu\nu\lambda\sigma}(q_1, q_2, q_3) = \sum_{i=1}^{54} T_i^{\mu\nu\lambda\sigma} \Pi_i(s, t, u), \tag{3}$$

with scalar functions Π_i depending on the Mandelstam variables $s = (q_1 + q_2)^2$, $t = (q_1 + q_3)^2$, $u = (q_2 + q_3)^2$ as well as the photon virtualities q_i^2, and Lorentz structures $T_i^{\mu\nu\lambda\sigma}$ [17, 18, 31]. This decomposition fulfills gauge invariance manifestly

$$\{q_1^\mu, q_2^\nu, q_3^\lambda, q_4^\sigma\} T^i_{\mu\nu\lambda\sigma} = 0, \tag{4}$$

is highly crossing symmetric (with only 7 distinct structures, all remaining 47 being related to these by crossing transformations), and ensures that the coefficient functions Π_i do not contain kinematic singularities and zeros. In addition, the BTT decomposition typically allows for a very economical representation of HLbL amplitudes, e.g. one of the structures coincides with the amplitude for a pseudoscalar pole, while even the sQED amplitude becomes very compact once expressed in terms of BTT functions.

3 HLbL contribution to $(g-2)_\mu$

The HLbL contribution to $a_\mu = (g-2)_\mu/2$ can be extracted with the help of well-known Dirac projector techniques [32]. With our decomposition of the HLbL tensor in 54 structures, this amounts to the calculation of the following two-loop integral:

$$\begin{aligned} a_\mu^{\mathrm{HLbL}} = &-\frac{e^6}{48 m_\mu} \int \frac{d^4 q_1}{(2\pi)^4} \frac{d^4 q_2}{(2\pi)^4} \frac{1}{q_1^2 q_2^2 (q_1+q_2)^2} \frac{1}{(p+q_1)^2 - m_\mu^2} \frac{1}{(p-q_2)^2 - m_\mu^2} \\ &\times \mathrm{Tr}\Big((\not{p} + m_\mu)[\gamma_\rho, \gamma_\sigma](\not{p} + m_\mu)\gamma_\mu(\not{p} + \not{q}_1 + m_\mu)\gamma_\lambda(\not{p} - \not{q}_2 + m_\mu)\gamma_\nu \Big) \\ &\times \sum_{i=1}^{54} \left(\frac{\partial}{\partial q_{4\rho}} T_i^{\mu\nu\lambda\sigma}(q_1, q_2, q_4 - q_1 - q_2) \right) \bigg|_{q_4=0} \Pi_i(q_1, q_2, -q_1 - q_2). \end{aligned} \tag{5}$$

After a Wick rotation of the momenta, five of the eight loop integrals can be carried out with the technique of Gegenbauer polynomials [33]. We have checked that this Wick rotation is justified even in the presence of anomalous thresholds in the scalar functions Π_i. In analogy to the pion-pole contribution [34], a three-dimensional integral representation for the full HLbL contribution to $(g-2)_\mu$ can be worked out [19] :

$$a_\mu^{\mathrm{HLbL}} = \frac{\alpha^3}{432\pi^2} \int_0^\infty d\Sigma\, \Sigma^3 \int_0^1 dr\, r\sqrt{1-r^2} \int_0^{2\pi} d\phi \sum_{i=1}^{12} T_i(Q_1, Q_2, Q_3) \bar{\Pi}_i(Q_1, Q_2, Q_3), \tag{6}$$

[3] In 4 space-time dimensions, there are two additional linear relations, hence a basis consists of 41 elements [29, 30].

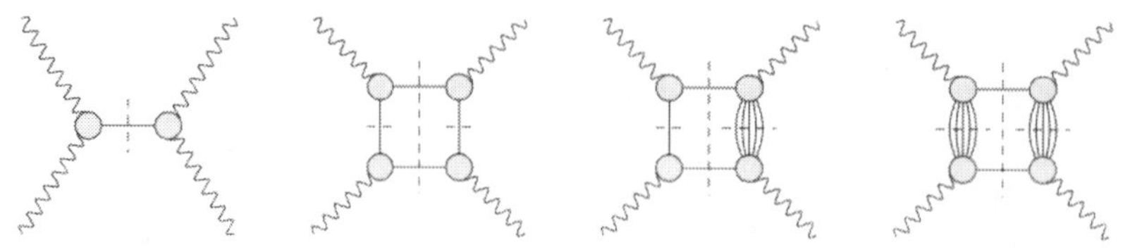

Figure 1. Unitarity diagrams according to the Mandelstam representation. Crossed diagrams are omitted.

where the T_i are known kernel functions, the $\bar{\Pi}_i$ suitable linear combinations of the BTT Π_i, and the Euclidean momenta are given by [30]

$$Q_{1,2}^2 = \frac{\Sigma}{3}\left(1 - \frac{r}{2}\cos\phi \mp \frac{r}{2}\sqrt{3}\sin\phi\right), \qquad Q_3^2 = \frac{\Sigma}{3}(1 + r\cos\phi). \tag{7}$$

There are only 6 distinct functions $\bar{\Pi}_i$, the remaining ones are again related to these by crossing symmetry. It suffices to calculate the $\bar{\Pi}_i$ in the kinematic limit where $q_4 \to 0$, the transition to $(g-2)_\mu$ then proceeds by means of Eq.(6).

4 Mandelstam representation

Although the scalar functions in the master formula Eq.(6) are needed only for the reduced kinematics where the limit $q_4 \to 0$ is taken, we define the dispersion relation in the Mandelstam variables of the four-point function with general kinematics and evaluate it only afterwards for the special case $q_4 \to 0$. This procedure has the following advantage: the HLbL contribution to $(g-2)_\mu$ splits into contributions from different topologies (shown in fig. 1), each of them linked to a specific sub-process, which is either data input or again a dispersively reconstructed quantity. These different contributions are discussed in the following.

Gauge invariance, encoded in the BTT decomposition, leads to Lorentz structures $T_i^{\mu\nu\lambda\sigma}$ of mass dimension 4, 6, and 8. Hence, we expect the scalar functions Π_i to be rather strongly suppressed at high energies. Thus we write down unsubtracted double-spectral (Mandelstam) representations for the Π_i [35], *i.e.* parameter-free dispersion relations. The input to the dispersion relation are the residues at poles (due to single-particle intermediate states) and the discontinuities along branch cuts (due to two-particle intermediate states). Both are defined by the unitarity relation, in which the intermediate states are always on-shell. We neglect contributions from intermediate states consisting of more than two particles in the primary cut. Heavier intermediate states are expected to be suppressed by higher thresholds and smaller phase space, in agreement with the outcome of model calculations.

The first topology in fig. 1 consists of the pion pole, *i.e.* the terms arising from a single pion intermediate state. This contribution is well-known [34] and given by

$$\begin{aligned}
\bar{\Pi}_1^{\pi^0\text{-pole}} &= -\frac{\mathcal{F}_{\pi^0\gamma^*\gamma^*}(-Q_1^2,-Q_2^2)\mathcal{F}_{\pi^0\gamma^*\gamma^*}(-Q_3^2,0)}{Q_3^2+M_\pi^2}, \\
\bar{\Pi}_2^{\pi^0\text{-pole}} &= -\frac{\mathcal{F}_{\pi^0\gamma^*\gamma^*}(-Q_1^2,-Q_3^2)\mathcal{F}_{\pi^0\gamma^*\gamma^*}(-Q_2^2,0)}{Q_2^2+M_\pi^2},
\end{aligned} \tag{8}$$

where $\mathcal{F}_{\pi^0\gamma^*\gamma^*}$ denotes the pion transition form factor (for off-shell photons but an on-shell pion).

The other topologies in fig. 1 are obtained by selecting two-pion intermediate states in the primary cut. The sub-process $\gamma^*\gamma^* \to \pi\pi$ is again cut in the crossed channel. If we single out the pion-pole contribution in both of the sub-processes, we obtain the box topologies for HLbL. For higher intermediate states in the crossed channel of $\gamma^*\gamma^* \to \pi\pi$, we obtain boxes with multi-particle cuts instead of poles in the sub-processes.

By explicitly constructing the Mandelstam representation, we have shown that the box topologies in the sense of unitarity have the same analytic structure as the one-loop sQED contribution, multiplied with pion electromagnetic form factors $F_\pi^V(q_i^2)$ for each of the off-shell photons (FsQED). The dispersion relation defines unambiguously this particular q_i^2 dependence. With the construction of the Mandelstam representation, we prove that FsQED and box topologies are the same. Note that the sQED loop contribution in terms of Feynman diagrams consists of boxes, triangles, and bulbs, but that the corresponding unitarity diagrams are just box topologies. This can be understood as follows: in sQED, the appearance of triangle and bulb diagrams is due to the seagull vertex, needed to ensure gauge invariance. In our formalism, gauge invariance is already encoded in the BTT tensor decomposition Eq.(3). Due to the high degree of crossing symmetry, the pion-box contribution can be expressed in terms of either fixed-s, -t, or -u dispersion relations, or in a symmetrized form

$$\Pi_i^{\pi\text{-box}}(s,t,u) = \frac{1}{3}\left[\frac{1}{\pi}\int_{4M_\pi^2}^{\infty} dt' \frac{\operatorname{Im}\Pi_i^{\pi\text{-box}}(s,t',u')}{t'-t} + \frac{1}{\pi}\int_{4M_\pi^2}^{\infty} du' \frac{\operatorname{Im}\Pi_i^{\pi\text{-box}}(s,t',u')}{u'-u} + \text{fixed-}t + \text{fixed-}u\right].$$

In this case the representation is exact.

Once heavier intermediate states are considered, see Fig. 1, a more detailed investigation of the double spectral functions is required. In practice, such contributions can be included using a partial-wave expansion, in which case the sub-process becomes a polynomial in the crossed variable and the crossed-channel cuts are neglected. Writing down all crossed versions of the unitarity diagrams in Fig. 1 with a two-particle primary cut, one sees that each double spectral region appears exactly twice in a symmetrized form as in (9), so that the prefactor changes according to $1/3 \to 1/2$ [31], with corrections suppressed by the mass scale of the neglected LHC. In particular, this representation becomes exact for $\pi\pi$-rescattering effects, which, by definition, are polynomial in the crossed Mandelstam variable.

For a numerical evaluation of the pion box contribution, the only input needed is the pion vector form factor in the space-like region. By fitting a dispersive representation which accounts for the prominent singularities in the low-energy region as suggested in [36, 37] to both space-like [38] and time-like form factor data [39–44], we obtain

$$a_\mu^{\pi\,\text{box}} = -15.9(2)\times 10^{-11} \tag{9}$$

where the uncertainty is determined from the differences between the time-like data sets as well as the details of the fit representation. We stress that previous evaluations of a "pion loop" [45, 46] had large model-dependent uncertainties, whereas our evaluation of an unambiguously-defined pion box has a negligible one.

5 Partial-wave expansion

Constraints from unitarity are most conveniently formulated in a partial-wave expansion for HLbL helicity amplitudes $h^J_{\lambda_1\lambda_2,\lambda_3\lambda_4}$ with angular momentum J and helicity labels λ_i. In this case the unitarity relation becomes diagonal

$$\operatorname{Im} h^J_{\lambda_1\lambda_2,\lambda_3\lambda_4}(s) = \frac{\sigma_\pi(s)}{16\pi S} h_{J,\lambda_1\lambda_2}(s) h^*_{J,\lambda_3\lambda_4}(s), \tag{10}$$

where $\sigma_\pi(s) = \sqrt{1-4M_\pi^2/s}$, $S = 2$ for indistinguishable particles, and $h_{J,\lambda_1\lambda_2}$ the helicity partial waves for $\gamma^*\gamma^* \to \pi\pi$, and once formulated in isospin basis, Watson's theorem guarantees that the phases on the right-hand side cancel to produce a well-defined imaginary part.

Table 1. Saturation of $a_\mu^{\pi\text{-box}}$ for maximal angular momentum $J_{\max}$.

$J_{\max}$	fixed-s	fixed-t	fixed-u	average
0	0.0%	106.2%	106.2%	70.8%
2	73.9%	102.3%	92.7%	89.6%
4	89.2%	101.5%	96.4%	95.7%
6	94.3%	100.7%	97.9%	97.6%
8	96.5%	100.4%	98.7%	98.5%

There are 41 independent helicity amplitudes for the full HLbL tensor, which reduce to 27 if one photon is taken on-shell. By overcoming several technical and conceptual challenges [19, 31], we were able to construct a set of 27 amplitudes $\check{\Pi}_i$ related to the 27 singly-on-shell helicity amplitudes by a basis change that we have derived in explicit analytic form. In the limit $q_4 \to 0$ a subset of the $\check{\Pi}_i$ includes all the scalar functions needed as input in (6) [31]. Moreover, this set of 27 amplitudes is manifestly free of Tarrach [28] or $d = 4$ ambiguities [30]. For singly-on-shell kinematics, there still exist 15 sum rules among the 27 helicity amplitudes, which we have exploited to optimize to a certain degree the representation with respect to the convergence of the partial-wave expansion.

The pion box provides an ideal test case for our partial-wave framework since the full result is known and explicit expressions for all BTT scalar functions are available. Our results, for simplicity obtained by using a vector-meson-dominance pion form factor $F_\pi^V(q^2) = M_\rho^2/(M_\rho^2 - q^2)$, with $a_\mu^{\pi\text{-box, VMD}} = -16.4 \times 10^{-11}$, are shown in Table 1, demonstrating that each fixed-s, -t, -u representation approaches the full result (going up to $J_{\max} = 20$, we checked that also the remaining differences disappear after partial-wave resummation). The vanishing S-wave contribution for fixed-s is well understood and partly a matter of convention [19]. The convergence pattern looks very reasonable.

6 $\pi\pi$ rescattering effects

The partial-wave decomposition of the pion box demonstrates that the traditional sQED pion loop can be understood as the result of resumming the Born-term contributions to the $\gamma^*\gamma^* \to \pi\pi$ partial waves $h_{J,\lambda_1\lambda_2}(s)$ introduced in the context of the unitarity relation (10). However, unitarity for the sub-process itself implies

$$\operatorname{Im} h^I_{J,\lambda_1\lambda_2}(s) = \sin\delta^I_J(s) e^{-i\delta^I_J(s)} h^I_{J,\lambda_1\lambda_2}(s), \tag{11}$$

with isospin labels I and $\pi\pi$ phase shifts δ^I_J. This relation is clearly violated for the (real) Born terms alone. With a partial-wave framework at our disposal, we are thus in the position to evaluate the corresponding unitarity corrections, more conventionally referred to as $\pi\pi$-rescattering effects, as a first important step to account for contributions beyond the pion box.

In contrast to the on-shell and singly-virtual case [20–22], the calculation of the $\gamma^*\gamma^* \to \pi\pi$ partial waves for two off-shell photons is complicated by the fact that even for S-waves two different helicity partial waves, $h_{0,++}$ and $h_{0,00}$, become coupled, including off-diagonal kernel functions required to eliminate kinematic singularities [6, 18]. We applied this framework to construct the $\gamma^*\gamma^* \to \pi\pi$ amplitudes that correspond to the rescattering corrections to the Born terms, whose solution can still be derived based on Muskhelishvili–Omnès methods [47, 48]. We used $\pi\pi$ phase shifts based on the modified inverse-amplitude method [49], for the main reason that it has a simple analytic expression, while at the same time it reproduces accurately the low-energy properties of the phase shifts as well as

Table 2. S-wave rescattering corrections to $a_\mu^{\pi\text{-box}}$, in units of 10^{-11}, for both isospin components and in total.

cutoff	1 GeV	1.5 GeV	2 GeV	∞
$I=0$	-9.2	-9.5	-9.3	-8.8
$I=2$	2.0	1.3	1.1	0.9
sum	-7.3	-8.3	-8.3	-7.9

pole position and couplings of the $f_0(500)$ resonance [19]. We also tested the sensitivity to the asymptotic part of the dispersive integrals by studying solutions with different cutoff values $\Lambda = [1\text{ GeV}, \infty)$, constructed with finite-matching-point techniques [21, 50–53]. Our results for the rescattering contribution, summarized in Table 2, turn out to be stable over a wide range of cutoffs, indicating that our input for the $\gamma^*\gamma^* \to \pi\pi$ partial waves reliably unitarizes the Born-term LHC, which should indeed dominate at low energies. The isospin-0 part of the result can be interpreted as a model-independent implementation of the contribution from the $f_0(500)$ of about -9×10^{-11} to HLbL scattering in $(g-2)_\mu$. In total, we obtain for the $\pi\pi$-rescattering effects related to the pion-pole LHC

$$a_{\mu,J=0}^{\pi\pi,\pi\text{-pole LHC}} = -8(1)\times10^{-11}, \tag{12}$$

where the error is dominated by the uncertainties related to the asymptotic parts of the integral.

Finally, it is instructive to consider the separate contributions not in the isospin, but in the charge basis. Our numerical analysis [19] shows that the rescattering effects in (12) are, as expected, dominated by the charged pion, with input for the $\gamma^*\gamma^* \to \pi\pi$ partial waves fully consistent both with the recent COMPASS measurement [54] of its dipole polarizability and the corresponding two-loop chiral prediction [55]. For this reason (12) can be considered a model-independent implementation of effects related to the low-energy constants L_9 and L_{10} in Chiral Perturbation Theory, which were suspected to produce large effects in [24]. Our calculation proves that this is not the case, and that the related rescattering corrections are indeed of much more reasonable size (a similar conclusion was reached within a model approach in [25]).

To summarize, we have shown that our framework allows us to estimate very accurately the combined effect of two-pion intermediate states generated by a pion-pole LHC and its S-wave unitarization

$$a_\mu^{\pi\text{-box}} + a_{\mu,J=0}^{\pi\pi,\pi\text{-pole LHC}} = -24(1)\times10^{-11}, \tag{13}$$

which is considered to be among the most important contributions after the dominant pseudoscalar poles, but was so far affected by significant uncertainties. This first numerical result based on our dispersive approach lays the foundation for extensions towards higher partial waves, an improved LHC in the $\gamma^*\gamma^* \to \pi\pi$ subamplitudes as well as higher-mass intermediate states, all important prerequisites for a model-independent data-driven evaluation of the complete HLbL contribution to $(g-2)_\mu$.

The speaker (MP) thanks the organizing committee of the *KLOE-2 Workshop on e^+e^- collider physics at 1 GeV* for their kind invitation. Financial support by the DFG (SFB/TR 16, "Subnuclear Structure of Matter," SFB/TR 110, "Symmetries and the Emergence of Structure in QCD"), the DOE (Grant No. DE-FG02-00ER41132), the National Science Foundation (Grant No. NSF PHY-1125915), the Swiss National Science Foundation (Project No. P300P2_167751) and the European Community's 7th Framework Programme (PIEF-GA-2013-622527) is gratefully acknowledged.

References

[1] G. W. Bennett *et al.* [Muon g-2 Collaboration], Phys. Rev. D **73** (2006) 072003 [hep-ex/0602035].

[2] T. Blum, A. Denig, I. Logashenko, E. de Rafael, B. Lee Roberts, T. Teubner and G. Venanzoni, arXiv:1311.2198 [hep-ph].

[3] J. Grange *et al.* [Muon g-2 Collaboration], arXiv:1501.06858 [physics.ins-det].

[4] N. Saito [J-PARC g-'2/EDM Collaboration], AIP Conf. Proc. **1467** (2012) 45.

[5] A. Kurz, T. Liu, P. Marquard and M. Steinhauser, Phys. Lett. B **734** (2014) 144 [arXiv:1403.6400 [hep-ph]].

[6] G. Colangelo, M. Hoferichter, A. Nyffeler, M. Passera and P. Stoffer, Phys. Lett. B **735** (2014) 90 [arXiv:1403.7512 [hep-ph]].

[7] M. Hayakawa, T. Blum, T. Izubuchi and N. Yamada, PoS LAT **2005** (2006) 353 [hep-lat/0509016].

[8] T. Blum, M. Hayakawa and T. Izubuchi, PoS LATTICE **2012** (2012) 022 [arXiv:1301.2607 [hep-lat]].

[9] T. Blum, S. Chowdhury, M. Hayakawa and T. Izubuchi, Phys. Rev. Lett. **114** (2015) 1, 012001 [arXiv:1407.2923 [hep-lat]].

[10] J. Green, O. Gryniuk, G. von Hippel, H. B. Meyer and V. Pascalutsa, arXiv:1507.01577 [hep-lat].

[11] T. Blum, N. Christ, M. Hayakawa, T. Izubuchi, L. Jin and C. Lehner, Phys. Rev. D **93**, 014503 (2016) [arXiv:1510.07100 [hep-lat]].

[12] T. Blum, N. Christ, M. Hayakawa, T. Izubuchi, L. Jin, C. Jung and C. Lehner, Phys. Rev. Lett. **118**, 022005 (2017) [arXiv:1610.04603 [hep-lat]].

[13] A. Gérardin, H. B. Meyer and A. Nyffeler, Phys. Rev. D **94**, 074507 (2016) [arXiv:1607.08174 [hep-lat]].

[14] G. Colangelo, M. Hoferichter, M. Procura and P. Stoffer, JHEP **1409** (2014) 091 [arXiv:1402.7081 [hep-ph]].

[15] G. Colangelo, M. Hoferichter, B. Kubis, M. Procura and P. Stoffer, Phys. Lett. B **738** (2014) 6 [arXiv:1408.2517 [hep-ph]].

[16] V. Pauk and M. Vanderhaeghen, Phys. Rev. D **90** (2014) 11, 113012 [arXiv:1409.0819 [hep-ph]].

[17] P. Stoffer, arXiv:1412.5171 [hep-ph].

[18] G. Colangelo, M. Hoferichter, M. Procura and P. Stoffer, JHEP **1509** (2015) 074 [arXiv:1506.01386 [hep-ph]].

[19] G. Colangelo, M. Hoferichter, M. Procura and P. Stoffer, arXiv:1701.06554 [hep-ph].

[20] R. García-Martín and B. Moussallam, Eur. Phys. J. C **70**, 155 (2010) [arXiv:1006.5373 [hep-ph]].

[21] M. Hoferichter, D. R. Phillips and C. Schat, Eur. Phys. J. C **71**, 1743 (2011) [arXiv:1106.4147 [hep-ph]].

[22] B. Moussallam, Eur. Phys. J. C **73**, 2539 (2013) [arXiv:1305.3143 [hep-ph]].

[23] K. T. Engel, H. H. Patel and M. J. Ramsey-Musolf, Phys. Rev. D **86**, 037502 (2012) [arXiv:1201.0809 [hep-ph]].

[24] K. T. Engel and M. J. Ramsey-Musolf, Phys. Lett. B **738**, 123 (2014) [arXiv:1309.2225 [hep-ph]].

[25] J. Bijnens and J. Relefors, JHEP **1609**, 113 (2016) [arXiv:1608.01454 [hep-ph]].

[26] R. A. Leo, A. Minguzzi and G. Soliani, Nuovo Cim. A **30** (1975) 270.

[27] W. A. Bardeen and W. K. Tung, Phys. Rev. **173** (1968) 1423 [Phys. Rev. D **4** (1971) 3229].

[28] R. Tarrach, Nuovo Cim. A **28** (1975) 409.

[29] G. Eichmann, C. S. Fischer, W. Heupel and R. Williams, arXiv:1411.7876 [hep-ph].
[30] G. Eichmann, C. S. Fischer and W. Heupel, Phys. Rev. D **92** (2015) 5, 056006 [arXiv:1505.06336 [hep-ph]].
[31] G. Colangelo, M. Hoferichter, M. Procura, P. Stoffer, in preparation.
[32] J. Aldins, T. Kinoshita, S. J. Brodsky and A. J. Dufner, Phys. Rev. D **1** (1970) 2378.
[33] J. L. Rosner, Annals Phys. **44** (1967) 11.
[34] M. Knecht and A. Nyffeler, Phys. Rev. D **65** (2002) 073034 [hep-ph/0111058].
[35] S. Mandelstam, Phys. Rev. **112** (1958) 1344.
[36] H. Leutwyler, hep-ph/0212324.
[37] G. Colangelo, Nucl. Phys. Proc. Suppl. **131** (2004) 185 [hep-ph/0312017].
[38] S. R. Amendolia *et al.* [NA7 Collaboration], Nucl. Phys. B **277**, 168 (1986).
[39] M. N. Achasov *et al.*, J. Exp. Theor. Phys. **103**, 380 (2006) [Zh. Eksp. Teor. Fiz. **130**, 437 (2006)] [hep-ex/0605013].
[40] R. R. Akhmetshin *et al.* [CMD-2 Collaboration], Phys. Lett. B **648**, 28 (2007) [hep-ex/0610021].
[41] B. Aubert *et al.* [BaBar Collaboration], Phys. Rev. Lett. **103**, 231801 (2009) [arXiv:0908.3589 [hep-ex]].
[42] F. Ambrosino *et al.* [KLOE Collaboration], Phys. Lett. B **700**, 102 (2011) [arXiv:1006.5313 [hep-ex]].
[43] D. Babusci *et al.* [KLOE Collaboration], Phys. Lett. B **720**, 336 (2013) [arXiv:1212.4524 [hep-ex]].
[44] M. Ablikim *et al.* [BESIII Collaboration], Phys. Lett. B **753**, 629 (2016) [arXiv:1507.08188 [hep-ex]].
[45] J. Bijnens, E. Pallante and J. Prades, Phys. Rev. Lett. **75**, 1447 (1995) Erratum: [Phys. Rev. Lett. **75**, 3781 (1995)] [hep-ph/9505251].
[46] M. Hayakawa, T. Kinoshita and A. I. Sanda, Phys. Rev. Lett. **75**, 790 (1995) [hep-ph/9503463].
[47] N. I. Muskhelishvili, *Singular Integral Equations*, Wolters-Noordhoff Publishing, Groningen, 1953 [Dover Publications, 2nd edition, 2008].
[48] R. Omnès, Nuovo Cim. **8**, 316 (1958).
[49] A. Gómez Nicola, J. R. Peláez and G. Ríos, Phys. Rev. D **77**, 056006 (2008) [arXiv:0712.2763 [hep-ph]].
[50] P. Büttiker, S. Descotes-Genon and B. Moussallam, Eur. Phys. J. C **33**, 409 (2004) [hep-ph/0310283].
[51] C. Ditsche, M. Hoferichter, B. Kubis and U.-G. Meißner, JHEP **1206**, 043 (2012) [arXiv:1203.4758 [hep-ph]].
[52] M. Hoferichter, C. Ditsche, B. Kubis and U.-G. Meißner, JHEP **1206**, 063 (2012) [arXiv:1204.6251 [hep-ph]].
[53] M. Hoferichter, J. Ruiz de Elvira, B. Kubis and U.-G. Meißner, Phys. Rept. **625**, 1 (2016) [arXiv:1510.06039 [hep-ph]].
[54] C. Adolph *et al.* [COMPASS Collaboration], Phys. Rev. Lett. **114**, 062002 (2015) [arXiv:1405.6377 [hep-ex]].
[55] J. Gasser, M. A. Ivanov and M. E. Sainio, Nucl. Phys. B **745**, 84 (2006) [hep-ph/0602234].

Recent Results on Meson Decays from A2

Patrik Adlarson for the A2 Collaboration[1,*]

[1] *Institut für Kernphysik, Johannes Gutenberg-Universität Mainz, D-55099 Mainz, Germany*

Abstract. Light meson decays are used to investigate topics related to fundamental aspects of particle physics. Precision measurements of meson Dalitz decays give input to theoretical evaluations of the Hadronic Light-by-Light contribution (HLbL) to the anomalous magnetic moment of the muon. The pseudoscalar η' decays allow for studies of topics like $\pi\pi$ scattering lengths, effective field theories and fundamental symmetries. The A2 collaboration, using the Crystal Ball/TAPS setup at MAMI, has recently published several high precision results on transition form factors which are related to HLbL. The value obtained for the slope parameter of the π^0 electromagnetic (e/m) transition form factor (TFF) is $a_\pi = 0.030 \pm 0.010_{tot}$. The slope parameters for the e/m TFFs of $\eta\gamma$ and $\omega\pi^0$ are $\Lambda_{\eta\gamma}^{-2} = (1.97 \pm 0.13)$ GeV^{-2} and $\Lambda_{\omega\pi^0}^{-2} = (1.99 \pm 0.22)$ GeV^{-2}, respectively. In 2014 the collaboration had a dedicated experimental campaign for the production of η' mesons. One of the main goals was to measure the dynamics of $\eta' \to \eta\pi^0\pi^0$ with high precision. An overview of the physics cases, the experimental setup and the results are given.

1 Introduction

Mesons and their properties continue to be of interest to the hadron and particles-physics communities. The electromagnetic transition form factors (TFF) are of interest for probing the precision frontier of the Standard Model (SM) but also for understanding the intrinsic properties of the hadrons themselves. In recent years, a renewed interest in the TFFs has in recent years surged as these enter as input to the theoretical calculations of the Hadronic Light-by-Light (HLbL) contributions to the anomalous magnetic moment of the muon, $(g-2)_\mu$. Also, the study of the pseudoscalar η' mesons along with its decay modes can give insight into our understanding of hadron and particles physics. Besides the η' Dalitz decay, which is related to the aforementioned $(g-2)_\mu$ determination, one can also study the decay properties and dynamics in $\eta' \to \eta\pi\pi$ and $\eta' \to \pi\pi\pi$; Pseudoscalar-Vector(PV) interactions of interest for Chiral Effective Field Theories can be studied via $\eta' \to \omega\gamma$; searches for forbidden or suppressed decays can be conducted on e.g. $\eta' \to 2\pi$, $\eta' \to 4\pi$ and $\eta' \to l^+l^-\pi^0$. In addition, photo-induced production of the η' mesons is a selective probe for studying nucleon resonances [1]. This proceeding will have, as its main focus, recent results from the A2 collaboration on meson TFF: $\pi^0 \to e^+e^-\gamma$ [2], $\eta \to e^+e^-\gamma$ and $\omega \to \pi^0e^+e^-$ [3], and the decay $\eta' \to \pi^0\pi^0\eta$. These proceedings are structured to, first, give motivations for the respective physics cases (section 2), an overview of the Crystal Ball-TAPS experimental setup at MAMI (section 3), a description of the major analysis steps (section 4), the results (section 5), and, finally a summary (section 6).

*e-mail: adlarson@kph.uni-mainz.de

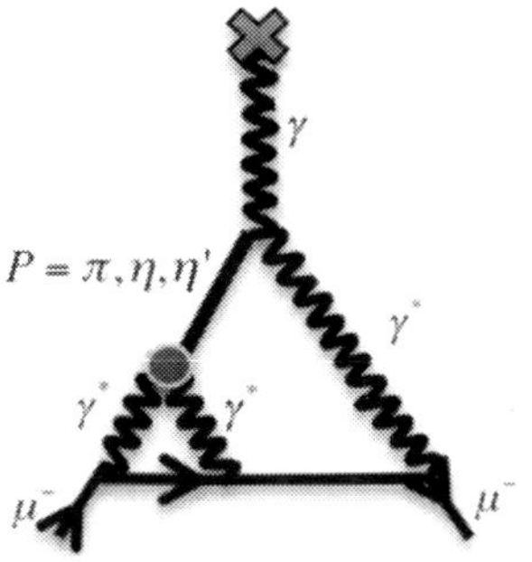

Figure 1. The Hadronic Light-by-Light contribution.

2 Motivation: a_μ and Meson TFF

Some of the most precisely known quantities in the SM are those of the anomalous magnetic moments of leptons, $a_l = (g_l - 2)/2$. Although a_e is more precisely known [4], the muon anomaly has greater sensitivity to heavier physics since the relative contribution of heavier virtual particles goes as $(m_\mu/m_e)^2 \sim 43000$ [5]. For the muon magnetic moment the most precisely measured value, which is known to the 10th decimal, comes from the E821 experiment at Brookhaven National Laboratory [6]. Comparing the E821 result to the SM calculation [7] gives a difference of $\Delta_\mu^{EXP-SM} = 28.7 \pm 8.0 \cdot 10^{-10}$, a deviation of 3.6σ. Another recent theoretical evaluation finds a deviation of 4.5σ [8]. This could be an indication of physics Beyond Standard Model, but for this to be confirmed, higher accuracy for both experiments and theory are still needed. The SM calculation to a_μ is split up into its electromagnetic (a_μ^{QED}), weak ($a_\mu^{W,Z}$) and hadronic components ($a_\mu^{hadr} = a_\mu^{HVP} + a_\mu^{HLbL}$). Although the highest contribution comes from QED, the greatest uncertainty comes from the hadronic part. New experiments are planned at Fermilab and J-PARC, with the goal to measure a_μ with a reduced uncertainty of a factor 4, $\delta a_\mu \sim 1.6 \cdot 10^{-10}$ [9, 10]. The accuracy probed by these experiments will be sensitive to mass scales of $m_\mu/(\delta a_\mu^{EXP})^{1/2} \sim 8$ TeV, thus competitive to LHC energy scales [11]. With the projected experimental accuracy, the limiting uncertainty will then be dominated by the hadronic contributions to the SM calculation of a_μ. The hadronic component of a_μ can be further sub-divided into a_μ^{HVP} and a_μ^{HLbL}, referring to the hadronic vacuum polarization (HVP) and HLbL contributions, respectively. The HVP term is dominant of the two and can be directly linked to hadronic cross-sections in e^+e^--scattering via dispersion relations. The HLbL contribution, which is the diagram shown in Fig. 1, is more difficult to estimate. Until recently, this contribution was mostly obtained from hadronic models, suffering from large and uncontrolled uncertainties [12]. Recently, however, dispersive approaches have been proposed with the goal to reduce the present day errors of the HLbL contribution. One theory group uses dispersion relations based on the analytic structure of the HLbL tensor [15], and one group based on the properties of the electromagnetic vertex function [13]. The first is also a data driven approach, which includes experimental data to evaluate the HLbL contribution in a systematical way. If the experimental data input is precise enough, this can help to constrain the dispersion relation results. Even if data are not precise enough to constrain the calculations, they can still be used to assess the predictive power of these approaches.

2.1 Motivation: Meson TFF

An important ingredient for a better understanding of the HLbL contributions comes from the study of meson TFFs [14]. The meson TFF refers to the vertex where a meson transforms to another state by emitting a virtual photon. Meson TFFs can be accessed in kinematical regions of four-momentum transfer q^2 through study of both space- and time-like processes. In Mainz, experimental activities are focused on both regions with the BESIII (space-like) and A2 (time-like) detectors [16]. Time-like TFFs can be studied in Dalitz decays and annihilation processes, while space-like TFFs are studied in photon-photon fusion processes. The TFFs with the lightest pseudo-scalar particles (π^0, η, η') are of greatest importance to the HLbL contribution as the expected importance of the contribution goes as $1/m_P^2$ [20]. The A2 collaboration can kinematically access and measure the $P = \pi^0$, η, η' Dalitz decays, i.e. $P \to \gamma^*\gamma \to e^+e^-\gamma$, as well as the decay $\omega \to \pi^0 e^+e^-$. Quantum Electro Dynamics (QED) predicts a specific meson decay dependence on the observable $m_{ll} = \sqrt{q^2}$. As QED describes the point-like behavior of the meson decay, any deviation is attributed to the properties of the electromagnetic TFF. For the Dalitz decays, the momentum transfer range available is given by $4m_l^2 < q^2 < m_P^2$, where m_P is the mass of the pseudoscalar meson, P. This applies to the case $P \to \gamma\gamma^*$. In general, however, $A \to B\gamma^*$ has the range $m_l^2 \leq q^2 \leq (m_A - m_B)^2$. The time-like TFF is obtained by dividing the differential decay distribution with its corresponding QED contribution, e.g. $QED_{P\gamma}$. For the Dalitz decays, this is given by

$$\frac{d\Gamma(P \to l^+l^-\gamma)}{dm_{ll} \cdot \Gamma(P \to \gamma\gamma)} = |QED_{P\gamma}||F_{P\gamma}(m_{ll})|^2, \tag{1}$$

where $F_P(m_{ll})$ is the meson TFF for the Pseudoscalar Dalitz decays. For the decay $\omega \to \pi^0 e^+e^-$, the $\omega \to \pi^0\gamma^*$ TFF is

$$\frac{d\Gamma(\omega \to l^+l^-\pi^0)}{dm_{ll} \cdot \Gamma(\omega \to \gamma\pi^0)} = |QED_{\omega\pi^0}||F_{\omega\pi^0}(m_{ll})|^2. \tag{2}$$

In the Vector Meson Dominance model (VMD), the virtual photon, γ^*, couples to an intermediate virtual-vector-meson state V(e.g. ρ, ω, ϕ). This mechanism becomes more enhanced in the time-like region for decays where q is allowed to approach close to the resonance region, $q^2 = m_V^2$. An observable often used to quantify the TFF dependence is the slope parameter, a_P, which reflects the slope value extracted at $q^2 = 0$,

$$a_P \equiv \frac{F_P(m_{ll})}{dq^2}|_{q=0}. \tag{3}$$

Related to this quantity is also the effective mass Λ which, under the VMD approximation, parametrizes the TFFs in a pole approximation,

$$F_{P\gamma/\omega\pi^0}(m_{ll}) = (1 - \frac{m_{ll}^2}{\Lambda^2})^{-1}, \ \Lambda^{-2} = a_P. \tag{4}$$

Due to the smallness of the momentum-transfer range for the π^0 Dalitz decay, it is parametrized as

$$F_{\pi^0\gamma}(m_{ll}) = (1 + a_\pi \frac{m_{ll}^2}{m_{\pi^0}^2}). \tag{5}$$

Experimentally, the PDG value of the slope parameter for the π^0 Dalitz decay, $a_{\pi^0} = 0.032 \pm 0.004$ [17] is dominated by the CELLO result, $a_{\pi^0} = 0.0326 \pm 0.0026_{\text{stat}} \pm 0.0026_{\text{syst}}$ [18]. It should be noted, however, that the CELLO result introduces a certain model dependence, as the slope parameter is extrapolated from the space-like region under the assumption of the validity of

VMD. Furthermore, the extrapolation is from the range of momentum transfers above 0.5 GeV^2, where the actual measurement took place, towards small momentum transfers. It is, therefore, also beneficial to have measurements of the slope parameter that does not rely on extrapolations, but are measured directly in the time-like region. Up until recently, the most accurate value of the slope parameter measured from π^0 Dalitz decay data is from the SINDRUM collaboration [19]. SINDRUM obtained $a_{\pi^0} = 0.025 \pm 0.014_{\text{stat}} \pm 0.026_{\text{syst}}$, and their result was based on the analysis of $54 \cdot 10^3$ events. More recently, and about the same time as the A2 result became available [2], the NA62 collaboration also released new results [21]. These are based on $1.05 \cdot 10^6$ Dalitz decays and the obtained value is $a_\pi = 0.0368 \pm 0.0051_{\text{stat}} \pm 0.0025_{\text{syst}}$. The most accurate theoretical descriptions of the slope-parameter are obtained from using Padé approximants [22] or dispersive theory [23], giving $a_{\pi^0} = 0.0324 \pm 0.0012_{\text{stat}} \pm 0.0019_{\text{syst}}$ and $a_\pi = 0.0307 \pm 0.0006_{\text{tot}}$, respectively.

For the η Dalitz decay, the best experimental result for the slope parameter of the Dalitz decays come from the NA60 collaboration from p–A collisions, $\Lambda_{\eta\gamma}^{-2} = (1.934 \pm 0.067_{stat} \pm 0.050_{syst})\ \text{GeV}^{-2}$ [24]. These are in good agreement with their first results, $\Lambda_{\eta\gamma}^{-2} = (1.95 \pm 0.17_{\text{stat}} 0.05_{\text{syst}})\ \text{GeV}^{-2}$ obtained in peripheral In–In data [25]. There are two previous results from the A2 collaboration at MAMI, based on $1.35 \cdot 10^3$ [26] and $2.2 \cdot 10^4$ [27] Dalitz decays. The higher-statistics result gave the value $\Lambda_\eta^{-2} = (1.95 \pm 0.15_{\text{stat}} \pm 0.10_{\text{syst}})\ \text{GeV}^{-2}$. A recent theoretical prediction comes from the Jülich group, using a dispersive approach [28, 29]. In addition, there are predictions based on Padé approximants [32, 33] and using a chiral Lagrangian approach [30, 31]. For the η' Dalitz decay, the best experimental result comes from the BESIII collaboration [34].

Considering also other TFFs, much attention has been given to the time-like $\omega\pi^0$ TFF in the q-range covered by the $\omega \to \pi^0 l^+ l^+$ decay. This is because the available experimental data at high m_{ll} masses are discrepant, so far, with every available theoretical approach. The experimental data from Lepton-G [69] and NA60 [24, 25] are, however, in agreement with each other. These high mass data points, $m_{ll} > 600$ MeV, are not compatible with theoretical approaches using also model-independent assumptions. One group uses analyticity and unitarity in the framework known as the method of unitary bounds [36] and finds inconsistency between their predictions and the NA60 data. Therefore, it is of importance to obtain new experimental input that can be used for further probing the high m_{ll} mass region.

2.2 Motivation: $\eta' \to \eta\pi\pi$

As mentioned above, one of the reasons to study η' decays is due to its connection to the HLbL contribution. In general, decays of the η' meson also play an important role for our understanding of low energy Quantum Chromo Dynamics (QCD). The low-energy effective-field theory of QCD, Chiral Perturbation Theory (ChPT) is constructed from the chiral $SU(3)_L \times SU(3)_R$ symmetry in the limit of vanishing quark masses (chiral limit), with the symmetry being spontaneously broken to $SU(3)_V$ [37]. ChPT works well below m_σ and describes the interaction of the pseudoscalar mesons π, K, and η. In ChPT η' is, however, not included explicitly due to the $U(1)_A$ anomaly, which renders it massive also in the chiral limit. To include η' in a Lagrangian together with π, K, and η, one can assume that the number of color charges is large (large-N_C). Under this assumption, the $SU(3)_L \times SU(3)_R$ symmetry extends to $U(3)_L \times U(3)_R$ [38–41, 48]. How well large-N_C ChPT works is not well established and, therefore, needs to be confirmed through precision experiments. A potential drawback with large-N_C ChPT concerns the fact that it does not include resonances like ρ, a_0, σ directly as external states. They are, however, taken into account through the low energy constants. The assumption is that the external states have higher masses compared to the center-of-mass total energy for any given process including η' and, therefore, does not need to be taken into account explicitly [41]. The validity of this assumption can, however, be questioned, and it can be argued that the resonance effects need to be considered directly. For this purpose, Resonance Chiral Theory (RChT) has been developed, where the pseudoscalars are allowed to interact with nonets of vectors, axials and scalars [42]. To test the effectiveness of the ChPT extensions, the decay $\eta' \to \eta\pi\pi$ can be used. In 2010 both frameworks were considered by using $\eta' \to \eta\pi\pi$ as a probe [41]. This decay is particularly suitable since final state interactions are mainly dominated by scalars as G-parity conservation suppresses vector contributions. In addition, as both η' and η participate in the process, the mixing properties of these two mesons can also be studied [41, 43].

There are also other motivations for studying $\eta' \to \eta\pi\pi$. In the isospin limit, the $\eta' \to \eta\pi^0\pi^0$ and $\eta' \to \eta\pi^+\pi^-$ decay amplitudes should be the same, allowing for a direct comparison of these two decay channels with theoretical predictions. Due to isospin violation, however, the virtual charge exchange reaction $\pi^+\pi^- \to \pi^0\pi^0$ gives rise to a cusp in the $m_{\pi^0\pi^0}$ spectrum visible at the mass below the $m_{\pi^+\pi^-}$ mass threshold. The effect of the cusp was first seen by NA48/2 in $K^+ \to \pi^+\pi^0\pi^0$ and it has been used to extract the S-wave $\pi\pi$ scattering length combination $a_0 - a_2$ [44–49]. For $\eta' \to \eta\pi^0\pi^0$, the cusp effect and its size has been predicted within the framework of the non-relativistic effective-field theory to be about 6%, compared to phase space [50]. Previously, also the $\eta' \to \eta\pi\pi$ Dalitz plot parameters were estimated within the framework of U(3) chiral effective-field theory in combination with a relativistic coupled-channels approach [51–53].

The first experimental data on the neutral decay mode, $\eta' \to \eta\pi^0\pi^0$, was taken by the GAMS-2000 spectrometer at the IHEP accelerator U-70 with $5.4 \cdot 10^3$ measured decays [54]. A higher statistics sample was made a few years later as the upgraded GAMS-2000 spectrometer, GAMS-4π, in 2009 measured $1.5 \cdot 10^4$ events [55]. The corresponding decay with charged pions $\eta' \to \eta\pi^+\pi^-$ has been measured with higher statistics by the VES [56] and BESIII [57] collaborations. The highest-statistics results come from the BESIII collaboration measuring $\sim 43.8 \cdot 10^3$ decays. An updated result from BESIII is foreseen in the near future with even higher statistics in the final event sample.

To describe the matrix element for the $\eta' \to \eta\pi_1\pi_2$ decay the Dalitz plot is used. Three body decays can be expressed in terms of the variables X and Y, defined as

$$X = \frac{\sqrt{3}}{Q}(T_{\pi_1} - T_{\pi_2}), \quad Y = \frac{T_\eta}{Q}(\frac{m_\eta}{m_\pi} + 2) - 1 \ . \tag{6}$$

The observables T_{π_1}, T_{π_2} denote the kinetic energies of the two final-state pions and T_η denotes the kinetic energy of η. All kinetic energies are calculated in the η' rest frame. The sum of the kinetic energies are given by Q, $Q = T_\eta + T_{\pi_1} + T_{\pi_2} = m_{\eta'} - m_\eta - 2m_\pi$. The Dalitz plot is expanded around $X = Y = 0$ and a polynomial is used to describe the matrix element. The polynomial parametrization used in the most recent measurements [55–57] is given in the form

$$|M|^2 \sim 1 + aY + bY^2 + cX + dX^2, \tag{7}$$

where a, b, c and d are Dalitz plot parameters. All Dalitz plot parameters of odd-powered X are expected to be consistent with 0 for symmetry reasons. For $\eta' \to \pi^0\pi^0\eta$ this is due to Bose-Einstein symmetry of the wave function, and for $\eta' \to \pi^+\pi^-\eta$ as a consequence of charge-conjugation symmetry. For the $\eta' \to \pi^0\pi^0\eta$, it is not possible to distinguish the two neutral pions from each other. By definition, $T_{\pi_1} > T_{\pi_2}$, which means that X is always greater or equal to 0. Therefore, any odd-powered dependence on X, e.g. the cX term, makes no sense in the neutral channel. Some

Experiment $\eta' \to \pi^0\pi^0\eta$	***a***	***b***	***d***
GAMS4π [55]	-0.067(16)(4)	-0.064(29)(5)	-0.067(20)(3)
Experiment $\eta' \to \pi^+\pi^-\eta$			
VES [56]	-0.127(16)(8)	-0.106(28)(14)	-0.082(17)(8)
BESIII [57]	-0.047(11)(3)	-0.069(19)(9)	-0.073(12)(3)
Theory $\eta' \to \pi\pi\eta$			
Large-Nc* [41]	-0.098(48)	-0.050(1)	-0.092(8)
RChT* [41]	-0.098(48)	-0.033(1)	-0.072(8)
U(3) Ch.EFT ($\eta' \to \pi^+\pi^-\eta/\eta' \to \pi^0\pi^0\eta$) [53]	-0.123/-0.109	-0.104/-0.087	-0.047/-0.036

Table 1. Dalitz plot-parameter results for $\eta' \to \pi\pi\eta$ measurements and calculations with errors given in the parentheses. For the experimental data, the first and second parentheses denote the statistical and systematical errors, respectively. *In ref [41] parameter values of $eY \cdot X^2$ and $f \cdot X^4$ are given. For Large-Nc, $e = 0.003(2)$ and $f = 0.002(1)$ while, for RChT, $e = -0.009(2)$ and $f = 0.001(1)$.

recent experimental determinations and theoretical estimates of the Dalitz plot parameters are given in Table 1. There is an ambiguity in the Dalitz plot parameters for the charged Dalitz plot parameters a and b when comparing the VES [56] and BESIII [57] result. Given the connection between the charged and neutral decays and the predicted cusp effect, a new high precision measurement would, therefore, provide valuable input into better understanding QCD dynamics.

3 Experimental Setup

The electron accelerator MAMI (Mainz Microtron) [58, 59] at the Institute for Nuclear Physics of the University of Mainz consists of a cascade of several accelerators, which allows the production of an unpolarized or polarized continuous wave electron beam with nominal energies up to 1.604 GeV. The photon beam were produced, for the experiments described here, when the electron beam interacted with a thin 10-μm Cu radiator producing bremsstrahlung electrons and photons. To reduce the width of the beam, the beam photons are collimated by a Pb collimator before entering the target region. The Pseudoscalar (P= π^0, η, η') and Vector (V=ω) particles are produced when the bremsstrahlung photons impinge on a 5 or 10 cm long Liquid Hydrogen target (LH_2), $\gamma p \to P/Vp$. The corresponding postbremsstrahlung electrons are detected in a tagger, providing the measurement of the energy, E_γ, of photons incident to the target. There are two different taggers, for energies covering ~0.08-1.5 GeV the Glasgow-Mainz tagger is used [60–62]. For photon beam energies needed to produce η' a special tagging device was built, the so-called end-point tagging spectrometer (EPT) [63] with the purpose of tagging postbremsstrahlung electrons with very low energy. The final state $e^\pm$- and γ-particles from the meson decays are measured by using the Crystal Ball (CB) [64] and TAPS [65, 66], being the central and forward calorimeters, respectively. The experimental setup is shown in figure 2. The CB detector consists of 672 NaI(Tl) crystals which are formed into a sphere. The sphere encapsulates the LH_2 target and covers 93% of 4π. Also covering the target is the Particle Identification Detector (PID), made of 24 scintillator bars. Its main purpose is to separate charged from neutral particles. The PID becomes important for the TFF analyses where one needs to suppress charge conversion events from e.g. $P \to \gamma\gamma$ events. The TAPS detector is installed 1.5 m downstream and consists of 384 hexagonal BaF_2 crystals. In the η' experiment, 18 of the BaF2 crystals closest to the beamline were replaced with 72 $PbWO_4$ crystals. This enabled running with a

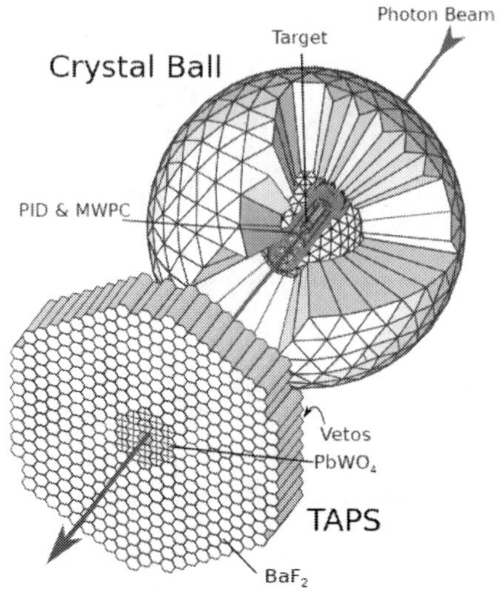

Figure 2. The Crystal Ball and TAPS experimental setup, indicating the direction of the photon beam as the red arrow. For the η' production run the 18 innermost TAPS BaF_2 crystals were replaced with 72 $PbWO_4$ crystals.

higher MAMI electron current, without decreasing the TAPS efficiency due to the higher count rate. The experimental trigger was set to exceed a given energy sum collected in the CB.

The results presented here are from several different run periods between the years 2007-2014, spanning different photon beam energy ranges. For the π^0 Dalitz decay two beam times were used: the first by tagging the bremsstrahlung electrons in the E_γ range 140-800 MeV and the second in the tagged E_γ range 220-1450 MeV. The η Dalitz and $\omega \to \pi^0 e^+ e^-$ decays were collected and analyzed using the two same runs. In the first run, the energies of the incident photons were analyzed up to 1402 MeV and in the second run up to 1448 MeV. The EPT run aimed at producing η' with the maximum electron-beam energy of MAMI, 1.604 GeV, which allowed to tag the E_γ range 1426-1577 MeV.

4 Analysis

4.1 $\pi^0 \to e^+ e^- \gamma$

The π^0 Dalitz decay was searched for in events containing 3 and 4 reconstructed clusters. For 3-cluster events it was assumed that the missing cluster is the proton, which can happen when the proton energy is low and fails to escape the target region. For the selection of the final event sample and determination of the kinematics, kinematic fit was used [67]. For details of the kinematic fit, its parametrizations and resolutions, the interested reader is referred to [68]. Kinematic fit was used both to select the signal and rejecting the main background contribution coming from $\pi^0 \to \gamma\gamma$ events. For signal selection, only events with a Confidence Level (CL)>1% for the hypothesis $\gamma p \to p 3\gamma$ were kept for further analysis. The electromagnetic showers for electrons/positrons are similar to γ, such that $\gamma p \to p 3\gamma$ can be tested to identify $\gamma p \to p e^+ e^- \gamma$. Kinematic-fit was also used as a veto rejecting events which fulfilled the hypothesis $\gamma p \to p\pi^0 \to p\gamma\gamma$ with CL>10^{-5}. The lepton pairs were distinguished from photons by selecting CB clusters which also had matching hits in the PID. To reduce background from conversion events of $\pi^0 \to \gamma\gamma$, an additional requirement was also that two different PID elements were fired. Both the signal and background contributions were simulated in Monte Carlo (MC) to reflect their actual behavior in the experiment, both in terms of the production cross sections and for the decay dynamics. The final event sample was split up into 18 bins with respect to $m(e^+e^-)$ in the range 15-120 MeV. In each bin, the actual Dalitz-decay-signal contents was found by fitting a polynomial on the remaining, but small, non-resonant background together with a Gaussian giving $\sim 4.0 \cdot 10^5$ π^0 Dalitz decays.

4.2 $\eta \to e^+e^-\gamma$ and $\omega \to e^+e^-\pi^0$

Similarly to the π^0 Dalitz decay, $\eta \to e^+e^-\gamma$ and $\omega \to e^+e^-\pi^0$ were also selected first by identifying the number of final state clusters. For the η Dalitz decay both 3- and 4-cluster events were selected, whereas, for the ω decay, only 5-cluster events was selected for further analysis. 4-cluster events were neglected in the ω analysis as the proton information missing for such events resulted in much larger background contributions. The kinematic fit was used to identify the signal channels as well as suppressing background. For the ω channel, the hypotheses $\gamma p \to 4\gamma p$ and $\gamma p \to \pi^0\gamma\gamma \to 4\gamma p$ were tested, while, for the η Dalitz decay the hypothesis $\gamma p \to 3\gamma p$ was tested. The main background for the η Dalitz decay with three cluster events comes from $\pi^0 \to \gamma\gamma$ and $\eta \to \gamma\gamma$. To reject these events, the kinematic-fit hypotheses tested also $\gamma p \to \pi^0 p$ and $\gamma p \to \eta p$. If the CL was found to be greater than 10^{-5} for the two latter hypotheses, these events were rejected. After kinematic-fit selection, the lepton pairs were identified by selecting clusters with matching hits in the PID. For the η Dalitz decay, the only serious background candidate was $\eta \to \gamma\gamma$ with one photon undergoing conversion in the material between the production vertex and the calorimeter material. For the ω, there are three major sources of background: $\gamma p \to \pi^0\pi^0 p$, $\gamma p \to \pi^0\eta p$ and $\omega \to \pi^+\pi^-\pi^0$. Suppressing the first two background channels required analysis of energy losses, dE/dx, in the PID elements, while to suppress $\omega \to \pi^+\pi^-\pi^0$, cuts on the kinematic-fit CL, vertex cuts and using the fact that nuclear-interaction and electromagnetic showers interact differently with the CB crystals were needed. After final event selection for the η Dalitz decay and $\omega \to \pi^0e^+e^-$, the sample was split up into bins with respect to $m(e^+e^-)$ and for each bin, the signal content was estimated by fitting a polynomial and Gaussian to the experimental data. In the remaining final event sample, $\sim 5.4 \cdot 10^4$ η Dalitz decays and $\sim 1.1 \cdot 10^3$ $\omega \to \pi^0e^+e^-$ decays were obtained.

4.3 $\eta' \to \eta\pi^0\pi^0$

The process $\gamma p \to \eta' p \to \pi^0\pi^0\eta p \to 6\gamma p$ was searched for in events reconstructed with 7 clusters detected in both CB and TAPS, assuming that one of the clusters was due to the recoil proton. In the beam energy range of when producing η', most of the recoil protons were produced within the polar angle covered by TAPS. Study of background reactions with MC simulations showed that the process $\gamma p \to 3\pi^0 p \to 6\gamma p$ can mimic $\gamma p \to \pi^0\pi^0\eta p \to 6\gamma p$ if the invariant mass is greater than $\sim$ 820 MeV. This can happen when invariant masses of the three $\gamma\gamma$ pairs, $m(\gamma\gamma)$, are found close to two masses of π^0 and a mass of η. Similarly, the process $\gamma p \to \pi^0\eta p \to 4\pi^0 p \to 8\gamma p$ can mimic the signal channel when two of the eight final-state photons are not detected. Suppression of the $\gamma p \to 3\pi^0 p \to 6\gamma p$ background was done by testing the corresponding kinematical-fit hypothesis and applying a CL cut. For the final event sample, the requirement on the signal channel was CL>0.02. It was also required that the hypothesis $\gamma p \to 3\pi^0 p \to 6\gamma p$ had to have a CL<0.08. The sample of remaining events is shown in the left frame of Fig. 3. It is seen that there is still a non-peaking background distribution to be accounted for in the final event sample. To estimate the signal content in the Dalitz plot, the data were, first, distributed into bins corresponding to different regions in X and Y of the Dalitz plot. Then the data were fitted with a polynomial of order 4 together with the MC η' signal line shape. The MC η' signal line shape came from a large statistics sample obtained under the same analysis condition as the experimental distribution. The estimated number of $\eta' \to \eta\pi^0\pi^0$ decays in the final event sample is $\sim 1.4 \cdot 10^5$.

5 Results

5.1 Meson TFF

The total number of π^0/η Dalitz decays and $\omega \to \pi^0 e^+e^-$ was corrected for the efficiency in each bin of $m(e^+e^-)$. To obtain the transition form factors $|F_{\pi^0\gamma}|$, $|F_{\eta\gamma}|$ and $|F_{\omega\pi^0}|$, the number of events in each bin was divided by the corresponding QED contributions and decay widths, $\Gamma(P \to \gamma\gamma)$, $\Gamma(\omega \to \pi^0\gamma)$. The form factor $|F_{\pi^0\gamma}|^2$, shown as a function of $m(l^+l^-)$ for the π^0 Dalitz decay, is seen in Fig. 4. Similarly, the $|F_{\eta\gamma}|^2$ and $|F_{\omega\pi^0}|^2$ are shown in Figs. 5 and 6, respectively. The statistical uncertainty in an individual bin of $|F_{P\gamma}|^2$ and $|F_{\omega\pi^0}|^2$ was given by the uncertainty in estimating the number of decays as determined by the Gaussian and polynomial fitting. The systematic uncertainties were estimated for each individual $m(e^+e^-)$ bin by repeating its fitting procedure several times after refilling the experimental $m(e^+e^-\gamma)/m(e^+e^-\pi^0)$ spectra under different selection criteria used in the analysis, e.g. cuts on the kinematical fit CL, different cuts on PID, dE/dx, cluster radius and z-vertex position. By testing several different fits, this showed the concistency of the results. The average of the results for all fits made for one bin was used to obtain the final form factor values. The systematic uncertainties were taken as the root mean square of all the results from each bin.

By fitting Eqs. (4) and (5) to the form factor data points, the pole approximation and its slope parameter were obtained. For the π^0 Dalitz decay, this gives $a_\pi = 0.030 \pm 0.010_{\rm tot}$, which is in good agreement with the PDG value [17], the recent NA62 result [21] and theoretical estimates [22, 23]. For the data points of the η Dalitz decay, the fit with the pole approximation, (Eq. 4), gives the fitted value $\Lambda_\eta^{-2} = (1.97 \pm 0.13)$ GeV^{-2}, which is in agreement with the latest NA60 result [24] and theoretical calculations [29, 30, 32, 33]. The fit value obtained for $|F_{\omega\pi^0}|$, $\Lambda_{\omega\pi^0}^{-2} = (1.99 \pm 0.21_{\rm tot})$ GeV^{-2}, is a bit lower compared to previous measurements from NA60, $\Lambda_{\omega\pi^0}^{-2} = (2.223 \pm 0.026_{\rm stat} \pm 0.037_{\rm syst})$ GeV^{-2} [24]. The new result from the A2 collaboration is in better agreement with theoretical calculations, compared to the data from other experiments. A firm conclusion may not be drawn presently as the accuracy of the present data points at large $m(e^+e^-)$ masses is not sufficient for it. More high-statistics measurements of the $\omega \to \pi^0e^+e^-$ decay, especially at large $m(e^+e^-)$ masses, are needed to solve the problem of the inconsistency remaining between the calculations and the experimental data.

5.2 $\eta' \to \eta\pi^0\pi^0$

The Dalitz plot for the final event sample is shown in the right frame of Fig. 3 for the bin width ΔX=ΔY =0.1. The fit function minimized in MINUIT [75] is

$$\chi^2 = \sum_{i=1}^{bins}\left(\frac{N_i - \sum_{i=1}^{nbins}\epsilon_i \cdot f(X_i, Yi)}{\sigma_i}\right)^2 \tag{8}$$

where N_i is the experimentally reconstructed number of events in bin i and ϵ_i is the efficiency for the same bin and $f(X_i, Y_i)$ is the fitting function given in Eq. 6. The bins are represented one-dimensionally and are calculated such that a bin number specifies a rectangular area in the Dalitz

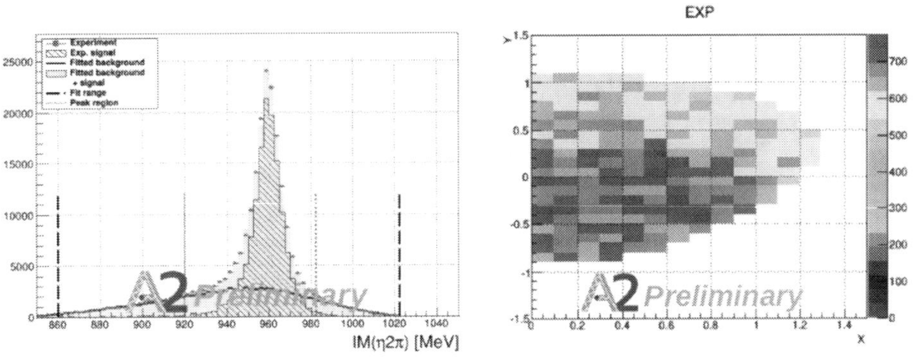

Figure 3. (left) The final event sample for the decay $\eta' \to \eta\pi^0\pi^0$. (right) The preliminary Dalitz plot with a bin width of 0.1.

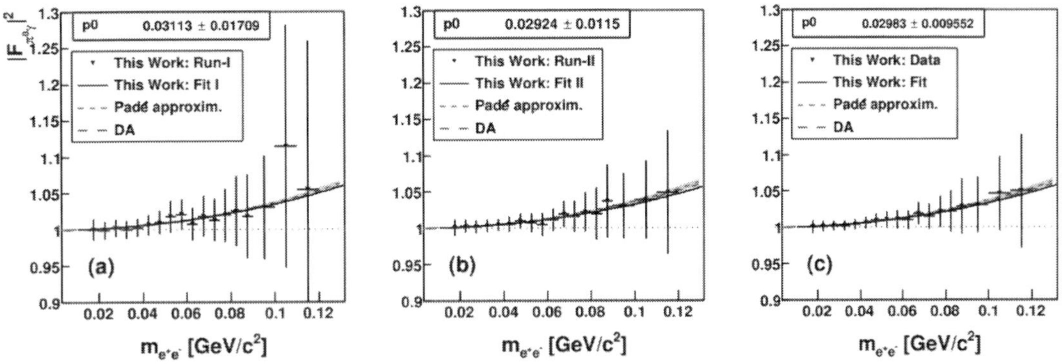

Figure 4. $|F_{\pi^0\gamma}|^2$ results with total uncertainties (black filled triangles) from two different run sets: Run 1 (a), Run 2 (b). The combined result is shown in (c). The fitted values to the points are shown by blue lines, with $p0$ being the slope parameter a_π. Comparison to calculations with Padé approximants [22] is seen as the short-dashed magenta line with a gray error band and to the dispersive analysis (DA) [23] shown as the red, long-dashed line. The error band for the latter analysis is by a factor of four narrower, compared [22], and is omitted.

plot region. The efficiency, ϵ_i, is obtained from a high statistics MC sample of $9 \cdot 10^6$ signal events. The preliminary values for the Dalitz-plot parameters are given in table 2. The preliminary results are in good agreement with the GAMS4π results, but have a better statistical accuracy. Investigation of systematic effects as well as a detailed study of the cusp region are in progress.

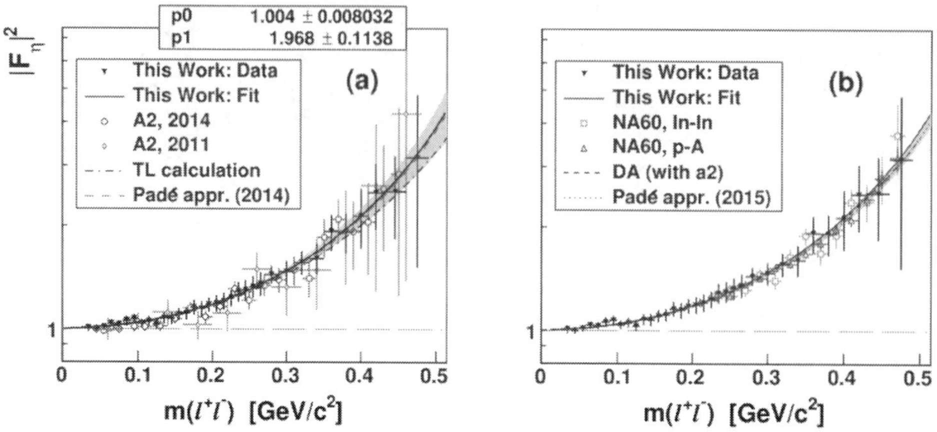

Figure 5. $|F_{\eta\gamma}(m_{e^+e^-})|^2$ results (black filled triangles) fitted with the pole approximation (black solid line). $p0$ and $p1$ denotes the normalization and slope parameter Λ^{-2}, respectively. The former results from the A2 Collaboration [27] (open magenta circles) and [26] (open green diamonds) are shown in panel (a). The results of NA60 from peripheral In–In data [25] and p–A collisions [24] are seen in (b). The calculation from Ref. [30] is seen in (a) as a blue dash-dotted line. The recent dispersive approach [29] is shown in (b) by a blue dashed line. The calculations with Padé approximants are shown in (a) for a previous solution [32] (red dashed line with an error band) and in (b) for the latest solution [33] (red dotted line with an error band).

Result	*a*	*b*	*d*	χ^2/dof
A2 Preliminary	-0.075(8)	-0.069(13)	-0.054(8)	247.6/195 = 1.27

Table 2. *Preliminary results of A2 for the Dalitz-plot parameters of $\eta' \to \eta\pi^0\pi^0$. Uncertainties (inside the parentheses) are statistical.*

6 Summary

In these proceedings, three new results from the A2 collaboration were presented for the TFFs $|F_{\pi^0\gamma}|$, $|F_{\eta\gamma}|$ and $|F_{\omega\pi^0}|$. The results were obtained by analyzing the decays $\pi^0 \to e^+e^-\gamma$, $\eta \to e^+e^-\gamma$ and $\omega \to \pi^0 e^+e^-$. The slope parameters for the e/m TFF of $\pi^0\gamma$, $\eta\gamma$ and $\omega\pi^0$ are $a_\pi = 0.030 \pm 0.010$, $\Lambda_{\eta\gamma}^{-2} = (1.97 \pm 0.13)$ GeV^{-2} and $\Lambda_{\omega\pi^0}^{-2} = (1.99 \pm 0.22)$ GeV^{-2}, respectively. While the results for the π^0 and η Dalitz decays are in agreement with other experimental results and theoretical calculations, the $\omega \to \pi^0 e^+e^-$ result from A2 are in a better agreement with theoretical calculations, compared to the available experimental data. However, because of limited experimental statistics, no firm conclusion can be drawn to rule out the previous results. For the $\eta' \to \eta\pi^0\pi^0$ the preliminary results of the Dalitz-plot parameters are presented and the systematic effect are currently under investigation. In addition to the analysis of $\eta' \to \eta\pi^0\pi^0$ analyses of $\eta' \to e^+e^-\gamma$, $\eta' \to \omega\gamma$ and $\omega \to \eta\gamma$ with the A2 data are in progress as well.

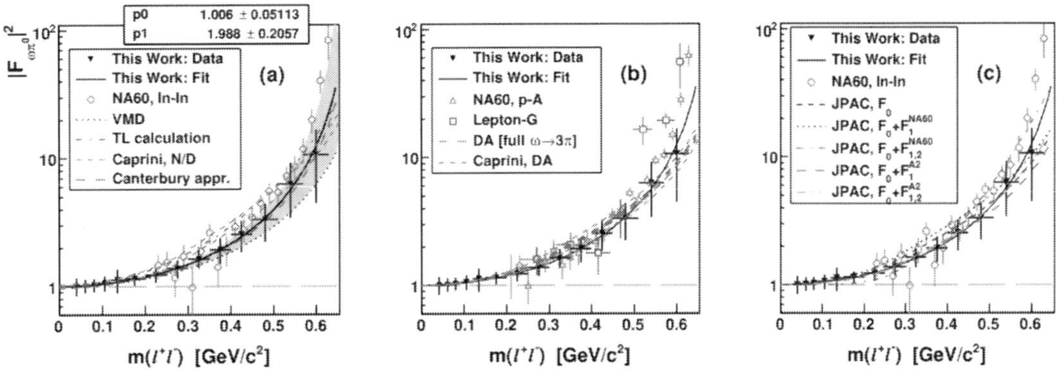

Figure 6. $|F_{\omega\pi^0}(m_{e^+e^-})|^2$ results (black filled triangles) fitted with the pole approximation (black solid line). $p0$ and $p1$ denotes the normalization and slope parameter Λ^{-2}, respectively. Earlier experimental results are Lepton-G [69] (open red squares in (b)); NA60 in peripheral In–In data [25] (open green circles in (a) and (c)); NA60 in p–A collisions [24] (open green triangles in (b)). There are several theoretical calculations: The VMD prediction (blue dashed line in (a)); calculation from Refs. [30, 31] is shown by a red dash-dotted line in (a); dispersive analysis calculation by the Bonn group [70] (shown by error-band borders, magenta dashed lines in (b)); upper and lower bounds by Caprini [71] (cyan dashed lines) for two cases of the discontinuity calculated with the partial-wave amplitude $f_1(t)$ based on the improved N/D model [72] (a), and from [70] (b); calculation based on a model-independent method using Canterbury approximants [73] (magenta long-dashed line with gray error band); basic calculation from JPAC [74] (blue dashed line in (c)) and inclusion of higher order terms of the inelastic contributions in the $\omega\pi^0$ TFF by a fit to the NA60 In–In data, for the solutions with adding one (black dotted line, in (c)) and two (red dash-dotted line in (c)) terms; similar effect from including higher order terms by fitting them to present $|F_{\omega\pi^0}(m_{\ell^+\ell^-})|^2$ results shown for solutions with one (magenta long-dashed line in (c)) and two terms (cyan dash-double-dotted line in (c)).

References

[1] V. L. Kashevarov *et al.* **(A2 Collaboration at MAMI)**, arXiv:1701.04809 [nucl-ex].

[2] P. Adlarson *et al.* **(A2 Collaboration at MAMI)**, arXiv:1611.04739 [hep-ex].

[3] P. Adlarson *et al.* **(A2 Collaboration at MAMI)**, arXiv:1609.04503 [hep-ex].

[4] D. Hanneke, S. Fogwell and G. Gabrielse Phys. Rev. Lett. **100**, 120801 (2008).

[5] T. Blum, A. Denig, I. Logashenko, E. de Rafael, B. L. Roberts, T. Teubner, G. Venanzoni, arXiv:1311.2198 [hep-ph].

[6] G. W. Bennett, *et al.* **(The g-2 Collaboration)**, Phys. Rev. D**73**, 072003 (2006).

[7] M. Davier, A. Hoecker, B. Malaescu, Z. Zhang. Eur. Phys. J. C**71**, 1515 (2011).

[8] M. Benayoun, P. David, L. DelBuono, F. Jegerlehner, Eur. Phys. J. C 75, 613 (2015).

[9] J. Grange *et al.* **(Muon g-2 Collaboration)**, FERMILAB-FN-0992-E, FERMILAB- DESIGN-2014-02.

[10] T. Mibe **(J-PARC g-2 Collaboration)**, Chinese Physics C **34**, 745 (2010).

[11] A. Czarnecki and W.J. Marciano, Phys. Rev. **D** 64, 013014 (2001).

[12] A. Nyffeler, Phys.Rev. D**94** no.5, 053006 (2016).

[13] V. Pauk, M. Vanderhaeghen, Phys. Rev. D 90, 113012 (2014).
[14] L. G. Landsberg, Phys. Rep. **128**, 301 (1985).
[15] G. Colangelo, M. Hoferichter, M. Procura, JHEP 1409, **091** (2014); G. Colangelo, M. Hoferichter, B. Kubis, M. Procura, P. Stoffer, Phys. Lett. B **738**, 6 (2014); G. Colangelo, M. Hoferichter, M. Procura and P. Stoffer JHEP **1509**, 074 (2015); see B. Kubis, S. Leupold, M. Procura, these proceedings.
[16] See C. Redmer, these proceedings.
[17] C. Patrignani *et al.* **(Particle Data Group)**, Chin. Phys. C **40**, 100001 (2016).
[18] H. J. Behrend *et al.* **CELLO Collaboration**, Z. Phys. C **49** (1991) 401 (1991).
[19] R. Meijer Drees *et al.*, **SINDRUM Collaboration**, Phys. Rev. D **45**, 1439 (1992).
[20] E. Czerwinski, S. Eidelman, C. Hanhart, B. Kubis, A. Kupść, S. Leupold, P. Moskal, S. Schadmand, arXiv:1207.6556 [hep-ph].
[21] C. Lazzeroni *et al.* **(NA62 Collaboration)**, arXiv:1612.08162 [hep-ex].
[22] P. Masjuan, Phys.Rev. D **86**, 094021 (2012).
[23] M. Hoferichter. B. Kubis, S. Leupold, F. Niecknig, S.P. Schneider, Eur. Phys. J. C **74**, 3180 (2014); See S. Leupold, these proceedings.
[24] R. Arnaldi *et al.*, Phys. Lett. B **757**, 47 (2016).
[25] R. Arnaldi *et al.*, Phys. Lett. B **677**, 260 (2009).
[26] H. Berghäuser *et al.*, Phys. Lett. B **701**, 562 (2011).
[27] P. Aguar-Bartolome *et al.* Phys. Rev. C **89**, 044608 (2014).
[28] C. Hanhart, A. Kupść, U.-G. Meißner, F. Stollenwerk, and A. Wirzba, Eur. Phys. J. C **73**, 2668 (2013); *ibid.* **75**, 242 (2015).
[29] C. W. Xiao, T. Dato, C. Hanhart, B. Kubis, U.-G. Meißner, A. Wirzba, arXiv:1509.02194 [hep-ph].
[30] C. Terschlüsen, Diploma Thesis, University of Gießen, 2010.
[31] C. Terschlüsen, S. Leupold, and M. F. M. Lutz, Eur. Phys. J. A **48**, 190 (2012).
[32] R. Escribano, P. Masjuan, and P. Sanchez-Puertas, Phys. Rev. D **89**, 034014 (2014).
[33] R. Escribano, P. Masjuan, and P. Sanchez-Puertas, Eur. Phys. J. C **75**, 414 (2015).
[34] M. Ablikim *et al.* **BESIII collaboration**, Phys. Rev. D 92, 012001 (2015).
[35] R. I. Dzhelyadin *et al.*, Phys. Lett. B **102**, 296 (1981).
[36] B. Ananthanarayan, I. Caprini, and B. Kubis, Eur. Phys. J. C **74**, 3209 (2014).
[37] J. Gasser, H. Leutwyler, Nucl. Phys., B **250**, 465 (1985).
[38] G. ´t Hooft, Nucl. Phys., B **72**, 461 (1974).
[39] E. Witten, Nucl. Phys., B **156**, 269 (1979).
[40] R. Kaiser, H. Leutwyler, Eur. Phys. J., C **17**, 623 (2000).
[41] R. Escribano, P. Masjuan, J. J. Sanz- Cillero. JHEP, **05**, 094 (2011).
[42] G. Ecker, J. Gasser, A. Pich, E. de Rafael, Nucl. Phys. B **321**, 311 (1989).
[43] P. Bickert, P. Masjuan, S. Scherer, arXiv:1612.05473.
[44] J. R. Batley et al. Phys. Lett., B**633**, 173 (2006).
[45] N Cabibbo. Phys. Rev. Lett., **93** 121801 (2004).
[46] N Cabibbo, G. Isidori, JHEP, **03** 021 (2005).
[47] G. Colangelo, J. Gasser, B. Kubis, A. Rusetsky, Phys. Lett. B **638**, 187, (2006).
[48] E. Gamiz, J. Prades, I. Scimemi, Eur. Phys. J., C **50**, 405 (2007).
[49] M. Bissegger, A. Fuhrer, J. Gasser, B. Kubis, A. Rusetsky, Nucl. Phys., B **806**, 178 (2009).

[50] B Kubis, S P. Schneider, Eur. Phys. J., C**62**, 511 (2009).
[51] N. Beisert, B. Borasoy, Nucl. Phys. A, **716** 186 (2003).
[52] B. Borasoy, R. Nissler, Eur. Phys. J. A, **26** 383 (2005).
[53] B. Borasoy, Ulf-G. Meissner, R. Nissler, Phys. Lett., B **643**, 41 (2006).
[54] D. Alde *et al.* **(LAPP collaboration)**, Phys. Lett., B **177** 115 (1986), [Yad. Fiz.45,117(1987)].
[55] A. M. Blik *et al.* **(GAMS4π collaboration)**, Phys. Atom. Nucl. **72** 231 (2009), [Yad. Fiz.72,258(2009)].
[56] V. Dorofeev *et al.* **(VES Collaboration)**, Phys. Lett., B **651** 22 (2007).
[57] M. Ablikim *et al.* **(BESIII Collaboration)**, Phys. Rev. D **83**, 012003 (2011).
[58] H. Herminghaus *et al.*, IEEE Trans. Nucl. Sci. **30**, 3274 (1983).
[59] K.-H. Kaiser *et al.*, Nucl. Instrum. Methods Phys. Res. A **593**, 159 (2008).
[60] I. Anthony *et al.*, Nucl. Instrum. Methods Phys. Res. A **301**, 230 (1991).
[61] S. J. Hall *et al.*, Nucl. Instrum. Methods Phys. Res. A **368**, 698 (1996).
[62] J. C. McGeorge *et al.*, Eur. Phys. J. A **37**, 129 (2008).
[63] P. Adlarson *et al.*, Phys. Rev. C 92, 024617 (2015).
[64] A. Starostin *et al.*, Phys. Rev. C **64**, 055205 (2001).
[65] R. Novotny, IEEE Trans. Nucl. Sci. **38**, 379 (1991).
[66] A. R. Gabler *et al.*, Nucl. Instrum. Methods Phys. Res. A **346**, 168 (1994).
[67] V. Blobel E. Lohrmann. Statistische Methoden der Datenanalyse. Teubner Studienbücher, Teubner, e-book http://www.desy.de/ blobel/eBuch.pdf, 1998.
[68] S. Prakhov *et al.*, Phys. Rev. C **79**, 035204 (2009).
[69] R. I. Dzhelyadin *et al.*, Phys. Lett. B **102**, 296 (1981).
[70] S. P. Schneider, B. Kubis, and F. Niecknig, Phys. Rev. D **86**, 054013 (2012).
[71] I. Caprini, Phys. Rev. D **92**, 014014 (2015).
[72] G. Köpp, Phys. Rev. D **10**, 932 (1974).
[73] P. Masjuan, private communication with Sergey Prakhov, 2016.
[74] I. V. Danilkin, C. Fernández-Ramírez, P. Guo, V. Mathieu, D. Schott, M. Shi, and A. P. Szczepaniak, Phys. Rev. D **91**, 094029 (2015).
[75] F. James, MINUIT Function Minimization and Error Analysis: Reference Manual Version 94.1, 1994.

Review of $e^+e^- \to \pi^+\pi^-\eta$

Olga Shekhovtsova[1,2,*]

[1] *NSC KIPT Akhiezer Institute for theoretical Physics, 61108 Kharkov, Ukraine*
[2] *Institute of Nuclear Physics PAN, 31-234 Cracow, Poland*

Abstract. In this talk we review the current experimental status of electron-positron annihilation into $\pi^+\pi^-\eta$ as well as the theoretical models to describe the data. Application of the CVC hypothesis to get the τ^- decay to $\pi^-\pi^0\eta$ branching fraction is discussed and its values from the different experiments is listed.

1 Introduction

The data on the $e + e^-$ annihilation in the energy region below 2. – 2.5 GeV is the main source of information about properties of the light vector meson and is of great interest for a better understanding of the light quark interactions. For the process

$$e^+ + e^- \to \pi^+ + \pi^- + \eta \tag{1}$$

the main mechanism of production proceeds through one intermediate state $\eta\rho(770)$ [1] which makes the precise measurements of the cross section of the reaction (1) of great importance for the determination of the $\rho(1400)$ and $\rho(1700)$ parameters. Also data on $e^+e^- \to \pi^+\pi^-\eta$ can be used to test the conservation of the vector current (CVC) hypothesis.

The reaction (1) was first observed at the end of 70's in the DM1 experiment [1] and since that time it was studied by Orsay [2], Novosibirsk [3–6] groups and by the BaBar collaboration [7, 8]. From a theoretical point of view the $\eta\pi\pi$ production in the e^+e^--annihilation and τ-lepton decay was discussed in [4, 9–11] using different approaches.

The paper is organized as follows. In Sec. 2 the theoretical approaches to describe hadronic currents $\gamma* \to \pi^+\pi^-\eta$ are described. The experimental data are reviewed in Sec. 3. Sec. 4 is devoted to the CVC study and comparison with the spectral function of the $\tau^- \to \pi^-\pi^0\eta\nu_\tau$ decay. The rare decay $\phi \to \pi^+\pi^-\eta$ is the subject of Sec 5.

2 Theoretical models

The description of hadronic interactions in the energy region populated by resonances is problematic. In fact, perturbative QCD does not work in the region below 2 GeV. At very low energies ($E << M_\rho$), Chiral Perturbation Theory (ChPT), based on the chiral symmetry of massless QCD, provides

*e-mail: oshekhov@ifj.edu.pl

[1] A production via $e^+e^- \to a2(1300)\pi$ can contribute to the cross section, however, this reaction has not been observed.

successful description of hadronic interactions. At higher energy there is no a similar framework and we have to rely on some phenomenological models. Three theoretical models have been applied to study the process (1) and they are shortly described below.

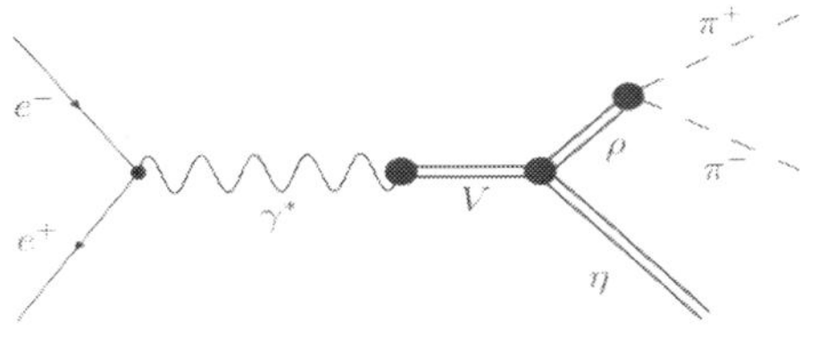

Figure 1. The diagram of the $e^+e^- \to \eta\pi^+\pi^-$ process in the framework of the VMD model.

The Vector Meson Dominance (VMD) model with three intermediate ρ meson states is a generally accepted scheme in the contemporary $\eta\pi\pi$ data analysis. It is characterized by the Feynman diagram shown in Fig. 1, where the intermediate resonance V represents the mesons $\rho(770), \rho(1450), \rho(1700)$. The corresponding cross section can be written as [5]

$$\sigma(s) = \frac{4\alpha^2}{3}\frac{1}{s\sqrt{s}}\int\limits_{4m_\pi^2}^{(\sqrt{s}-m_\eta)^2} \frac{\sqrt{q^2}\Gamma_\rho(q^2)P_\eta^3(s,q^2)}{(q^2-m_\rho^2)^2+(\sqrt{q^2}\Gamma_\rho(q^2))^2}|F(s)|^2 dq^2, \tag{2}$$

where q is the 4-momentum of the $\pi^+\pi^-$ system and $P_\eta^2(s,q^2) = [(s-m_\eta^2-q^2)^2 - 4m_\eta^2 q^2]/4s$.

The transition form factor $F(s)$ describes the vertex $\gamma^* \to \eta\rho$ and takes the form:

$$F(s) = \sum_V \frac{g_V e^{i\phi_V} m_V^2}{s - m_V^2 + i\sqrt{s}\Gamma_V(s)},$$

where $g_V e^{i\phi_V} = g_{V\rho\eta}/g_{V\gamma}$, $g_{V\rho\eta}$ and $g_{V\gamma}$ are the coupling constants for the transitions $V \to \rho\eta$ and $V \to \gamma^*$, respectively. Here the energy dependence of the ρ meson widths is driven by two-pion loops $\Gamma_V(q^2) = \Gamma_V \frac{m_V^2}{q^2}[p_\pi(q^2)/p_\pi(m_V^2)]^3$, $p_\pi(q^2) = \sqrt{q^2-4m_\pi^2}$.

Since the coupling constants are not expected to have sizable imaginary parts, the fit to the experimental data is done assuming the phases $\phi_{\rho(1450)}$ and $\phi_{\rho(1700)}$ equal to zero or π (if the reference $\rho(770)$ phase is put equal to zero). The SND Collaboration analysis [5] favours $\phi_{\rho(1450)} = \phi_{\rho(1700)} = \pi$. Notice that the fit with $\rho(1700)$ is slightly better than without it; however, the value obtained for $g_{\rho(1700)}$ deviates from zero by only 2σ.

The description of the process (1) within the framework of the Resonance Chiral Theory (RChT) is presented in [9, 10]. RChT is based on the ChPT approach with the explicit inclusion of the vector and axial–vector resonances. Its main features can be summarized in two points: 1) the RChT Lagrangian contains the ChPT one and hence the RChT amplitudes satisfy chiral symmetry and match the chiral amplitudes at low energies; 2) the high-energy amplitude of the hadronic form factors provided by RChT is suppressed, as enforced by QCD.

The $\eta\pi\pi$ production in the RChT framework is described by the diagrams in Fig. 3. Leaving only the diagram 4d), we obtain the VMD approach. In Ref. [9], in addition to the photon propagation, only the lightest $\rho(770)$ resonance was considered. Comparison with experimental data [1–4, 7] demonstrates fair agreement in the energy region from the $\eta\rho$ threshold to 1.4 GeV (Fig. 4, left

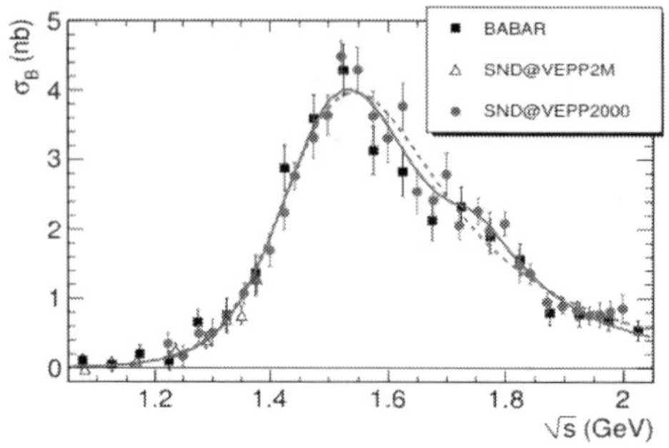

Figure 2. The cross section $e^+e^- \to \pi^+\pi^-\eta$ measured in the SND experiment: the solid line is the VMD parametrization including $\rho(1700)$, the dashed line without it. The figure is taken from [5].

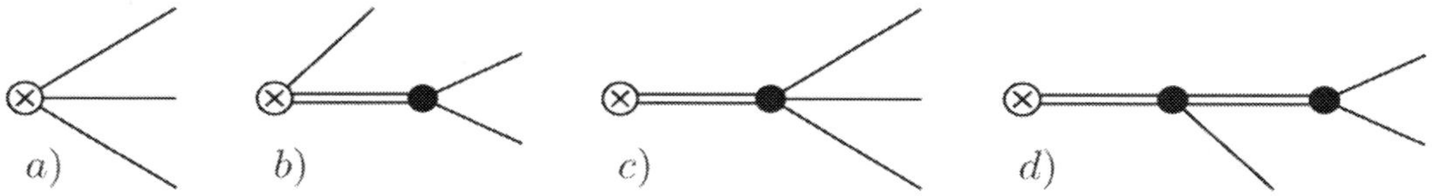

Figure 3. Hadronization of the vector current in $e^+e^- \to \pi^+\pi^-\eta$ within the RChT approach. Single lines represent pseudoscalar mesons (π, η) and double lines are for the ρ resonances.The diagram a) corresponds to the Wess-Zumino-Witten interaction Lagrangian, the diagrams b) c) and d) to the effective Lagrangian describing interactions with resonances.

panel). An extended approach to study the reaction $e^+e^- \to \pi^+\pi^-\eta(\pi^0)$, including the excited $\rho(1450)$ and $\rho(1700)$ mesons as well as tiny isospin symmetry breaking effects, was proposed in [10]. In order to obtain the values of the model parameter, fits based on different sets of $e^+e^- \to \pi^+\pi^-\pi^0$ and $e^+e^- \to \pi^+\pi^-\eta$ data were carried out. For details, see Section VII of [10]. Comparison between the theoretical prediction for the cross section of process (1) and the experimental data is presented in Fig. 4, right panel. Notice, in addition, that the fit favors an higher mass for the $\rho(1450)$ resonance ($M_{\rho'} = 1.55$ GeV instead of its PDG value $M_{\rho'} = 1.465(25)$ GeV) and lower width ($\Gamma_{\rho'} = 0.238$ GeV instead of its PDG value $\Gamma_{\rho'} = 0.400(60)$ GeV). Also the contribution of the $\rho(1700)$ is one half that of the $\rho(1450)$.

The reaction (1) within the extended nonlocal version of the Nambu-Jona-Lasinio (NJL) model studied in [11, 12] and the corresponding Feynman diagrams are shown in Fig. 5. The standard NJL model describes the physical processes via an intermediate states of a virtual photon and ρ and ω mesons in its ground state. The contribution of the first radially excited meson states is included in the extended NJL model [13] and provides a reasonable good description of a wide range of physical process for energy region up to about 1.5 GeV [12].

Hadronization mechanism to produce $\eta\pi\pi$ is presented in Fig. 6. The VMD approach could be directly obtained from the NJL model by taking into account only the diagram with the triangle loops. Precise formulae for the cross section can be found in Section III of [11]. Numerical values of the model parameters are fixed in Eq.(4) of [11] whereas for the resonance masses and widths their PDG values are chosen. The comparison with the BaBar data is presented in Fig. 7.

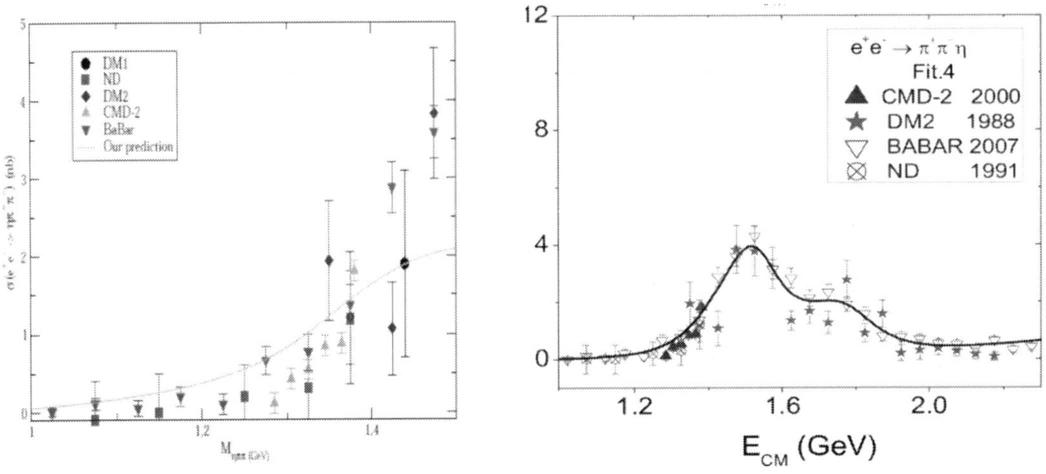

Figure 4. Comparison between the experimental data for $e^+e^- \to \pi^+\pi^-\eta$ and the theoretical prediction based on the RChT approach. The left panel corresponds to the propagation via $\gamma + \rho(770)$, the right one to the elaborated model used in Ref. [10]. The figures are taken from [9, 10].

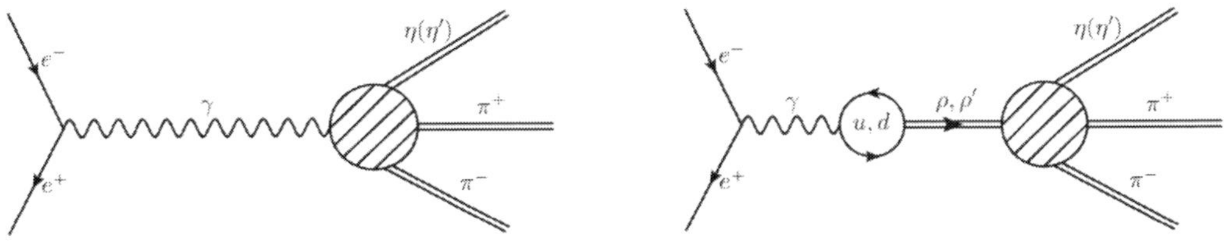

Figure 5. Electron-positron annihilation with $\eta\pi\pi$ production within the NJL approach. The left diagram corresponds to the production via virtual photons, the right one is via intermediate $\rho(770)$, $\rho(1450)$ states. The dashed circles are the sum of two subdiagrams from Fig. 6.

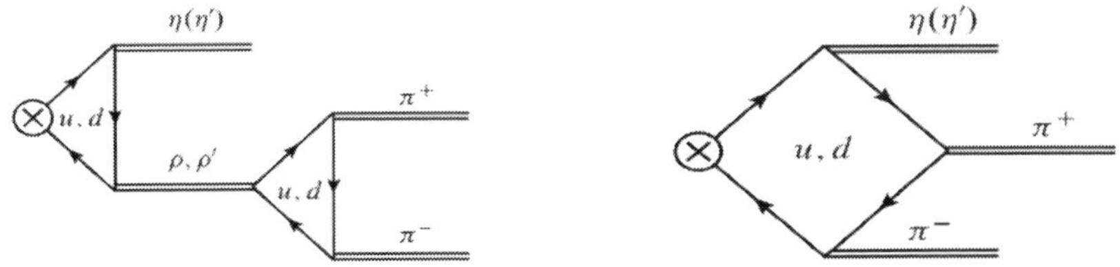

Figure 6. Vertex subdiagrams $V\eta\pi\pi$ with a quark box loop (left) and with two quark-triangle loops (right).

3 Experimental data

As the η meson is a short lifetime particle it is observed decaying either in $\pi^+\pi^-\pi^0$ or via its $\gamma\gamma$ decay. Therefore the process (1) contributes to the reactions: $e^+e^- \to \pi^+\pi^-\pi^0\pi^+\pi^-$ and $e^+e^- \to \pi^+\pi^-\gamma\gamma$.

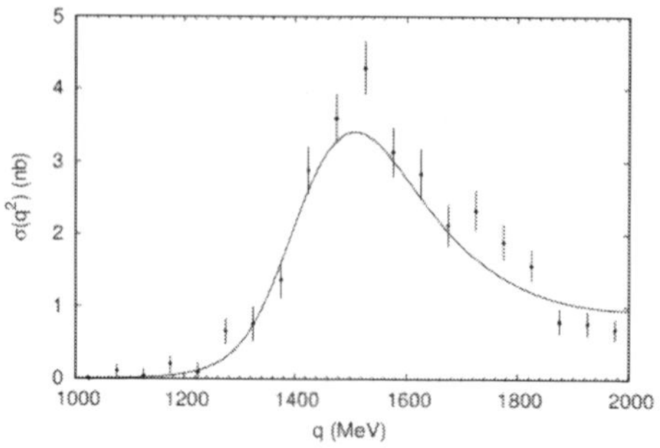

Figure 7. Comparison of the extended NJL model prediction with the BaBar experiment data [7]. The figure is taken from [11].

The reaction (1) was first observed at the end of the 70's in the DM1 experiment [1] [2] for the center-of-mass (CM) energy region 1.4 – 2.8 GeV. The DM1 experiment data analysis demonstrated that the $\rho\eta$ production only via the intermediate $\rho(770)$ resonance [15], predicts a cross section of about 0.5 nb, an order of magnitude below the experimental data. To solve the problem, the production via the intermediate $\rho'(1600)$ was added. The ND Collaboration [3] studied the process in the CM energy region 1.05 – 1.40 GeV and showed the same excess of production compared with the theoretical prediction [15] for energies above 1.35 GeV, see Fig. 8, confirming the hypothesis that the main contribution in this energy region comes from the intermediate $\rho'(1600)$ meson.

Further comparison of the $\eta\rho$ mass spectrum in diffractive photoproduction [16] and e^+e^--annihilation has brought to the idea of the existence of two ρ' resonances with a mass of 1.47 GeV and 1.70 GeV [17]. This parametrization has been described in Sec. 2 and since the DM2 experiment it is the generally accepted pattern in the $\eta\pi\pi$ data analysis.

Later the process was studied by Orsay (DM2 Collaboration) [2] and Novosibirsk groups (SND, CMD groups) [3, 4]. The most recent measurements were performed at the VEPP-2000 collider by the SND group [5] for the CM energy range from 1.22 to 2.00 GeV (the η meson is recontructed via the two-photon mode) and at B-factory with the BaBar detector using the initial state radiation technique [7] in the (1.025; 2.975) GeV CM energy region (the η meson is recontructed via three-pion mode). Systematic uncertainties of about 8% at BaBar [7] and below 8.3% by the SND experiment were reported. The preliminary results from BaBar using the $\eta \to \gamma\gamma$ channel [8] and from the CMD-3 collaboration using both $\eta \to \pi^+\pi^-\pi^0$ and $\eta \to \gamma\gamma$ decay modes have been presented recently [6, 8].

As seen from Figs. 2, 9 all measurements are consistent. The cross section shows a steep rise starting from the $\eta\rho(770)$ threshold, peaking at about 1.5 GeV, and then decreases with increasing energy. The measured $e^+e^- \to \eta\pi^+\pi^-$ cross section was fitted using the VMD model (see Section 2) with and without $\rho(1700)$ contribution. Based on the fit results one cannot draw a definite conclusion that at the current statistics the $\rho(1700)$ contribution is needed for the description of the data.

4 Check of the CVC hypothesis

The data on electron-positron annihilation with $\eta\pi\pi$ production can be used to estimate the branching fraction of the $\tau^- \to \eta\pi^-\pi^0\eta\nu_\tau$ decay [19] and test of the CVC hypothesis.

[2]In fact the e^+e^--annihilation into five pions was observed by the M3N collaboration [14], however, the intermediate mechanisms producing the final state were not studied.

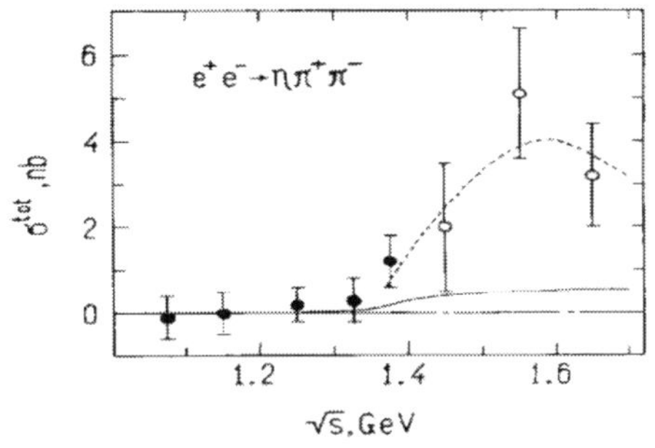

Figure 8. Energy dependence of the $e^+e^- \to \eta\pi^+\pi^-$ cross section. In the VMD model, the solid line includes only the $\rho(770)$ resonance, whereas the dashed line takes into account also the $\rho(1600)$. The ND data are represented by open circles, the DM1 ones by black circles. The figure is taken from [3].

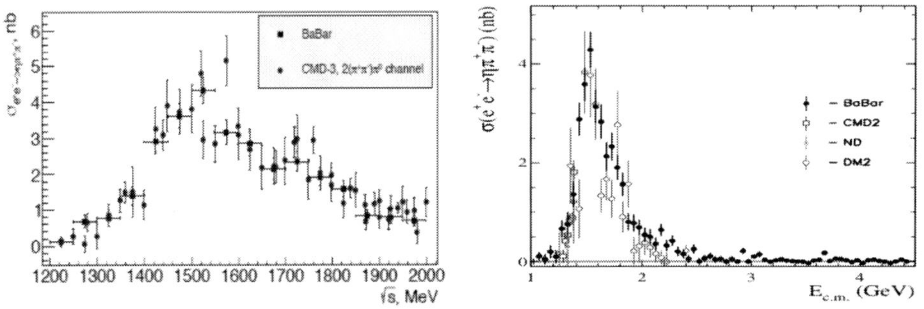

Figure 9. Comparison between experimental data for the process $e^+e^- \to \eta\pi\pi$ done by CMD-3(left panel) and BaBar Collaborations (right panel). The figures are taken from [6, 7].

The hypothesis of conserved vector current [18] and isospin symmetry is based on two assumptions: first, it assumes that the isovector part of the electromagnetic current and the vector part of the hadronic weak current are only different components of the same isospin current and, second, that the three components of this isospin current are conserved. As a consequence, hadronic currents describing vector τ decays and hadronic low energy (up to the τ lepton mass) e^+e^--annihilation are related and can be obtained one from the other.

Applying CVC for the $e^+e^- \to \pi^+\pi^-\eta$ data one obtains the branching fraction of the $\tau^- \to \pi^-\pi^0\eta\nu_\tau$ decay:

$$\frac{Br(\tau^- \to \pi^-\pi^0\eta\nu_\tau)}{Br(\tau^- \to e^-\nu_\tau)} = \frac{3|V_{ud}^2|S_{EW}}{2\pi\alpha^2}\int_{4m_\pi^2}^{\infty} dq^2 \frac{q^2}{m_\tau^2}\left(1-\frac{q^2}{m_\tau^2}\right)^2\left(1+\frac{2q^2}{m_\tau^2}\right)\sigma_{e^+e^-\to\eta\pi\pi}(q^2) \quad (3)$$

where S_{EW} is an electroweak correction and the V_{ud} is the ud element of the Cabibbo-Kabayashi-Maskawa matrix. The estimate for $Br((\tau^- \to \pi^-\pi^0\eta\nu_\tau)$ is summarized in Table 1. As one can see the branching ratio derived from the e^+e^- CMD-2 and old data data is in agreement with the world average experimental value whereas the predicted BaBar value is slightly above it. A possible source of this difference might be an yet unknown isospin breaking corrections.

Group	$Br,\%$	Ref.
Old e^+e^- data	0.13 ± 0.015	[19]
BaBar 2007	0.165 ± 0.015	[19]
Old e^+e^- +BaBar 2007	0.153 ± 0.018	[19]
BaBar 2016	0.160 ± 0.009	[8]
BaBar 2007+2016	0.162 ± 0.008	[8]
CMD-2	$0.156 \pm 0.004 \pm 0.01$	[4]
PDG	0.139 ± 0.007	[20]

Table 1. $Br(\tau^- \to \pi^-\pi^0\eta\nu_\tau)$ value obtained on the base of the CVC hypothesis from e^+e^- data and its world average experimental value.

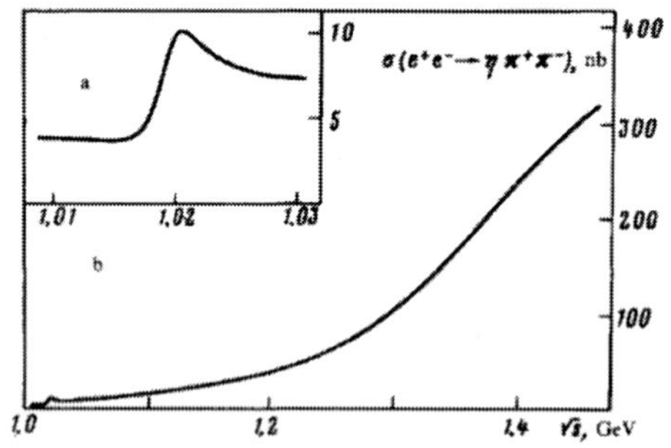

Figure 10. The VMD predicted cross section for $e^+e^- \to \eta\pi\pi$ and the influence related with production $\phi \to \eta\pi\pi$. The figure is taken from [21].

5 Rare decay $e + e^- \to \phi \to \eta\pi^+\pi^-$

Investigation of the energy dependence of the process (1) cross section for the CM energy near the ϕ meson mass is necessary for the search of the rare $\phi \to \eta\pi^+\pi^-$ decay. This latter is both OZI suppressed and G-parity violating. Its branching ratio has been theoretically estimated within VMD: $Br(\phi \to \eta\pi^+\pi^-) = 0.35 \cdot 10^{-6}$ [21] and its influence on the $\eta\pi^+\pi^-$ cross section is shown in Fig. 10. Experimentally the upper limit was measured by the CMD-2 Collaboration and is $Br(\phi \to \eta\pi^+\pi^-) < 1.8 \cdot 10^{-5}$ 90% CL [22]. Therefore the current experimental upper limit is about 50 times above than its theoretical prediction.

6 Conclusion

Summarizing, during the last decade the process of $e^+e^- \to \eta\pi^+\pi^-$ has been studied both by theoreticians, using the description based on the Resonance Chiral Lagrangian or the extended Nambu-Jona-Lasinio model, and by experimental collaborations. The data sets at energies 1. – 2. GeV have been collected in e^+e^--annihilation by the CMD and SND Collaborations (Novosibirsk, VEPP-2M collider) and using the radiative return technique at the SLAC B-factory with the BaBar detector. The systematic uncertainty has been decreased substantially compared with the old experiments and reached the level below 8% (the preliminary analysis by CMD-3 claims even a 5% level). The data of the three experiments are in agreement. However, a more detailed study based on the cross section approximation proposed in Ref. [19] might be applied for a better comparison between the data in whole energy region.

To fit the measured cross section, the VMD model with three isovector states $\rho(770)$, $\rho(1450)$ and $\rho(1700)$ has been used. However, the currents statistics does not allow to conclude whether the $\rho(1700)$ is necessary to reproduce the data or new more precise measurements at energies above 1.4 GeV would be necessary to study the region in detail. This study will require an estimate of a background related with the production via $a_2(1320)$, $\rho(1400;1700) \to a_2\pi$, but to our knowledge the corresponding process has not yet been calculated theoretically.

Also new measurements at low energies near the ϕ meson region are important to improve our knowledge about the rare decay $\phi \to \eta\pi^+\pi^-$ (the upper experimental limit is 50 times higher than the theoretical prediction). This is planned to be done at the KLOE-2 experiment.

Assuming the isospin invariance the $e^+e^- \to \eta\pi\pi$ cross section via the CVC relation predicts the branching ratio $\tau^- \to \eta\pi^-\pi^-\nu_\tau$. Therefore comparison between the predicted value and its measured value would allow to test the CVC hypothesis. Currently the branching ratio derived from the CMD-2 and old e^+e^- data is in agreement with the average experimental value whereas the predicted value based on BaBar data is slightly above it. A possible source of this difference might be due to not yet calculated isospin-breaking corrections.

Last but not least, the $\eta\pi\pi$ contribution gives about 5% to the total hadronic cross section at energy $\sqrt{s} = 1.5$ GeV and therefore gives a non-negligible contribution to the anomalous magnetic moment of muon, $(g-2)_\mu$.

References

[1] B. Delcourt, D. Bisello, J. C. Bizot, J. Buon, A. Cordier and F. Mane, Phys. Lett. **113B**, 93 (1982) Erratum: [Phys. Lett. **115B**, 503 (1982)].
[2] A. Antonelli *et al.* [DM2 Collaboration], Phys. Lett. B **212**, 133 (1988).
[3] V. P. Druzhinin *et al.*, Phys. Lett. B **174**, 115 (1986).
[4] R. R. Akhmetshin *et al.* [CMD-2 Collaboration], Phys. Lett. B **489**, 125 (2000).
[5] V. M. Aulchenko *et al.* [SND Collaboration], Phys. Rev. D **91**, 052013 (2015).
[6] E. P. Solodov *et al.*, AIP Conf. Proc. **1735**, 020005 (2016).
[7] B. Aubert *et al.* [BaBar Collaboration], Phys. Rev. D **76**, 092005 (2007) Erratum: [Phys. Rev. D **77**, 119902 (2008)].
[8] R. Waldi [BaBar Collaboration], Talk atICHEP 2016, *http://indico.cern.ch/event/432527/contributions/1072438/*.
[9] D. Gomez Dumm and P. Roig, Phys. Rev. D **86**, 076009 (2012).
[10] L. Y. Dai, J. Portoles and O. Shekhovtsova, Phys. Rev. D **88**, 056001 (2013).
[11] M. K. Volkov, A. B. Arbuzov and D. G. Kostunin, Phys. Rev. C **89**, 015202 (2014).
[12] M. K. Volkov and A. B. Arbuzov, Phys. Part. Nucl. **47**, 489 (2016).
[13] M. K. Volkov and V. L. Yudichev, Phys. Part. Nucl. **31**,282 (2000); M. K. Volkov, D. Ebert and M. Nagy, Int. J. Mod. Phys. A **13**, 5443 (1998).
[14] G. Cosme *et al.*, Nucl. Phys. B **152**, 215 (1979).
[15] J. Layssac and F. M. Renard, Nuovo Cim. A **6**, 134 (1971).
[16] M. Atkinson *et al.* [Omega Photon Collaboration], Nucl. Phys. B **242**,269 (1984); M. Atkinson *et al.* [Omega Photon Collaboration], Z. Phys. C **30**, 531 (1986).

[17] A. Donnachie and A. B. Clegg, Z. Phys. C **34**, 257 (1987).
[18] H. B. Thacker and J. J. Sakurai, Phys. Lett. **36B**, 103 (1971);
Y. S. Tsai, Phys. Rev. D **4**, 2821 (1971).
[19] V. Cherepanov and S. Eidelman, Nucl. Phys. Proc. Suppl. **218**, 231 (2011).
[20] C. Patrignani *et al.* [Particle Data Group], Chin. Phys. C **40**, 100001 (2016).
[21] N. N. Achasov and V. A. Karnakov, JETP Lett. **39**, 342 (1984).
[22] R. R. Akhmetshin *et al.* [CMD-2 Collaboration], Phys. Lett. B **491**, 81 (2000).

The $\gamma\gamma$ Physics Program at BESIII

C.F. Redmer for the BESIII Collaboration[1,★]

[1] *Institut für Kernphysik, Johannes Gutenberg-Universität Mainz, J.-J.-Becher Weg 45, 55128 Mainz, Germany*

Abstract. A key motivation for the two-photon physics program of the BESIII collaboration is the need of high precision data on electromagnetic transition form factors as input to the calculations of the contribution of hadronic Light-by-light scattering to the anomalous magnetic moment of the muon. The data collected with the BESIII detector allow to study the momentum dependence of the form factors at small momentum transfers, which is of special relevance for a_μ. In this presentation the ongoing measurements of the transition form factors of π^0, η and η' mesons, as well as pion pairs, are discussed, and the potential for first double-tagged measurements at BESIII are pointed out.

1 Introduction

The production and investigation of hadrons at electron-positron colliders mostly relies on the annihilation of the beam particles. All states produced in this reaction are bound to the quantum numbers of the virtual photon, which is exchanged: $J^{PC} = 1^{--}$. In order to study hadrons with other quantum numbers usually the different decay modes of the produced resonances are exploited.

Final states with quantum numbers different from $J^{PC} = 1^{--}$ can be produced directly at e^+e^- colliders in two-photon collisions. Instead of annihilating, the leptons scatter and exchange two photons in the process, which fuse into a new, hadronic state. The investigation of these processes provides a clean laboratory for the spectroscopy of light hadronic $J^{PC} = 0, 2^{\pm+}$ states. In contrast to hadronic systems produced in e^+e^- annihilations, the mass of these hadrons is well below the center-of-mass energy of the collider. As they are produced in a purely electromagnetic environment, studying these light hadrons provides important input to the search for gluonium or gluon-rich states.

The cross section of hadronic final states in $\gamma\gamma$ collisions is proportional to their respective radiative width $\Gamma_{\gamma\gamma}$ and the squared transition form factor (TFF) $|F(Q_1^2, Q_2^2)|^2$, with $Q_i = -q_i$ being the momentum transfer of the scattered leptons and, thus, the virtuality of the colliding photons [1]. Therefore $\gamma\gamma$ physics also provides access to structural information of the produced hadronic states. This presentation focuses on the measurement of the TFF of light mesons at the BESIII experiment, which is motivated by the ongoing efforts in theory and experiment to determine the anomalous magnetic moment of the muon, a_μ.

The anomalous magnetic moment of the muon is one of the most precisely known observables in particle physics. Over the recent decades enormous efforts have been made to reach a precision in the order of 10^{-10} in the direct experimental measurement [2], as well as in the Standard Model prediction

★e-mail: redmer@uni-mainz.de

of a_μ. However, there remains a discrepancy of 3 to 4 standard deviations between the measured and the calculated values of a_μ [3, 4]. The potential origin of this difference has triggered a lot of activity, since, even if not being significant enough to claim a discovery, it could be considered as a hint for New Physics.

New experiments have been proposed to remeasure a_μ with a four times better precision [5, 6]. These proposals have triggered numerous efforts to improve the Standard Model prediction on the same level. While the dominating QED contribution has been calculated including corrections up to the tenth order [7], and the weak contribution also being well understood by means of perturbation theory [8], the error of the prediction of a_μ is completely dominated by the hadronic contributions. The main challenge is that QCD cannot be handled by perturbative methods at the relevant energy regime.

The largest hadronic contribution a_μ^{hVP} is due to the hadronic vacuum polarization. It can be handled within a dispersive framework, which requires hadronic cross sections measured at e^+e^- colliders as experimental input [4]. Various laboratories world wide have measured $\sigma(e^+e^- \to \text{hadrons})$ in scan experiments and exploiting the technique of Initial State Radiation (ISR). These results can be used to significantly reduce the uncertainty of a_μ related to hadronic vacuum polarization in the near future.

Another hadronic contributions comes from hadronic Light-by-Light scattering (hLbL), a_μ^{hLbL}. It cannot be as easily related to experimental observables as a_μ^{hVP}. Calculations of the hLbL process are possible at lowest energies by means of chiral perturbation theory, or at high energies, using perturbative QCD. The energy scale relevant for the magnetic moment of the muon is the intermediate regime, which cannot be addressed by perturbative means. In order to evaluate the hLbL, different hadronic models have been developed and are used for the calculation of a_μ^{hLbL}. Two more frequently quoted examples for these models are provided by Prades, de Rafael and Vainshtain [9], which is referred to as the "Glasgow Consensus", and by Jegerlehner and Nyffeler [3]. The different models agree on the dominating processes in their evaluation of hLbL, which comes from the transitions of the pseudoscalar mesons π^0, η and η' [10]. The validity of the models can, thus, be tested with measurements of the momentum transfer dependence of the respective TFF. The uncertainties of the resulting values of a_μ^{hLbL} are, however, strongly model dependent.

In order to reduce the model dependency of the prediction of a_μ^{hLbL}, data driven approaches are being developed. Their goal is to provide a more reliable estimate of the uncertainty of a_μ^{hLbL} by describing the dominating processes based on dispersive analyses [11, 12] and experimental information. Important inputs to these calculations are the TFF of π^0, of two pions and the pion polarizabilities.

Only few experiments investigated the momentum dependence of space-like TFF of light mesons. At e^+e^- machines the information can be accessed via the meson production in two-photon collision. However, the kinematics of two-photon reactions result in a maximum of the differential cross section at smallest scattering angles of the two leptons. Detectors designed to study the annihilation reactions of the beam particles do not cover smallest scattering angles. The design of the beam pipes and the need for magnets close to the interaction point to bring the beams to collision prevents the installation of detectors. Instead, a technique referred to as "single-tag measurement" is applied. Only one of the two scattered leptons and the complete hadronic system, which was produced, is reconstructed in the detector volume. The other lepton is assumed to have escaped detection at smallest scattering angles, i.e. it has been scattered along the beam axis. The assumption can be tested with the polar angle of the missing momentum of the detected particles. Since the momentum transfer of the undetected lepton is small, the exchanged photon is quasi-real and the TFF depends only on a single momentum transfer: $F(Q_1^2, Q_2^2 \approx 0) = F(Q^2)$.

The most recent results on the momentum dependence of TFF are provided by the B-factories [13–16]. Their studies of the π^0, η, η', and more recently the $\pi^0\pi^0$ system cover momentum transfers Q^2

from 4 GeV^2 to 40 GeV^2. The results, especially for the TFF of the π^0 at $Q^2 > 10\ \mathrm{GeV}^2$, triggered a discussion on the applicability of pQCD in this energy range, which is often referred to as the Belle-BaBar-puzzle. For the determination of a_μ^{hLbL}, however, these data are less important. Here, the Q^2-dependence of the TFF at approximately 1.5 GeV^2 is most relevant [17]. At these energies only measurements from the CELLO and CLEO experiments [18, 19] are available, which have rather low statistics.

The need for additional precision data on the momentum dependence of TFF, in order to improve the calculations of hLbL and, thus, to improve the prediction of a_μ, is a key motivation of the $\gamma\gamma$ physics program at BESIII. The BESIII experiment can contribute in the region of low momentum transfer and can provide data on the TFF dependence with high statistical accuracy.

2 The BESIII detector at BEPCII

The BESIII experiment [20] is located at the symmetric e^+e^- collider BEPCII, which is operated at the IHEP in Beijing (China). Data can be collected at center-of-mass energies from 2.0 GeV to 4.6 GeV. The BESIII detector setup consists of a large drift chamber surrounding the Beryllium beam pipe, a time-of-flight system, and an electromagnetic calorimeter, which are placed inside the bore of a superconducting solenoid, providing a 1.0 T magnetic field. The flux-return yoke is instrumented and serves as muon counter. The detector covers 93% of the solid angle. Momenta of charged particles and photon energies are measured with a resolution of 0.5% and 2.5% at 1 GeV, respectively.

The BESIII collaboration has collected $1.31\times10^9 e^+e^- \to J/\psi$ events, more than $500\times10^6 e^+e^- \to \psi(2S)$ events, 2.9 fb^{-1} at the $\psi(3770)$ peak, and more than 10 fb^{-1} in the center-of-mass region above 4 GeV, which are devoted to studies of the charmonia and charmonium-like states [21–23]. Additional data have been acquired for a τ-mass scan and a high statistics R-scan. Based on these data, the BESIII collaboration pursues a physics program [24], which focuses on charmonium spectroscopy, charm physics, light hadron spectroscopy, τ physics, and R-measurements.

The investigations in the field of $\gamma\gamma$ physics make use of all larger data sets, whereas the data taken at $\sqrt{s} = 3.773$ GeV constitute one of the largest individual samples. It is used to study the TFF of π^0, η and η'.

3 Electromagnetic Transition Form Factors of π^0, η and η'

The analysis of the TFF of π^0, η and η' is based on a single-tag technique. The two-photon reaction $e^+e^- \to \mathcal{P}e^+e^-$ with $\mathcal{P} = \pi^0, \eta, \eta'$ is reconstructed by measuring only one of the two scattered leptons and the decay products of the produced mesons $\mathcal{P}$ in the detector. The missing momentum of the measured particles is taken as the four-momentum of the undetected lepton. As it is undetected, its scattering angle smaller is required to be less than 8°, which corresponds to a distance from the beam axis equal to the radius of the beam pipe at the end of the detector.

The π^0 meson is reconstructed from its dominating decay into two photons. The main source of background in the event selection are QED processes, such as radiative Bhabha scattering. Hard radiative photons detected along with soft photons from any secondary process can reproduce the invariant mass of a pion. A set of three conditions has been worked out to successfully reduce the background contribution. First, a limit on the helicity angle of the photons is applied, which is defined as the angle between one of the photons in the rest frame of the pion and the direction of motion of the pion in the laboratory frame. Another condition to suppress radiative Bhabha scattering is imposed by setting a threshold on the scattering angle of the pion candidate. Finally, a minimum polar angle difference of the two photons is required in the laboratory frame to suppress contributions from cluster splitting. Another source of background comes from incompletely detected hadronic final states. Decays of $\psi(3770)$ into pairs of D mesons, the radiative return to $\psi(2S)$ and J/ψ, and, most dominantly, states from the from the $q\bar{q}$ continuum contribute. A condition, designed by the BaBar collaboration [13] to suppress initial state radiation (ISR) in two-photon production, turned out also to suppress this kind of background. Making use of four-momentum conservation to calculate the energy of a potential, unregistered ISR photon, it is also sensitive to the energy imbalance induced by incompletely reconstructed hadronic final states. Event candidates fulfilling all selection criteria show clear signal peaks, not only of the π^0, but also of the η meson in the invariant mass distribution of the two photons.

The differential distribution of the momentum transfer of the tagged lepton is used to extract the TFF of the π^0 and study its momentum dependence. In a first step, the remaining background is subtracted. The background contribution in each bin in Q^2 is estimated from the corresponding invariant mass spectrum of the two photon system. The pion signal content is determined from a fit to the spectrum and the number of signal events above a continuous background are considered as the background free content of the individual bin in Q^2. The background free Q^2 distribution is converted to the distribution of the differential cross section by normalizing it to the reconstruction efficiency and the integrated luminosity. Finally, the TFF is extracted by dividing out the point-like cross section, which is obtained from a Monte Carlo simulation, based on the Wess-Zumino-Witten term [25, 26].

The data taken by the BESIII collaboration at the $\psi(3770)$ peak allow to measure the TFF of the π^0 at momentum transfers in the region between 0.3 GeV^2 and 3.1 GeV^2. The interval matches the kinematic region, which has been identified as most relevant input for the calculations on hadronic Light-by-Light scattering [17]. The statistical accuracy obtained with the analysis scheme described above is unprecedented for $Q^2 \leq 1.5$ GeV^2; for larger momentum transfers it is compatible with the accuracy of the CLEO [19] result. The evaluation of the systematic uncertainties is being finalized. The method applied for background subtraction has been identified as the largest source of uncertainty. The publication of the final result is expected soon.

An improved understanding of the contribution of the π^0 TFF to a_μ^{hLbL} is, however, only sufficient on the level of the current experimental precision of the direct measurement of a_μ. In view of the announced measurements and their planned four-fold improvement in accuracy, it is necessary to include and to improve the understanding of the contributions of η and η' as well [17].

A dedicated measurement of the respective TFF has been started at BESIII, based on the decay modes $\eta \to \pi^0\pi^+\pi^-$ and $\eta' \to \eta\pi^+\pi^-$. Taking into account the subsequent decays of π^0 and η into two photons, both decay modes yield the same final state. Analogously to the investigation of the π^0 TFF, analysis strategy follows the single-tag technique in order to study the momentum dependence of the TFF. The dominating background contributions involve on the one hand decays of ω and ϕ mesons into three pions in the analysis of the η and η' TFF and on the other hand incompletely reconstructed decays of η' mesons in the analysis of the η TFF. Both kinds of background contributions are efficiently suppressed by kinematically fitting the decay systems to the masses of η and η', respectively.

Similar to the investigation of the momentum dependence of the TFF of the π^0, the TFF of η and η' can be studied based on the data taken at $\sqrt{s} = 3.773$ GeV, for momentum transfers between $0.3 \leq Q^2\,[\mathrm{GeV}]^2 \leq 3.5$. The statistical accuracy, which is achieved with the analysis strategy described above, is compatible with the published results of the CELLO and CLEO experiments [18, 19]. However, in contrast to these published results, the current analysis of the BESIII data makes only use of a single decay channel. The accuracy achieved by the CELLO and CLEO collaborations combines the results of several different decay channels. Extending the analysis by adding more decay channels, as well as analyzing the remaining data sets will improve the BESIII result significantly.

4 Electromagnetic Transition Form Factors of $\pi\pi$

In addition to the production of single, pseudoscalar mesons, the BESIII collaboration also started the investigation of multi-meson production in two-photon collisions. The aim is to gain additional information on scalar and tensor states, which decay into $\pi\pi$, $\pi\eta$, and $\eta\eta$ systems. The motivation to study the two pion system is twofold. From point of view of spectroscopy, it is possible to extract resonance parameters and also to study pion rescattering effects at low invariant masses of the $\pi\pi$ system. Additionally, the investigation is also motivated by its relevance as input for the data-driven, dispersive calculations for the contribution of hLbL to a_μ.

Previous measurements of the production of two pseudoscalars in two-photon collisions focused on the spectroscopy of intermediate states [27–31]. The published results are based on the untagged technique, where both leptons are scattered close to the beam axis and, thus, escape detection. Due to the behavior of the differential cross section in two-photon collisions, the event rate is largest in this kinematic configuration, which makes it preferable for spectroscopy. The mass range covered by most of the different experiments is restricted to invariant masses larger than approximately 500 MeV/c^2. A few data points at lower masses have been provided by the MARKII experiment.

Recently, the Belle collaboration published the first investigation of the momentum dependence of the TFF in the two-pion system [16]. The $\pi^0\pi^0$ system was studied for momentum transfers between $3 \leq Q^2[\mathrm{GeV}^2] \leq 30$, for invariant masses of the pion pairs between $0.5 \leq m_{\pi^0\pi^0}[\mathrm{GeV/c}^2] \leq 2.1$, and with complete coverage of the helicity angle. In a partial wave analysis the $|D_0|^2$, $|D_2|^2$, and $|S|^2$ wave amplitudes have been determined, and the TFF of the $f_0(980)$ resonance and the three possible helicity states of the $f_2(1270)$ resonance were measured and compared to different predictions [32, 33].

At BESIII a measurement of the $\pi^+\pi^-$ pair production in two-photon collisions is performed. The analysis follows the single-tag strategy, successfully used in the analysis of single pseudoscalar meson production. Further analysis conditions are designed to efficiently suppress the two dominating sources of background. On the one hand, the two-photon production of muon pairs require a special focus on the μ/π separation, and on the other hand there is a contribution of time-like pion pairs production, which stems from radiative Bhabha scattering, where the photon is virtual and couples to a ρ-meson, which in turn decays into two pions.

The QED background of muon production is well understood from the $\gamma\gamma$ physics studies at LEP [34, 35]. Reliable Monte Carlo generators exist, which are currently used to subtract the QED contributions from data. At the same time, the application of an artificial neural network is prepared. Being trained based on the Monte Carlo distributions, it will allow to suppress the background from muon pairs production with an improved statistical accuracy. The method particle identification has already been successfully applied in the measurement of hadronic cross sections [36]. The background involving the ρ-meson is subtracted by fitting the clearly visible peak in the $\pi^+\pi^-$ invariant mass with the Kühn-Santamaria parameterization [37]. The remaining events allow to study the production of $\pi^+\pi^-$ in bins of the momentum transfer Q^2, the pion invariant mass, and the pion helicity angle $\cos\theta^*$. This is the first measurement in the invariant mass region of $2m_\pi \leq M_{\pi\pi}[\mathrm{GeV}] \leq 2.0$ and the momentum transfer region of $0.2 \leq Q^2[\mathrm{GeV}^2] \leq 2.0$ with a full coverage of the helicity angle.

5 Double-tagged Measurements

In double-tagged measurements both of the scattered leptons are registered in the detector. Since the design of common detector setups at e^+e^- colliders focuses on the investigation of final states produced in the annihilation of the beam particles, in a double-tagged measurement the leptons can be registered at large angles only, usually at angles larger 20° with respect to the beam axis. Thus, the momentum transfer of each lepton is large. The cross section of two-photon reactions, however, becomes vanishingly small for two large virtualities. Currently, experimental information on TFFs depending on two virtualities $F(Q_1^2, Q_2^2)$ is not available. The knowledge of TFF with arbitrary virtuality is vital for a reliable error estimate of a_μ^{hLbL}.

First exploratory studies for a double-tagged measurement of the π^0 TFF have been started at BESIII. The aim is to exploit the large data sets, originally collected for charm physics and charmonium-like spectroscopy, which in total correspond to more than 10 fb^{-1}. Based on an analysis chain using only simple conditions on energy and momentum conservation, a part the data have been studied and a π^0 signal has been observed. The signal-to-background ratio is very low, but more sophisticated analysis techniques should be able to improve it. Efficiency studies based on Monte Carlo distributions show that a direct measurement of $F(Q_1^2, Q_2^2)$ is possible for $0.5 \leq Q_i^2$ [GeV]$^2 \leq 1.5$, with $i = 1, 2$. Due to the limited data set, the result of this analysis, regardless of the reconstruction efficiency, which can be achieved in a final, optimized analysis, can provide only limited information on the doubly-virtual TFF. Nevertheless, it is expected to have a significant impact on the uncertainty of a_μ^{hLbL} [38].

The statistics, which has been collected by the BESIII collaboration should provide sufficient accuracy for a meaningful confrontation with hadronic models. As pointed out recently by Nyffeler [38], certain models differ significantly in their Q^2 behavior of the TFF. For example, the momentum transfer dependence of the TFF in VMD and in the LMD+V model [17] differs by a damping factor of Q^2. The relative difference between the two models could be as large as 25% for $F(1\text{GeV}^2, 1\text{GeV}^2)$. This sizable effect could be resolved by a measurement at BESIII.

Another aspect of double-tagged measurements is related to the production mechanism in two-photon collisions. The general term for the cross section can be separated into individual terms depending on the transverse or longitudinal polarization of the individual photons, the relative parallel or perpendicular polarization of the two photons and the helicity of the two-photon system [1]. Another parameter is the angle of the plane of the leptons in the rest frame of the two photons. It can be used to separate multi-meson and tensor contributions to the TFF. Its determination requires good knowledge of all four-momenta in the final state, especially of both scattered leptons. In single-tag measurements there is, however, a large uncertainty on the azimuthal angle of the untagged lepton, which is propagated when determining the momentum transfer, i.e. the virtuality of the photons. Double-tagged measurements, in turn, suffer from a vanishing cross section, due to the limitations of conventional detector setups.

A remedy for the efficiency gaps at small angles is the installation of special tagging detectors close to the beam pipe. In this way, double-tagged measurements are not longer restricted to events with two large virtualities of the exchanged photons. Tagging detectors have already been installed at different facilities [39, 40]. Also at BESIII, a first tagging detector has been installed. It is made from spare parts of the KLOE calorimeter [41, 42], which is a sampling calorimeter made from lead an scintillating fibers. Due to spatial limitations, the tagger only covers polar angles from 1 to 10 mrad. First data have been taken and demonstrate that the tagging detector can also be used to measure photons emitted by initial state radiation. Motivated by the additional use and by the expected aging effects of the scintillating fibers in the high rate environment close to the collision region of the accelerator, it has been decided to replace the current detector design by crystal calorimeters, which will be installed in both hemispheres of the BESIII setup. These calorimeters will consist of two arrays of three by four LYSO crystals. The length of the arrays corresponds to approximately 12.2 radiation lengths. Monte Carlo simulations have been performed to optimize the setup. It was found that the energy resolution is mostly limited by the shower leakages due to the small volume of the detector. Currently, the crystals and the readout electronics are tested at the electron accelerator MAMI located at the Institute for Nuclear Physics of Mainz University, Germany. The installation of the detectors is planned for 2019.

6 Summary and Outlook

The need for new, high precision data on electromagnetic transition form factors of pseudoscalar mesons as input to the calculations of the contribution of hadronic Light-by-light scattering to the anomalous magnetic moment of the muon is one of the main motivations of the two-photon physics program at BESIII. Single-tag measurements are performed to study the momentum dependence of the respective TFF. The investigation of the π^0 TFF provides results for momentum transfers between 0.3 GeV^2 and 3.1 GeV^2, where the statistical accuracy is unprecedented for $0.3 \leq Q^2$ [GeV]$^2 \leq 1.5$ and compatible with previous measurements for larger values of Q^2.

The TFF of η and η' has been studied reconstructing only a single decay mode. The result covers the same kinematic region as the studies of the π^0 TFF, however the statistical accuracy does not exceed previous measurements. Considering additional decays of η and η' will significantly improve the final result.

The analysis of $\pi^+\pi^-$ in two-photon collisions will provide the first single-tag measurement at low invariant masses and small momentum transfers, with the full coverage of the pion helicity angle. In addition to the investigation of the TFF for different helicities, also rescattering effects, expected at small invariant masses of the pion pairs, can be studied. The analysis is being extended to the neutral channels $\pi^0\pi^0, \pi^0\eta$ and $\eta\eta$.

In addition to the single-tag studies, a first investigation on the possibility of double-tagged measurements has been started. The outcome of first precursor studies demonstrates the feasibility and gives a hint on the impact of a final result, which will be the first direct and model independent parameterization of the TFF of π^0.

The great potential of the BESIII experiment, to contribute valuable information on the field of $\gamma\gamma$ physics is currently being extended by the installation of tagging detectors. These add new prospects to the physics program, allowing to measure the scattered leptons with small momentum transfer, which, so far, escaped detection.

References

[1] V.M. Budnev, I.F. Ginzburg, G.V. Meledin, V.G. Serbo, Phys. Rept. **15**, 181 (1975)
[2] G.W. Bennett et al. (Muon g-2), Phys. Rev. **D73**, 072003 (2006), hep-ex/0602035
[3] F. Jegerlehner, A. Nyffeler, Phys. Rept. **477**, 1 (2009), 0902.3360
[4] M. Davier, A. Hoecker, B. Malaescu, Z. Zhang, Eur. Phys. J. **C71**, 1515 (2011), [Erratum: Eur. Phys. J.C72,1874(2012)], 1010.4180
[5] G. Venanzoni (Muon g-2), Nucl. Phys. Proc. Suppl. **225-227**, 277 (2012)
[6] T. Mibe (J-PARC g-2), Nucl. Phys. Proc. Suppl. **218**, 242 (2011)
[7] T. Aoyama, M. Hayakawa, T. Kinoshita, M. Nio, Phys. Rev. Lett. **109**, 111808 (2012), 1205.5370
[8] C. Gnendiger, D. Stöckinger, H. Stöckinger-Kim, Phys. Rev. **D88**, 053005 (2013), 1306.5546
[9] J. Prades, E. de Rafael, A. Vainshtein, Adv. Ser. Direct. High Energy Phys. **20**, 303 (2009), 0901.0306
[10] E. de Rafael, Phys. Lett. **B322**, 239 (1994), hep-ph/9311316
[11] G. Colangelo, M. Hoferichter, M. Procura, P. Stoffer, JHEP **09**, 074 (2015), 1506.01386

[12] V. Pauk, M. Vanderhaeghen, Phys. Rev. **D90**, 113012 (2014), 1409.0819
[13] B. Aubert et al. (BaBar), Phys. Rev. **D80**, 052002 (2009), 0905.4778
[14] P. del Amo Sanchez et al. (BaBar), Phys. Rev. **D84**, 052001 (2011), 1101.1142
[15] S. Uehara et al. (Belle), Phys. Rev. **D86**, 092007 (2012), 1205.3249
[16] M. Masuda et al. (Belle), Phys. Rev. **D93**, 032003 (2016), 1508.06757
[17] M. Knecht, A. Nyffeler, Phys. Rev. **D65**, 073034 (2002), hep-ph/0111058
[18] H. Behrend et al. (CELLO), Z. Phys. **C49**, 401 (1991)
[19] J. Gronberg et al. (CLEO), Phys. Rev. **D57**, 33 (1998), hep-ex/9707031
[20] M. Ablikim et al. (BESIII), Nucl. Instrum. Meth. **A614**, 345 (2010), 0911.4960
[21] M. Ablikim et al. (BESIII), Chin. Phys. **C37**, 123001 (2013), 1307.2022
[22] M. Ablikim et al. (BESIII), Chin. Phys. **C39**, 093001 (2015), 1503.03408
[23] M. Ablikim et al. (BESIII), accepted for publication by Chin. Phys. C, (2016), 1607.00738
[24] D.M. Asner et al., Int. J. Mod. Phys. **A24**, S1 (2009), 0809.1869
[25] J. Wess, B. Zumino, Phys. Lett. **B37**, 95 (1971)
[26] E. Witten, Nucl. Phys. **B223**, 422 (1983)
[27] J. Boyer, F. Butler, G. Gidal, G. Abrams, D. Amidei et al., Phys. Rev. **D42**, 1350 (1990)
[28] H. Behrend et al. (CELLO), Z. Phys. **C56**, 381 (1992)
[29] T. Mori et al. (Belle), Phys. Rev. **D75**, 051101 (2007), hep-ex/0610038
[30] S. Uehara et al. (Belle), Phys. Rev. **D78**, 052004 (2008), 0805.3387
[31] S. Uehara et al. (Belle), Phys. Rev. **D80**, 032001 (2009), 0906.1464
[32] G.A. Schuler, F.A. Berends, R. van Gulik, Nucl. Phys. **B523**, 423 (1998), hep-ph/9710462
[33] V. Pascalutsa, V. Pauk, M. Vanderhaeghen, Phys. Rev. **D85**, 116001 (2012), 1204.0740
[34] F.A. Berends, P. Daverveldt, R. Kleiss, Comput. Phys. Commun. **40**, 271 (1986)
[35] F.A. Berends, P. Daverveldt, R. Kleiss, Comput. Phys. Commun. **40**, 285 (1986)
[36] M. Ablikim et al. (BESIII), Phys. Lett. **B753**, 629 (2016), 1507.08188
[37] J.H. Kuhn, A. Santamaria, Z. Phys. **C48**, 445 (1990)
[38] A. Nyffeler, Phys. Rev. **D94**, 053006 (2016), 1602.03398
[39] D. Babusci, H. Czyz, F. Gonnella, S. Ivashyn, M. Mascolo, R. Messi, D. Moricciani, A. Nyffeler, G. Venanzoni, Eur. Phys. J. **EPJC72p1917.pdf**, 1917 (2012), 1109.2461
[40] V.V. Anashin et al., Phys. Part. Nucl. **44**, 657 (2013)
[41] M. Adinolfi et al., Nucl. Instrum. Meth. **A494**, 326 (2002)
[42] M. Adinolfi et al., Nucl. Instrum. Meth. **A482**, 364 (2002)

Latest studies on the reaction $e^+e^- \to K^+K^-\gamma$

Leonard Leśniak[1,⋆] and *Michał* Silarski[1,⋆⋆]

[1] *Institute of Physics, Jagiellonian University, 30-348 Kraków, Poland*

Abstract. Recent theoretical studies of the $e^+e^- \to K^+K^-\gamma$ process are described. Three main reaction mechanisms are considered: the initial state radiation, the final state radiation and the strong interaction between the outgoing K^+K^- mesons. The K^+K^- effective mass distributions are derived for three different models which in past have been used for a description of the $e^+e^- \to \pi^+\pi^-\gamma$ data. Also the numerical results for the angular photon and kaon distributions are presented. A new model of the $e^+e^- \to M_1M_2\gamma$ reactions is outlined which can serve for multichannel analyses of the radiative processes with a production of two pseudoscalar mesons M_1 and M_2.

1 Introduction

Strong interactions of the strange mesons K^+ and K^-at low energies are not well known. Since the K^+K^- threshold lies quite close to the DAΦNE accelerator energy, the low energy K^+K^- interactions can be studied using data collected by the KLOE experiment. Masses of the scalar resonances $f_0(980)$ and $a_0(980)$ are close to 1 GeV. There are, however, large uncertainties in their values. According to the Particle Data Group estimations [1] the $f_0(980)$ mass equals to 990 ± 20 MeV and its width lies between 10 and 100 MeV, while the $a_0(980)$ mass is 980 ± 20 MeV and the width value varies between 50 and 100 MeV. Let us stress here that the parameters of the scalar resonances found in experimental analyses are very much model dependent.

As an example let us consider the reaction $e^+e^- \to \pi^+\pi^-\gamma$ studied by the KLOE Collaboration in 2006 [2]. Fits to the experimental data done using the following two models: the kaon-loop model [3] and the so-called no-structure model [4], gave quite different values of the $f_0(980)$ mass ranges: 980-987 MeV in the first case and 973-981 MeV in the second case. The intervals of the maximal variations of the $f_0(980)$ coupling constants to K^+K^- were: 5.0-6.3 GeV for the first model and 1.6-2.3 GeV for the second one. The corresponding numbers for the $f_0(980)$ coupling constants to $\pi^+\pi^-$ were: 3.0-4.2 GeV and 0.9-1.1 GeV, respectively. One can attribute the differences between the resonance parameters to different parameterizations of the production amplitude P(m) which describes formation of the $f_0(980)$ resonance in the first step transition $e^+e^- \to f_0(980)\gamma$ (m being the $\pi^+\pi^-$ effective mass). Let us assume that the full reaction amplitude A(m) is schematically written as a product A(m)= P(m) · D(m), where D(m) is the resonant $f_0(980) \to \pi^+\pi^-$ decay amplitude. If the fits to data are performed with two different production functions P(m) then the fitted parameters of the resonant amplitude D(m) could be different as well. This is one possible source of the model dependence found in experimental analyses.

⋆e-mail: Leonard.Lesniak@ifj.edu.pl
⋆⋆e-mail: Michal.Silarski@uj.edu.pl

Let us enumerate some other problems frequently encountered in the data analyses of the reactions in which the scalar mesons are produced:

1. application of the Breit-Wigner formulae with constant widths of scalar mesons,
2. reduction of a number of scalar resonances to one (for example, assuming that near the $K\bar{K}$ threshold only the $f_0(980)$ resonance is present),
3. too simplified treatment of the final-state meson-meson interactions,
4. existence of two closed thresholds K^+K^- and $K^0\bar{K}^0$.

1.1 Motivation to study the the $e^+e^- \to K^+K^-\gamma$ reaction

Below we give some additional arguments in favour of further studies on the $e^+e^- \to K^+K^-\gamma$ reaction. The branching fraction of the $\Phi(1020)$ meson decay into $\pi^+\pi^-\gamma$ has been measured [5] but the branching fraction for the $\Phi(1020) \to K^+K^-\gamma$ channel is yet unknown. There are data for the radiative decay of the $\Phi(1020)$ meson into two scalar resonances. The corresponding ratios of the widths are: $\Gamma(f_0(980)\gamma)/\Gamma_{total} = (3.22 \pm 0.19) \cdot 10^{-4}$ and $\Gamma(a_0(980)\gamma)/\Gamma_{total} = (7.6 \pm 0.6) \cdot 10^{-5}$ [1]. Since both scalar resonances decay into K^+K^- mesons, one should experimentally observe the reaction $e^+e^- \to K^+K^-\gamma$. For the decay $\Phi \to K^0\bar{K}^0\gamma$ only the upper limit $1.9 \cdot 10^{-8}$ is known [6]. In the final state of the transition process $e^+e^- \to \Phi \to K^+K^-\gamma$ the K^+ and K^- mesons can interact so a new information about the K^+K^- strong interactions could be obtained from data. Thus the proposed experimental analysis could provide us with potentially interesting results.

2 Reaction mechanisms

There are several transition mechanisms processes which can lead to the same $K^+K^-\gamma$ final state. In the first one, called the initial state radiation (ISR) a photon is emitted from an electron or a positron in the e^+e^- collision associated with the production of strange mesons K^+ and K^-. As shown in Fig. 1 the two mesons are emitted from an intermediate photon γ^*.

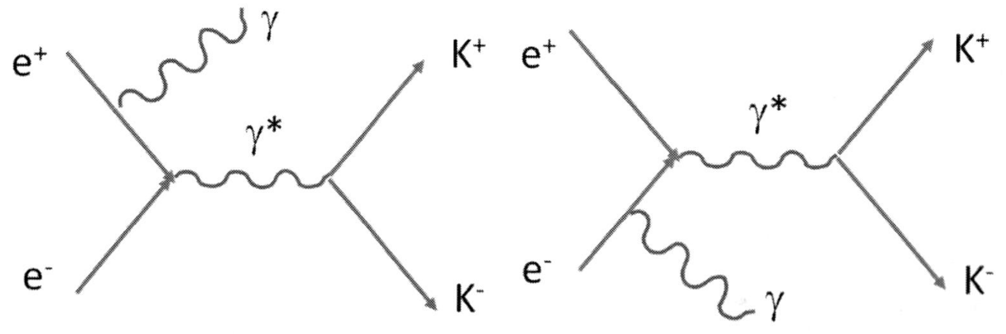

Figure 1. Diagrams corresponding to the initial state radiation in the $e^+e^- \to K^+K^-\gamma$ reaction

In other process, called the final state radiation (FSR), the outgoing photon is emitted from the K^+ or from the K^- meson, or directly from the same vertex connecting K^+ and K^- with γ^*. In Fig. 2 the third diagram of the FSR process is called the contact term. Both the ISR and FSR amplitudes can be calculated using the methods of quantum electrodynamics.

The mesons K^+ and K^- can interact strongly and in a case where the K^+K^- effective mass is close to 1 GeV this interaction has a resonant character. Two scalar resonances can be formed: isoscalar

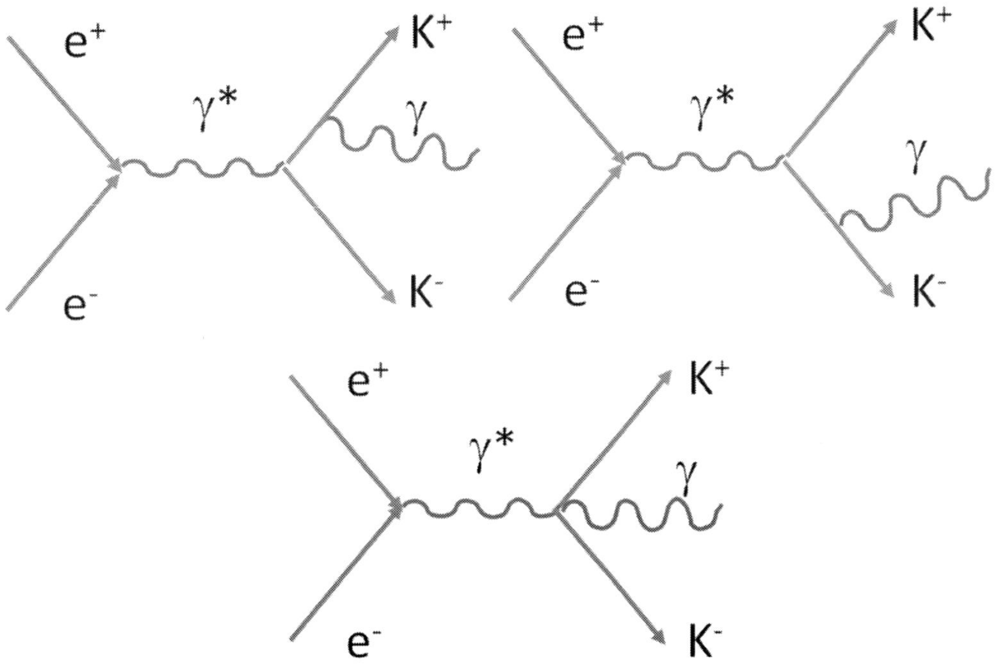

Figure 2. Diagrams contributing to final state radiation in the $e^+e^- \to K^+K^-\gamma$ reaction

$f_0(980)$ and isovector $a_0(980)$. The $f_0(980)$ resonance can decay into π^+ and π^- mesons. For a description of the reaction $e^+e^- \to \pi^+\pi^-\gamma$ the no-structure model (NS) has been formulated by Isidori, Maiani, Nicolaci and Pacetti [4]. In this model an essential role is played by the $\Phi(1020)$ meson point-like coupling to $f_0(980)\gamma$ (see Fig. 3). The model can be extended to describe the $e^+e^- \to K^+K^-\gamma$

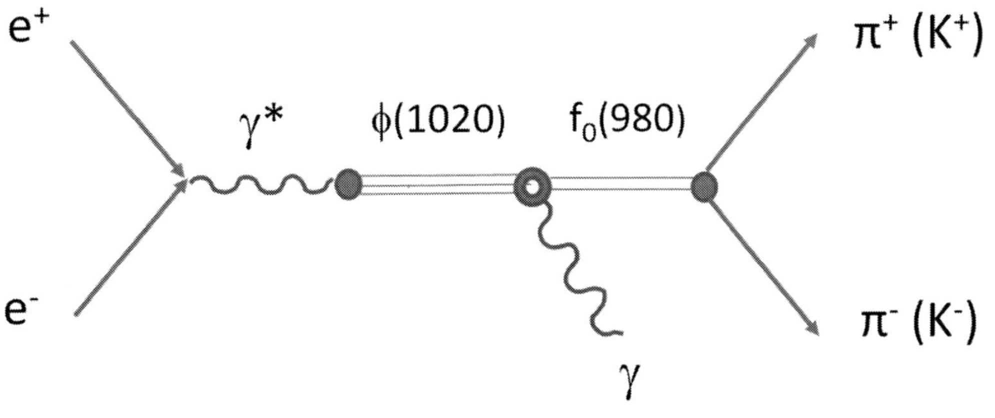

Figure 3. Diagram that corresponds to the no-structure model [4]

reaction since the $f_0(980)$ resonance decays also to the K^+K^- state.

Another model of the radiative $\Phi(1020)$ meson decays with a formation of kaon pairs has been developed by Achasov, Gubin and Shevchenko in Ref. [3]. In this model one calculates the amplitudes related to three diagrams shown in Fig. 4. In each of these diagrams the Φ meson is coupled to the scalar mesons $f_0(980)$ or $a_0(980)$ through the charged kaon-loop (KL model). Both no-structure and kaon-loop models have been used in analysis of the KLOE Collaboration for the $e^+e^- \rightarrow \pi^+\pi^-\gamma$ reaction [2].

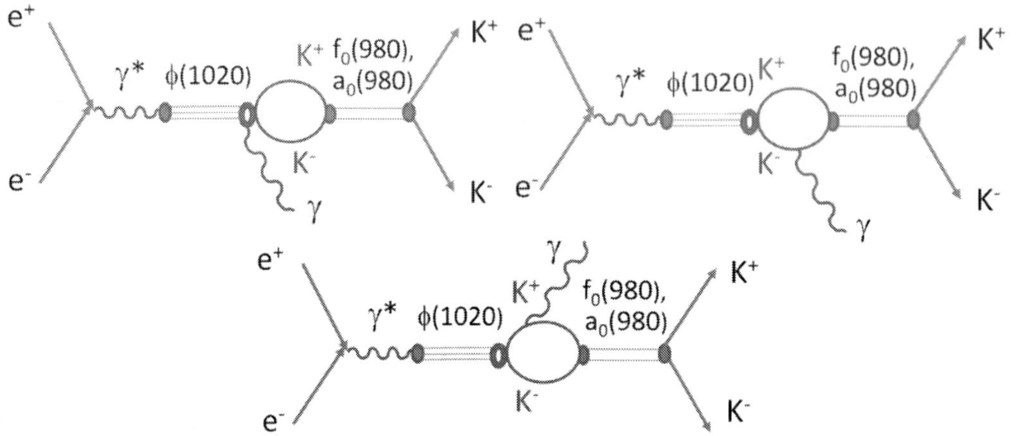

Figure 4. Diagrams of the kaon-loop model [3]

2.1 Reaction amplitudes

We have just discussed a few reaction mechanisms leading to the same final state of $K^+K^-\gamma$. Therefore the corresponding amplitudes have to be added in the total reaction amplitude M:

$$M = A(ISR) + A(FSR) + A(KL), \tag{1}$$

where $A(ISR)$ is the initial state radiation amplitude, $A(FSR)$ is the final state radiation amplitude and as an example we have added the amplitude $A(KL)$ describing the final state interaction between kaons in the kaon-loop approach. The modulus of the total amplitude squared reads:

$$|M|^2 = |A(ISR)|^2 + |A(FSR)|^2 + |A(KL)|^2 + 2Re[A^*(ISR)\,A(FSR)] \\ +2Re[A^*(ISR)\,A(KL)] + 2Re[A^*(FSR)\,A(KL)]. \tag{2}$$

Therefore the total differential cross-section can be witten as a sum of six terms:

$$d\sigma(total) = d\sigma(ISR) + d\sigma(FSR) + d\sigma(KL) + Int(ISR - FSR) + Int(ISR - KL) + Int(FSR - KL). \tag{3}$$

The first three terms are the cross-sections proportional to the moduli of the amplitudes squared like $d\sigma(ISR) \propto |A(ISR)|^2$. The three inteference terms are also present, for example, Int(FSR-KL) is the term corresponding to the interference of the ISR amplitude with the kaon-loop model amplitude. Let us notice that if the experimental cuts are chosen symmetrically with respect to interchange of the K^+ and K^- mesons, then the two interference terms vanish: Int(ISR-FSR)=Int(ISR-KL)=0.

2.2 Differential cross-section

Let us consider the differential cross-section of the reaction $e^+(p_{e^+})e^-(p_{e^-}) \to K^+(p_{K^+})K^-(p_{K^-})\gamma(q)$, where the particle four momenta are indicated by p or q:

$$d\sigma = \frac{(2\pi)^4}{2\sqrt{s(s-4m_e^2)}}|M|^2 d\Phi_3. \tag{4}$$

In the formula above s is the Mandelstam invariant $s = (p_{e^+} + p_{e^-})^2$, m_e is the electron mass, M denotes the total reaction amplitude and Φ_3 is the final state tree-body phase space. Next we define the K^+K^- and $K^-\gamma$ effective masses squared

$$m^2 = (p_{K^+} + p_{K^-})^2, \qquad m^2_{K^-\gamma} = (p_{K^-} + q)^2 \tag{5}$$

and two momentum transfers squared

$$t = (p_{e^-} - q)^2, \qquad t_1 = (p_{e^-} - p_{K^-})^2. \tag{6}$$

Then the four-fold differential cross-section can be written as

$$\frac{d\sigma}{dm^2 dm^2_{K^-\gamma} dt dt_1} = \frac{|M|^2}{(2\pi)^4 16 s(s-4m_e^2)(s-m^2) r(t_1)}. \tag{7}$$

Here

$$r(t_1) = \sqrt{-(t_1 - t_{1min})(t_1 - t_{1max})}, \tag{8}$$

where t_{1min} and t_{1max} are the lower and upper limits of t_1 which depend on m, $m_{K^-\gamma}$ and t in the following way:

$$t_{1min} = r_0 + r_1 t - b\sqrt{-t(t+s-m^2)}, \quad t_{1max} = r_0 + r_1 t + b\sqrt{-t(t+s-m^2)}. \tag{9}$$

The coefficients r_0, r_1 and b are expressed as

$$r_0 = m_K^2 - \frac{1}{2}s(1 - v\,z), \quad r_1 = -\frac{1}{2} + \frac{(s+m^2)\,v\,z}{2(s-m^2)}, \quad b = \frac{m\sqrt{s}\,v\sqrt{1-z^2}}{s-m^2}. \tag{10}$$

In the above equations m_K is the charged kaon mass, $v = \sqrt{1 - 4m_K^2/m^2}$ is the K^- velocity in the K^+K^- center-of-mass frame and z is the cosine of the angle between K^- momentum and the photon momentum in the same frame. In Eq. (10) we have neglected a small value of the electron mass squared. The variable z is directly related to the $K^-\gamma$ effective mass:

$$m^2_{K^-\gamma} = m_K^2 + \frac{1}{2}(s-m^2)(1 - v\,z). \tag{11}$$

3 K^+K^- effective mass distributions

As mentioned in the previous section, the KLOE data for the $e^+e^- \to \pi^+\pi^-\gamma$ process [2] have been described using the no-structure [4] and the kaon-loop models [3]. We have extended these models and calculated the K^+K^- effective mass distributions in the $e^+e^- \to K^+K^-\gamma$ reaction. The distributions we are going to discuss now correspond to the third term written in Eq. (2) and denoted for the KL model as $|A(KL)|^2$. The parameters of the kaon-loop model (KL) and the no-structure model (NS) have been taken from Table 1 of Ref. [2]. In this reference only the K^+K^- coupling to one scalar resonance $f_0(980)$ is present. We have also done calculations using somewhat different parameters of the two scalar resonances $f_0(980)$ and $a_0(980)$ as used earlier in Ref. [3]. In this case the masses of the $f_0(980)$ and $a_0(980)$ were equal to 980 MeV and the coupling constants to the K^+K^- system were equal. The corresponding results are shown in Fig. 5. The differential cross-section calculated for the model of Achasov, Gubin and Shevchenko is larger than the two other cross-sections. This can be simply explained by the fact that adding coherently two equal amplitudes related to the $f_0(980)$ and $a_0(980)$ resonances with the same parameters leads to an enhancement by a factor four in comparison with models that have only one scalar resonance.

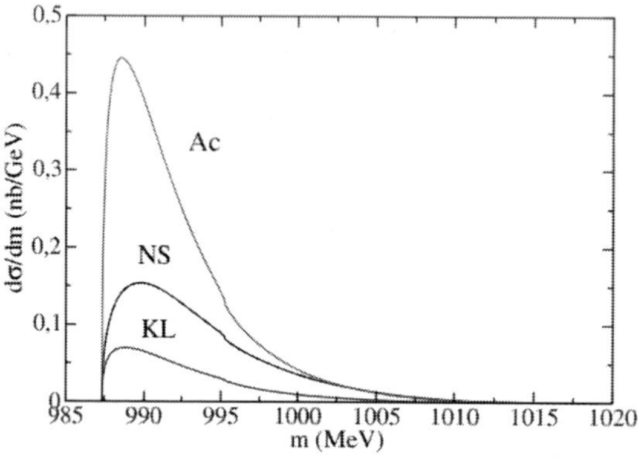

Figure 5. Comparison of the K^+K^- effective mass distributions calculated using three models with parameters taken from Refs. [3] (line labelled Ac) and [2] (labels NS and KL).

Now let us discuss other terms giving non-zero contributions to the differential cross-section $d\sigma/dm$. Here we have chosen symmetric cuts on the photon emission angle θ_γ in the centre-of-mass e^+e^- frame ($45^0 < \theta_\gamma < 135^0$). At the beginning let us take the NS model and make a comparison of its K^+K^- effective mass distribution with the ISR and FSR ones. Apart of the NS, ISR and FSR distributions we show in Fig. 6 lines of the NS-FSR interference term and the total differential cross-section. One sees that in the range of the effective mass m limited to 1000 MeV the FSR contribution dominates while the ISR cross-section is largely suppressed by the cuts put on the angle θ_γ. The NS contribution is quite small but the interference term of the NS model amplitude with the FSR amplitude is not negligible in comparison with the FSR one which gives some hope to be measured in experiment. Similar results are shown in Fig. 7 for the kaon-loop model with parameters fixed in Ref. [2]. It is interesting to see the negative contribution of the interference term in a part of the spectrum where m is smaller than about 993 MeV.

4 K^- angular distributions

In this section we first pass to a study of the double differential cross-section $d\sigma/[dm^2 dm^2_{K^-\gamma}]$ at fixed m. As seen in Eq. (11) the variation of the $K^-\gamma$ effective mass m while keeping the K^+K^- effective mass fixed, is equivalent to a variation of z which is the cosine of the angle between K^- momentum and the photon momentum in the K^+K^- center-of-mass frame. The z distributions for two m values are presented in Fig. 8. Here one sees maxima at $z = 0$ for the FSR and interference terms. The ISR function has a minimum at $z = 0$ for the θ_γ range between 45^0 and 135^0. The distribution (NS) of the

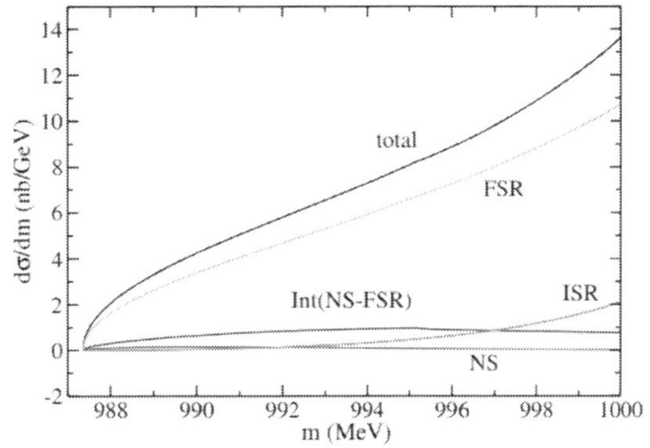

Figure 6. K^+K^- effective mass distributions calculated for $45^0 < \theta_\gamma < 135^0$. Four contributions to the total result are labelled FSR (final state radiation), ISR (initial state radiation), NS (NS model) and Int(NS-FSR) (interference of the NS and the FSR amplitudes).

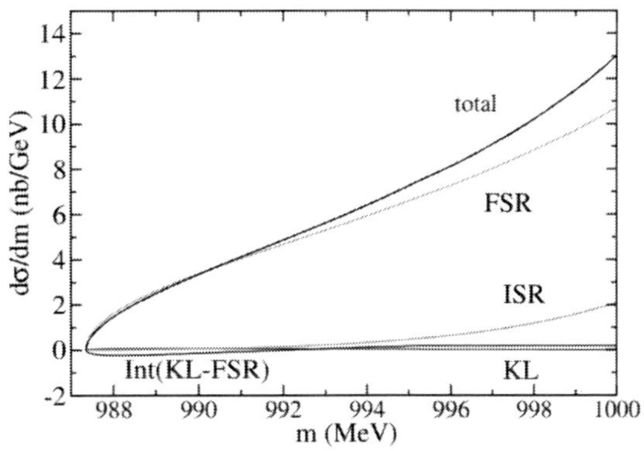

Figure 7. K^+K^- effective mass distributions with $45^0 < \theta_\gamma < 135^0$. Four contributions to the total result are labelled FSR (final state radiation), ISR (initial state radiation), KL (kaon-loop model) and Int(KL-FSR) (interference of the KL and the FSR amplitudes).

no-structure model is flat. This feature is common to all the models in which the K^+ and K^- mesons interact in the S-wave.

Angular distributions of the K^- mesons with respect to the electron momentum in the e^-e^+ center-of-mass frame have also been studied. The relevant K^- angle is denoted by θ_1. Once again using the no-structure model we have calculated all the six contributions to the double differential cross-section $d\sigma/[dm\ dcos\theta_1]$ seen in Eq. (3). So in Fig. 9 we notice two small interference terms ISR-FSR and NS-ISR in addition to other five lines labelled similarly as in Fig. 8. These two terms are asymmetric as they change sign when the angle θ_1 is changed into $180^0 - \theta_1$. Therefore they vanish after the integration over the full range of θ_1. Let us also notice that the shape of curves changes when one increases the K^+K^- effective mass from 990 MeV to 998 MeV.

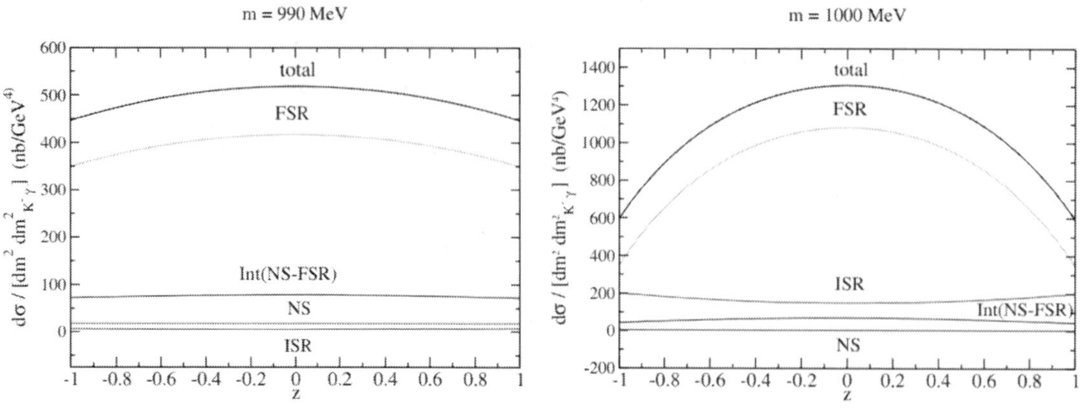

Figure 8. Distributions of the polar angle of K^- with respect to the photon momentum in the K^+K^- center-of-mass frame at fixed m equal to 990 and 1000 MeV. The curve labelled FSR corresponds to the final state radiation, the line labelled ISR to the initial state radiation, the line NS to the no-structure model, the interference term is labelled Int(NS-FSR), and the line labelled *total* is a sum of all the contributions.

5 Photon angular distributions

Finally we consider the photon angular distributions with respect to the e^- momentum in the e^-e^+ center-of-mass frame. We calculate the double differential cross-section $d\sigma/[dm\ dcos\ \theta_\gamma]$ at fixed values of m. As previously we use the no-structure model. Fig. 10 shows the five lines for two masses m = 990 MeV and m = 1000 MeV. Its labelling is that as in Fig. 8 caption. The most spectacular behaviour is a rise of the ISR term when $cos\ \theta_\gamma$ approaches to 1 or to -1. At these two values the ISR cross-section goes to infinity and therefore in experimental studies the cuts on the values of $cos\ \theta_\gamma$ are put in order to diminish the ISR background. For example, in Figs.6-8 we have cut the range of the photon angles from 0^0 to 45^0 and from 135^0 to 180^0.

6 Multichannel model of the $e^+e^- \to M_1M_2\gamma$ reactions

In this chapter we briefly outline basic properties of a special model which can be formulated for a simultaneous description of the reactions $e^+e^- \to M_1M_2\gamma$ in which M_1 and M_2 are the pseudoscalar mesons coupled to the K^+K^- mesons. Here one can enumerate the following set of reactions:
1. $e^+e^- \to \pi^+\pi^-\gamma$,
2. $e^+e^- \to \pi^0\pi^0\gamma$,
3 $e^+e^- \to \pi^0\eta\gamma$,
4. $e^+e^- \to K_S^0K_S^0\gamma$,
5. $e^+e^- \to K^+K^-\gamma$.

The diagrams corresponding to the proposed model are shown in Fig. 11. In the first step of the production process two charged kaons K^+ and K^- and a photon are created. Then in the second reaction stage the kaons interact forming a sytem of two mesons M_1 and M_2. We denote by T the corresponding set of transition amplitudes. A specific model of these $K^+K^- \to M_1M_2$ transitions should be unitary. Its desirable feature is analyticity of the transition amplitudes which relate different coupled channels. For a practical future description of the data all the transition amplitudes should have the same poles corresponding to the relevant scalar mesons present in the energy range close to 1 GeV.

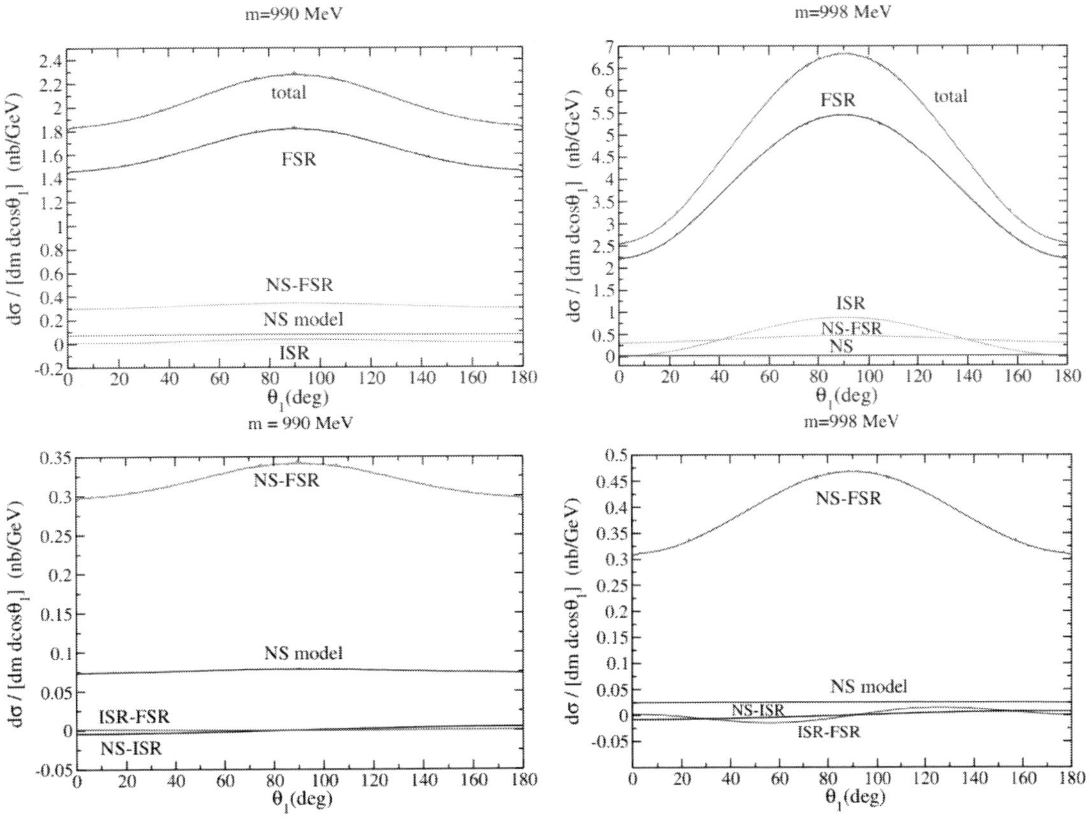

Figure 9. K^- angular distributions at fixed m for $45^0 < \theta_\gamma < 135^0$. The meaning of the labels total, FSR, ISR and NS is the same as in Fig.8. NS-FSR, ISR-FSR and NS-ISR denote the interference terms of the NS and FSR amplitudes, the ISR and FSR amplitudes, and the NS and ISR amplitudes, respectively.

7 Experimental implications

The differential cross-sections which have been calculated for different reaction mechanisms leading to the same final state $K^+K^-\gamma$ can be integrated within some experimental limits put on the photon minimum energy and on the photon polar angle defined with respect to the electron beam axis. If the minimum photon energy is equal to 10 MeV in the e^+e^- center-of-mass frame then the maximum efective K^+K^- mass is close to 1009 MeV. In Table 1 the reaction cross sections are given. Abbreviations for the reaction mechanisms are the same as in Fig. 6 caption.

Assuming integrated luminosity of 1.7 fb^{-1} one can obtain expected numbers of events. In Table 2 we show values of the expected number of events obtained for two different cuts on the photon angles θ_γ and for the minimum photon energy of 10 MeV in the e^+e^- center-of-mass frame. These numbers are not yet corrected for the experimental efficiency.

8 Summary

Three theoretical models have been extended in order to make predictions for the reaction $e^+e^- \to K^+K^-\gamma$. The resulting effective mass and angular distributions can be used in future experimental data

https://doi.org/10.1051/epjconf/201816600018

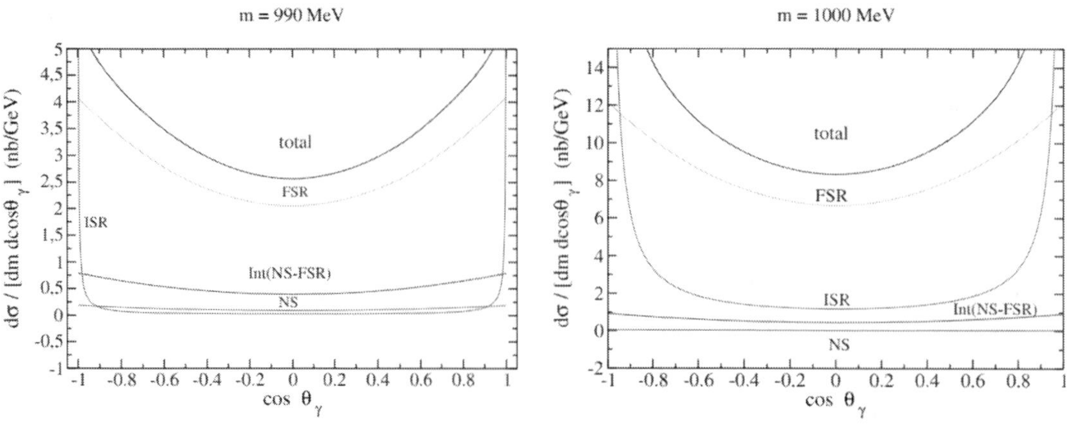

Figure 10. Photon angular distributions at fixed m=990 and 1000 MeV. For a description of lines see Fig.8 caption.

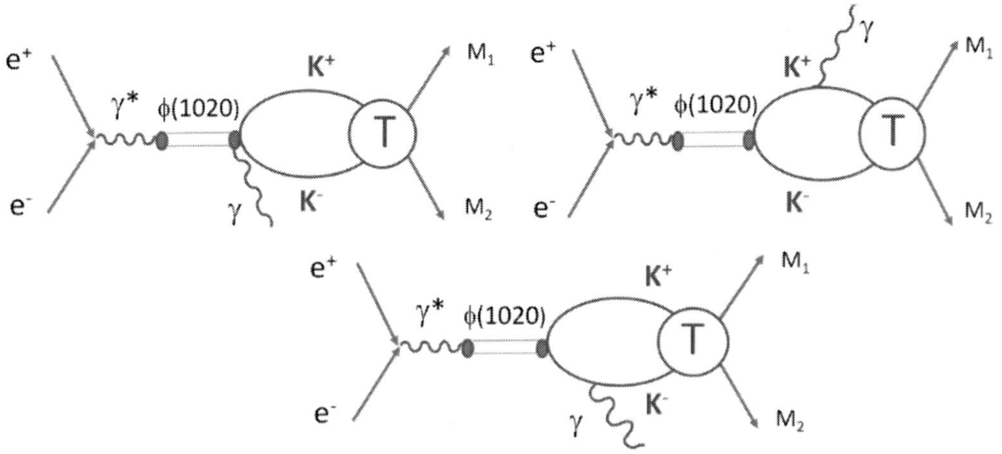

Figure 11. Diagrams corresponding to the proposed model of the reactions $e^+e^- \to M_1M_2\gamma$

analyses. Some features of the new model of the multichannel coupled reactions $e^+e^- \to M_1M_2\gamma$, where M_1 and M_2 are pseudoscalar mesons, have been outlined. This model can be applied in a combined analysis of the radiative $\Phi(1020)$ resonance decays into two mesons. It can also serve in determination of the threshold parameters of the K^+K^- strong interaction amplitudes as well as in a better specification of the properties of the scalar meson resonances $f_0(980)$ and $a_0(980)$.

This work has been supported by the Polish National Science Centre (grant no 2013/11/B/ST2/04245).

Table 1. The cross sections integrated over the K^+K^- effective mass from the threshold up to 1009 MeV for two ranges of the photon emission angle θ_γ.

reaction mechanism	$24^0 < \theta_\gamma < 156^0$	$45^0 < \theta_\gamma < 135^0$
FSR	0.330 nb	0.238 nb
NS	0.0020 nb	0.0014 nb
Int(NS-FSR)	0.021 nb	0.015 nb
ISR	0.183 nb	0.104 nb
total	0.536 nb	0.358 nb

Table 2. Numbers of events for the K^+K^- effective mass up to 1009 MeV and for two ranges of θ_γ for two photon angle ranges.

reaction mechanism	$24^0 < \theta_\gamma < 156^0$	$45^0 < \theta_\gamma < 135^0$
FSR	$5.6 \cdot 10^5$	$4.0 \cdot 10^5$
NS	$3.4 \cdot 10^3$	$2.4 \cdot 10^3$
Int(NS-FSR)	$3.6 \cdot 10^4$	$2.5 \cdot 10^4$
ISR	$3.1 \cdot 10^5$	$1.8 \cdot 10^5$
total	$9.1 \cdot 10^5$	$6.1 \cdot 10^5$

References

[1] C. Patrignani *et al.* (Particle Data Group), Chin. Phys. **C 40**, 1000001 (2016)
[2] F. Ambrosino *et al.* (KLOE Collaboration), Phys. Lett. **B 634**, 148 (2006)
[3] N. N. Achasov, V. V. Gubin and V. I. Shevchenko, Phys. Rev. **D 56**, 203 (1997)
[4] G. Isidori, L. Maiani, M. Nicolaci, S. Pacetti, JHEP **0605**, 049 (2006)
[5] R. R. Akhmetshin *et al.* (CMD-2 Collaboration), Phys. Lett. **B 462**, 371 (1999)
[6] F. Ambrosino *et al.* (KLOE Collaboration), Phys. Lett. **B 679**, 10 (2009)

Recent results at the ϕ-meson region from the CMD-3 detector at the VEPP-2000 collider

Vyacheslav Ivanov*[1], *Evgeny* Solodov[1], *Evgeny* Kozyrev[1], and *Georgiy* Razuvaev[1]

[1]Budker Institute of Nuclear Physics, 11, akademika Lavrentieva prospect, Novosibirsk, 630090 Russia

Abstract. This report is devoted to the recent results of the study of the processes $e^+e^- \to K_SK_L$, K^+K^-, $\pi^+\pi^-2\pi^0$, $2\pi^+2\pi^-$, $\eta\gamma$ at the ϕ-meson region with the CMD-3 detector at the VEPP-2000 collider. From the combined analysis of K_SK_L and K^+K^- final states the ϕ-meson parameters were measured, and the obtained ratio of $g_{\phi K^+K^-}$ and $g_{\phi K_SK_L}$ couplings was found to be compatible with the isospin symmetry. From the analyzes of the $\pi^+\pi^-2\pi^0$ and $2\pi^+2\pi^-$ final states the branching ratios $\mathcal{B}(\phi \to \pi^+\pi^-2\pi^0)$ and $\mathcal{B}(\phi \to 2\pi^+2\pi^-)$ were determined.

1 Introduction

A high precision measurement of the inclusive $e^+e^- \to hadrons$ cross section is required for a calculation of the hadronic contribution to the muon anomalous magnetic moment $(g-2)_\mu$ in the frame of the Standard Model. To confirm or deny the observed difference between the calculated [1] and measured [2] $(g-2)_\mu$ values more precise measurements of the exclusive channels of $e^+e^- \to hadrons$ are necessary. On the other hand, the study of the exclusive final states can provide useful information about the precision of the isospin symmetry, particle parameters, branching ratios and coupling constants.

2 VEPP-2000 collider and CMD-3 detector

The VEPP-2000 e^+e^- collider [3] at the Budker Institute of Nuclear Physics (Novosibirsk, Russia) covers the center-of-mass energy ($E_{c.m.}$) range from 0.32 to 2.0 GeV and employs a novel technique of round beams to reach luminosity up to 10^{32} cm^{-2}s^{-1} at 2.0 GeV. The Cryogenic Magnetic Detector (CMD-3), described in [4], is installed in one of the two beams interaction regions. The detector tracking system consists of the cylindrical drift chamber (DC) and double-layer cylindrical multiwire proportional Z-chamber, installed inside a thin superconducting solenoid with 1.0-1.3 T magnetic field. Both subsystems are also used to provide the trigger signals. DC contains 1218 hexagonal cells in 18 layers and allows one to measure charged particle momentum with 1.5-4.5% accuracy in the 40-1000 MeV/c range, and the polar θ and azimuthal φ angles with 20 mrad and 3.5-8.0 mrad accuracy, respectively. Amplitude information from the DC wires is used to measure the

* Corresponding author: vyacheslav_lvovich_ivanov@mail.ru

ionization losses dE/dx of charged particles. The barrel electromagnetic calorimeters based on liquid xenon (LXe) (5.4 X_0) and CsI crystals (8.1 X_0) are placed outside the solenoid [5]. The total amount of material in front of the barrel calorimeter is 0.13 X_0 that includes the solenoid as well as the radiation shield and vacuum vessel walls. BGO crystals of 13.4 X_0 thicknesses are used in the endcap calorimeter. The magnetic flux return yoke is surrounded by scintillation counters which are used to tag cosmic events.

The energy range $E_{c.m.} = 920 - 1070$ MeV was scanned in 2013 year, and 9.8 pb^{-1} of integral luminosity in 22 energy points was collected. The integral luminosity at each $E_{c.m.}$ point was determined using the events of the processes $e^+e^- \to e^+e^-$ and $e^+e^- \to 2\gamma$ with about 1% accuracy [6]. The beam energy was monitored by using the Back-Scattering-Laser-Light system [7].

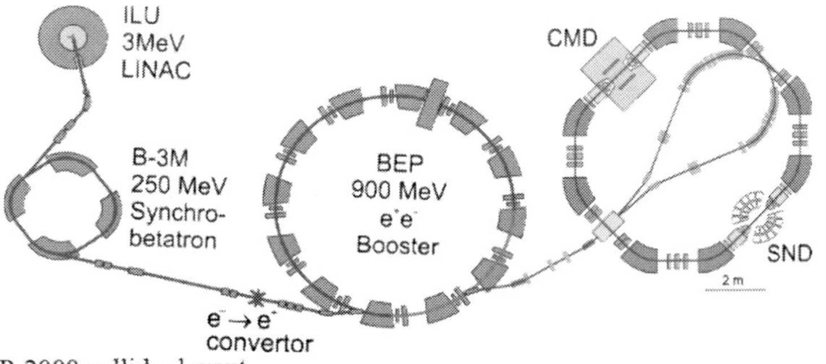

Fig. 1. VEPP-2000 collider layout.

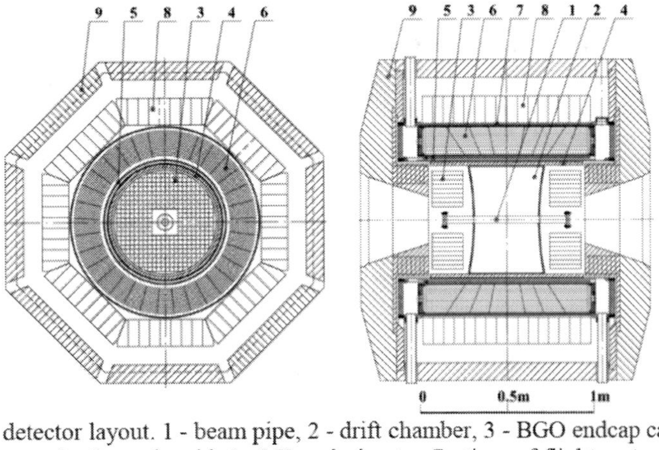

Fig. 2. CMD-3 detector layout. 1 - beam pipe, 2 - drift chamber, 3 - BGO endcap calorimeter, 4 - Z-chamber, 5 - superconducting solenoid, 6 - LXe calorimeter, 7 - time-of-flight system, 8 - CsI calorimeter, 9 - yoke.

The main goals of the VEPP-2000 complex are: to achieve <1% systematic error of the cross section measurement for the major exclusive $e^+e^- \to hadrons$ channels; to study other exclusive processes, especially those which were not measured before and were taken into account in the calculation of $(g-2)_\mu$ via the isotopic relations; to measure the parameters of "excited" vector mesons like ρ', ρ'', ω', ϕ' etc.; to test the conservation of vector current; to study the nucleon electromagnetic form factors; to study some of the diphoton processes (e.g. $e^+e^- \to \eta'$).

It should be noted that the study of ϕ-meson energy range is not a priority of CMD-3, but ϕ is very important for studying of photon, pion and kaon detection efficiencies and for studying the $\pi^\pm$ energy depositions in the calorimeters (necessary for proper π/μ

separation). On the other hand, CMD-3 can provide interesting results on some problems, where the systematic uncertainty is decisive.

3 Study of the $e^+e^- \to K_SK_L$, K^+K^- processes at ϕ

To select the charged kaons, we demand the event to have 2 tracks in the DC, passing from the beams interaction region and satisfying the collinearity conditions $||\varphi_1 - \varphi_2| - \pi| <$ 0.3 rad and $|\theta_1 + \theta_2 - \pi| <$ 0.3 rad. Then the almost background-free sample of signal K^+K^- events with charged kaons are extracted using dE/dx losses (see Fig. 3).

The signal identification for the K_SK_L final state is based on the detection of two pions from the $K_S \to \pi^+\pi^-$ decay. The longitudinal distance z_v and the transverse coordinate ρ_v of the vertex should be less than 10 and 6 cm correspondingly. Pions from K_S decay are required to have polar angles 1 rad$< \theta_\pi < (\pi - 1)$ rad. Then, the dE/dx vs. particle momenta distribution is used to select the signal events (see Fig. 4). The signal/background separation is performed by the fitting of the distribution of invariant mass of two pions $m_{inv}(\pi^+,\pi^-)$ (see Fig. 5). The major background process was found to be $e^+e^- \to \pi^+\pi^-2\pi^0$.

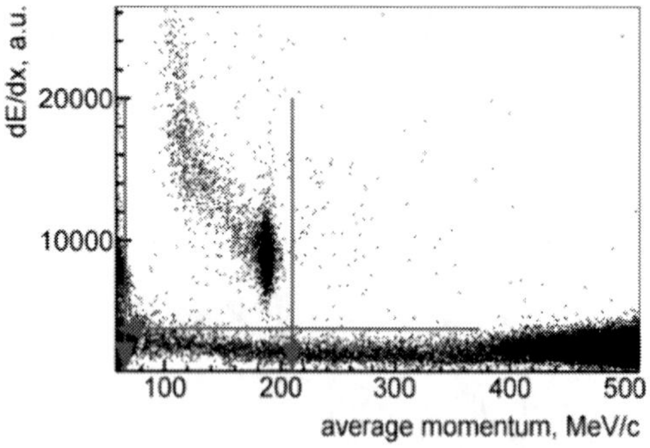

Fig. 3. The ionization losses vs momentum for positive tracks for data at $E_{c.m.} = 1.06$ GeV. Line shows selection of signal koans.

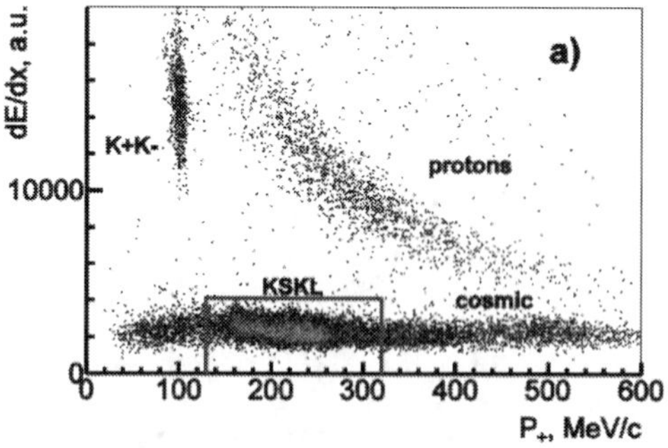

Fig. 4. The ionisation losses in DC vs momentum for positively charged pions for data at energy $E_{c.m.} = 1.01$ GeV. The line shows selections of pions from the K_S decay.

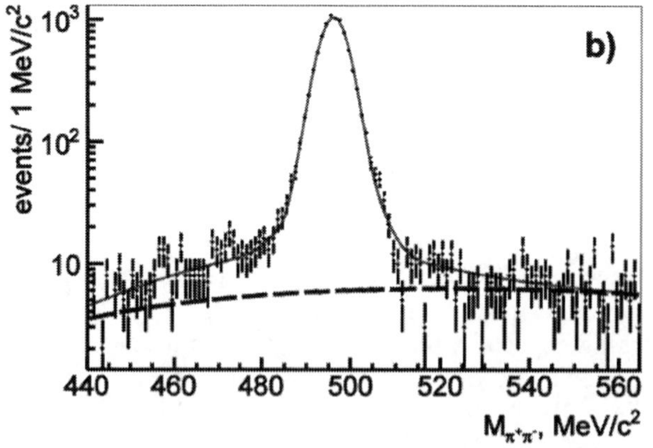

Fig. 5. Signal/background separation at $E_{c.m.} = 1.01$ GeV by the approximation of the $m_{inv}(\pi^+,\pi^-)$ distribution. Dashed line corresponds to the background, solid line – to the signal with background.

The measured $e^+e^- \to K_SK_L$, K^+K^- cross sections together with their approximation are shown in Figs. 6-7 respectively. Systematic uncertainty is estimated to be 3% in charged channel and 1.8% in neutral channel. The final results for K_SK_L are published in [8]. The cross sections were approximated according to the Vector Dominance Model:

$$\sigma_{KK} = \frac{8\pi\alpha}{3s^{5/2}} p^3{}_K \left| \frac{g_{\phi\gamma} g_{\phi KK}}{D_\phi(s)} \pm A_\rho + A_\omega + A \right|^2, \quad (1)$$

where $D_\phi(s)$ is the denominator of the ϕ propagator, $A_{\rho/\omega}$ are the functions describing contributions of $\rho(770)$ and $\omega(782)$ resonances, constant A describes the contribution of higher resonances. The sign «+» before the A_ρ in formula (1) corresponds to K^+K^- channel, «-» - to K_SK_L channel. Further, the couplings $g_{\phi\gamma}$ and $g_{\phi KK}$ are equal

$$g_{\phi\gamma} = \sqrt{\frac{3m^3{}_\phi \Gamma_\phi \mathcal{B}(\phi \to e^+e^-)}{4\pi\alpha}}, \quad (2)$$

$$g_{\phi KK} = \sqrt{\frac{6\pi m^2{}_\phi \Gamma_\phi \mathcal{B}(\phi \to K^+K^-)}{p^3{}_K(m_\phi)}}. \quad (3)$$

The results of the approximation procedure in comparison with the PDG values (see [9]) are shown in Table 1. The contributions from the $\rho(770)$ and $\omega(782)$ were fixed from quark model and SU(3) symmetry of u-, d- and s-quarks. But if we introduce in the fit the additional floating parameter $r_{\rho,\omega}$, describing the relative deviation of the $g_{\rho KK}$ and $g_{\omega KK}$ from their SU(3)-predicted values, we obtain a compatible with unity result $r_{\rho,\omega} = 0.76 \pm 0.11$. It is seen, that systematic errors dominate the uncertainty, and we still are working on the improvements.

Table 1. The results of the approximation of the cross sections of $e^+e^- \to K_SK_L$, K^+K^- processes measured with the CMD-3 detector in comparison with PDG values.

Parameter	CMD-3	PDG
m_ϕ, MeV	$1019.468 \pm 0.005_{stat} \pm 0.06_{syst} \pm 0.02_{model}$	1019.461 ± 0.019
Γ_ϕ, MeV	$4.254 \pm 0.009_{stat} \pm 0.016_{syst} \pm 0.012_{model}$	4.266 ± 0.031
$\mathcal{B}(\phi \to e^+e^-) \times \mathcal{B}(\phi \to K_SK_L) \times 10^{-5}$	$10.06 \pm 0.03_{stat} \pm 0.22_{syst} \pm 0.02_{model}$	10.10 ± 0.13
$\mathcal{B}(\phi \to e^+e^-) \times \mathcal{B}(\phi \to K^+K^-) \times 10^{-5}$	$15.83 \pm 0.04_{stat} \pm 0.47_{syst} \pm 0.1_{model}$	14.46 ± 0.23
$\Gamma_{\phi\to e^+e^-} \mathcal{B}(\phi \to K^+K^-) \times 10^{-1}$ keV	$6.735 \pm 0.014_{stat} \pm 0.199_{syst} \pm 0.042_{model}$	6.340 ± 0.08

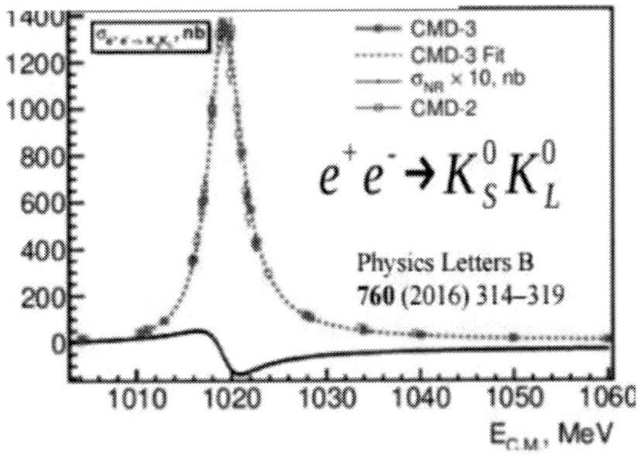

Fig. 6. Cross section of $e^+e^- \to K_SK_L$ process,

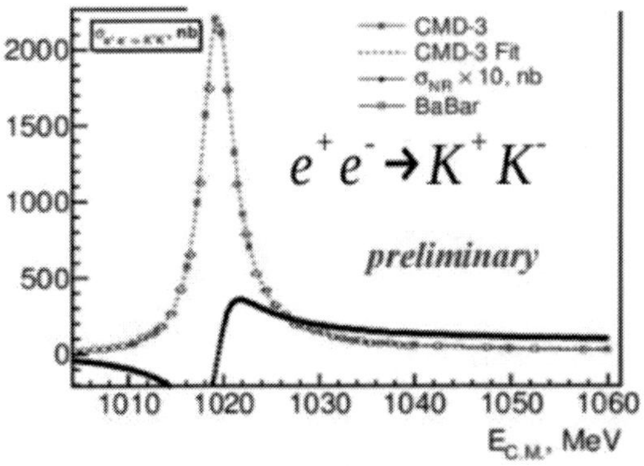

Fig. 7. Cross section of $e^+e^- \to K_SK_L$ process

measured with the CMD-3 detector, together with its approximation.

(preliminary results), measured with the CMD-3 detector, together with its approximation.

The combined analysis of $e^+e^- \to K_SK_L$, K^+K^- processes provides a possibility to check the precision of the isospin invariance of the u- and d-quarks, which predicts the value of unity to the factor

$$R = \frac{g_{\phi K^+K^-}}{g_{\phi K_SK_L}\sqrt{Z(m_\phi)}} = \sqrt{\frac{\mathcal{B}(\phi\to K^+K^-)p^3{}_{K_S}}{\mathcal{B}(\phi\to K_SK_L)p^3{}_{K^+}}\frac{1}{Z(m_\phi)}}, \quad (4)$$

where p_{K^+} and p_{K_S} are the momenta of charged and neutral kaons in $\phi \to K\bar{K}$ decays at $E_{c.m.} = m_\phi$, and

$$Z(m_\phi) = \frac{\pi\alpha/\beta}{1-e^{-\pi\alpha/\beta}}\left(1+\frac{\alpha^2}{4\beta^2}\right) \quad (5)$$

is the Sommerfeld-Gamov-Sakharov factor, which takes into account the Coulomb interaction of charged kaons in the final state (α is the fine structure constant, β is the velocity of charged kaons over the speed of light). The previous measurements gave $R_{SND} = 0.92 \pm 0.03$, $R_{CMD-2} = 0.943 \pm 0.013$, $R_{BaBar} = 0.972 \pm 0.017$ (see [9]) and one can suspect a significant deviation of R from unity. Our measurement gives $R_{CMD-3} = 0.984 \pm 0.029$, and thus we do not see statistically significant deviation of R from unity.

4 Study of the $e^+e^- \to \pi^+\pi^-2\pi^0$, $2\pi^+2\pi^-$ processes at ϕ

4.1 Study of the $e^+e^- \to \pi^+\pi^-2\pi^0$ process

To select the $\pi^+\pi^-2\pi^0$ final state events, we perform the 5C kinematic fit with all the combinations of 4 photons and 2 DC-tracks with opposite curvatures in the event, demanding the energy and momentum conservation, as well as the invariant mass of one pair of photons to be fixed at π^0 mass. Having selected the combination with the smallest $\chi^2{}_{5C}$, we apply the cut $\chi^2{}_{5C}$<30 to suppress the background (see Fig. 8). For the signal/background separation the approximation of the invariant mass of the second pair of photons (the invariant mass of which was not fixed at m_{π^0} in the kinematic fit) is used (see Fig. 9). The signal and background shapes were taken from simulation with additional smearing.

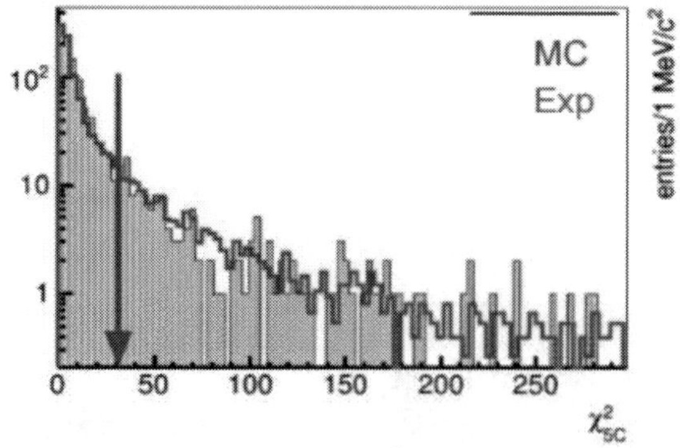

Fig. 8. The comparison of $\chi^2{}_{5C}$ distributions for data (filled histogram) and simulation (open histogram) at $E_{c.m.} = 1.01$ GeV.

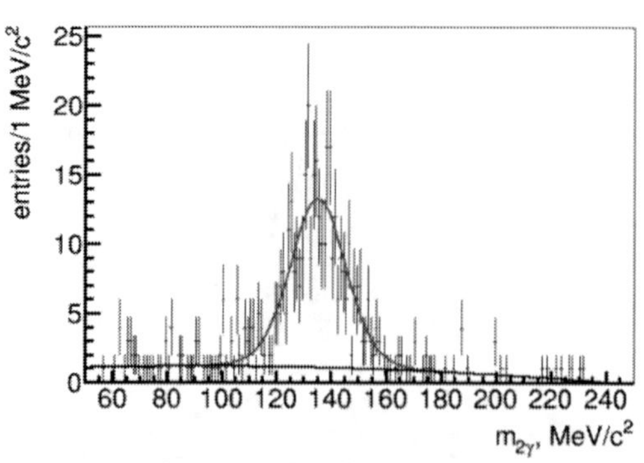

Fig. 9. The approximation of the invariant mass of the second pair of photons at $E_{c.m.} = 1.01$ GeV.

The calculated $e^+e^- \to \pi^+\pi^-2\pi^0$ cross section together with its approximation is shown in Fig. 10. For the approximation we use the function

$$\sigma_{\pi^+\pi^-2\pi^0}(\sqrt{s}) = \sigma_0 e^{A(\sqrt{s}-m_\phi)} \left|1 - Z\frac{m_\phi\Gamma_\phi}{m^2{}_\phi - s - i\sqrt{s}\Gamma_\phi}\right|^2, \quad (6)$$

and the fit results are $\sigma_0 = 7.54 \pm 0.15$ nb, $\mathrm{Re(Z)} = 0.136 \pm 0.025$, $\mathrm{Im(Z)} = -0.084 \pm 0.023$. From these results one can obtain the value

$$\mathcal{B}(\phi \to \pi^+\pi^-2\pi^0) = \frac{\sigma_0|Z|^2m^2{}_\phi}{12\pi\mathcal{B}(\phi\to e^+e^-)} = (5.3 \pm 1.6 \pm 0.8)\cdot 10^{-5}, \quad (7)$$

compatible with that in PDG [9]: $\mathcal{B}(\phi \to \pi^+\pi^-2\pi^0) = (4.7 \pm 0.5)\cdot 10^{-5}$.
Finally, it should be noted, that the major intermediate mechanisms of this process in the ϕ energy range are $e^+e^- \to \omega\pi^0$ and $e^+e^- \to a_1\pi$, and the determination of their amplitudes and relative phase via the partial wave analysis can be performed.

4.2 Study of the $e^+e^- \to 2\pi^+2\pi^-$ process

To extract the signal, we select the events with 3 or 4 DC-tracks, passing from the beams interaction region and having dE/dx, typical for pions. The major background processes were found to be $e^+e^- \to \phi \to \pi^+\pi^-\pi^0$, K^+K^-, K_SK_L and $e^+e^- \to 2e^+2e^-$. The $\pi^+\pi^-\pi^0$ final state is suppressed by the cut on the recoil masses of all the pairs of detected pions: $m_{recoil}(\pi_i,\pi_j) > 2m_\pi$. The K^+K^- contribution is suppressed by the cuts on dE/dx. The background from K_SK_L final state is suppressed by the cuts on K_S momentum and the invariant mass of two oppositely charged pions (see Fig. 11). Finally, Figs. 12-13 show the distributions of the energy disbalance in the events with four and three tracks, respectively. It is seen, that in the 4-track class we obtain almost background-free sample of signal events, whereas in 3-track class the signal/background separation is performed via the approximation of the energy disbalance distribution.

We determine the detection efficiency using the simulation of the signal process according to $e^+e^- \to a^\pm{}_1\pi^\mp$ model, as this intermediate mechanism is known to dominate the process in the region close to reaction threshold. From the Figs. 14a-b one can see good agreement between the polar angle distributions of charged pions in the experiment and simulation for 4-track and 3-track events, respectively.

The measured $e^+e^- \to 2\pi^+2\pi^-$ cross section together with its approximation using the formula (6) is shown in Fig. 15. The overall systematic uncertainty of cross section measurement is estimated to be 3.6%. The fit results are: $\sigma_0 = 1.263 \pm 0.027$ nb, $\mathrm{Re(Z)} = 0.146 \pm 0.030$, $\mathrm{Im(Z)} = -0.002 \pm 0.024$. From these one can obtain using the formula analogous to (7) the value

$$\mathcal{B}(\phi \to 2\pi^+2\pi^-) = (6.5 \pm 2.7 \pm 1.7)\cdot 10^{-6},$$

which is in agreement with previous CMD-2 measurement $\mathcal{B}_{CMD-2}(\phi \to 2\pi^+2\pi^-) = (4.0^{+2.8}_{-2.0})\cdot 10^{-6}$ [10], as well as with the estimation according to the vacuum polarization (which assumes Im(Z)=0):

$$\mathcal{B}(\phi \to \gamma^* \to 2\pi^+2\pi^-) = \frac{9\sigma_0(\mathcal{B}(\phi\to e^+e^-))^2m^2{}_\phi}{12\pi\mathcal{B}(\phi\to e^+e^-)} = 4.8\cdot 10^{-6}.$$

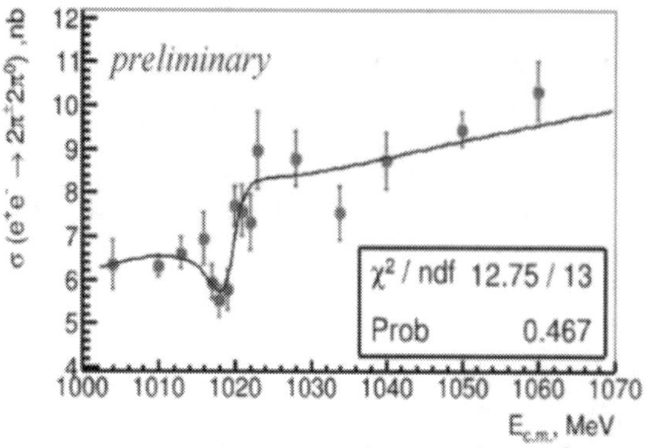

Fig. 10. Cross section of $e^+e^- \to \pi^+\pi^-2\pi^0$ process, measured with the CMD-3 detector, together with its approximation.

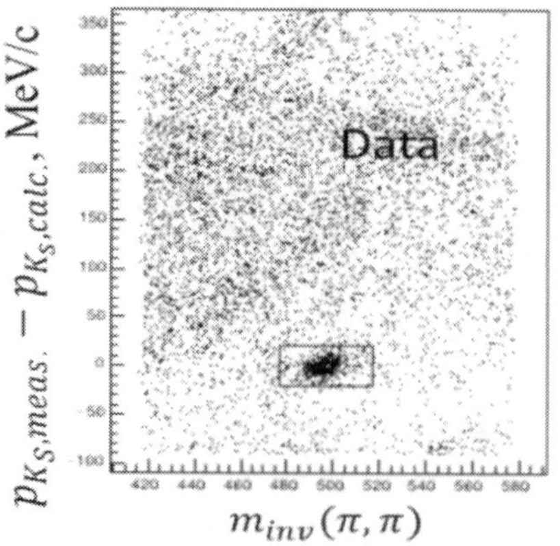

Fig. 11. The distribution of the $(p_{K_{S,measured}} - p_{K_L,calculated})$ versus the invariant mass of two oppositely charged pions in the experiment (data from all $E_{c.m.}$ are used). The events in the frame (mainly K_SK_L) are rejected.

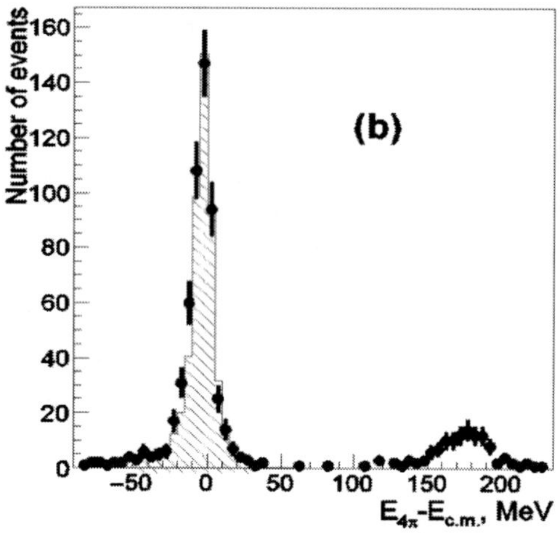

Fig. 12. The distribution of the energy disbalance for the events with four tracks at $E_{c.m.} = 958$ MeV.

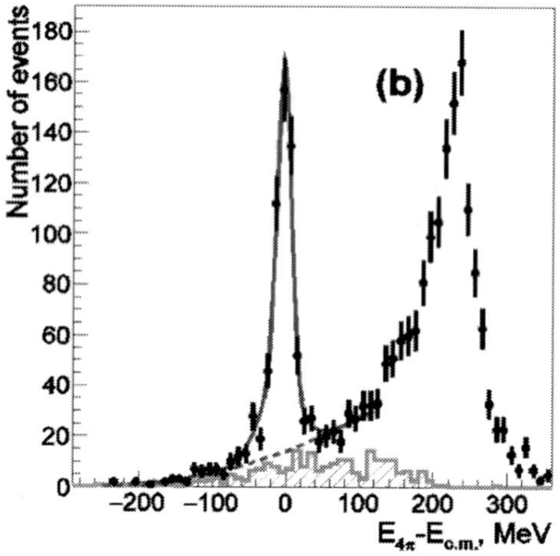

Fig. 13. The distribution of the energy disbalance for the events with three tracks at $E_{c.m.} = 958$ MeV. The dashed histogram represents the contribution of the $\pi^+\pi^-\pi^0$ final state. The peak at ~ 250 MeV is due to events of $e^+e^- \to 2e^+2e^-$ process.

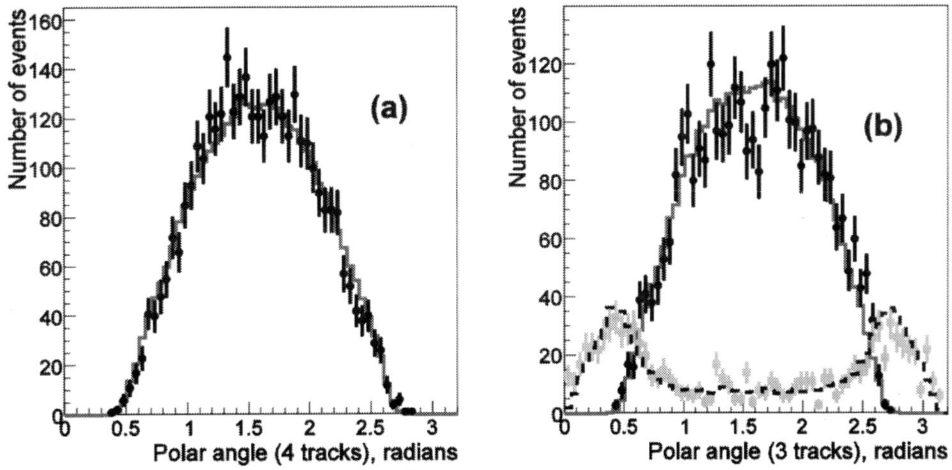

Fig. 14. The polar angle distributions of charged pions in the experiment (circle markers) and simulation (red open histogram) for 4-track (a) and 3-track (b) events, respectively (data from all $E_{c.m.}$ are used). For 3-track events also the distributions of the polar angle of missed pion in the experiment (square markers) and simulation (dashed line) are shown.

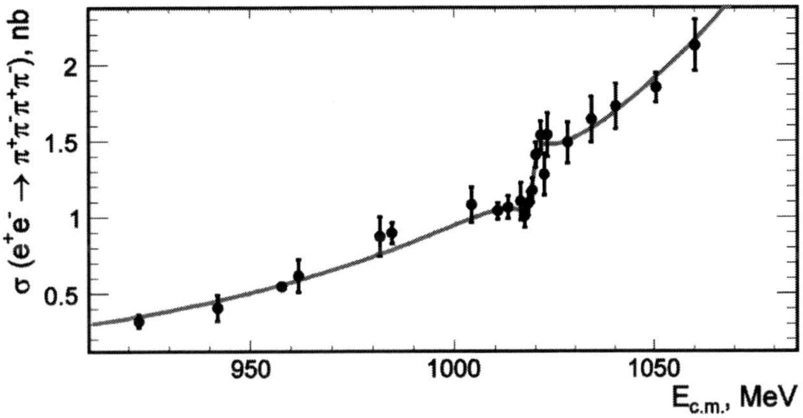

Fig. 15. Cross section of the $e^+e^- \to 2\pi^+2\pi^-$ process, measured with the CMD-3 detector, together with its approximation.

5 Study of the $e^+e^- \to \eta\gamma$ process at ϕ

The $e^+e^- \to \eta\gamma$ process was studied in $\eta \to 2\gamma$ decay mode in the $E_{c.m.}$ range from 1.0 to 1.03 GeV on the base of 5.5 pb^{-1} of integrated luminosity. To select the signal, we consider the events, having at least 3 photons with the polar angles in the range from 0.42 to $\pi - 0.42$ rad. Then the kinematic cuts, as shown in Fig. 16, are applied. After that the major background processes are $e^+e^- \to \pi^0\gamma$ and $e^+e^- \to 3\gamma$. We perform the kinematic fit, requiring the energy and momentum conservation and considering the common vertex of three photons as a floating parameter. After kinematic fit we arrange 3 photons according to their energies: $E_{\gamma_1} > E_{\gamma_2} > E_{\gamma_3}$ (see $m_{inv}(\gamma_1,\gamma_3)$ vs. $m_{inv}(\gamma_2,\gamma_3)$ distribution in Fig, 17). As it follows from the kinematics, in the ϕ-meson energy region the monochromatic photon in $\eta\gamma \to 3\gamma$ reaction has, in the majority of cases, the medium energy E_{γ_2}.

Signal/background separation is performed using the approximation of the $m_{inv}(\gamma_1, \gamma_3)$ distribution in the experiment (see Fig. 18). The preliminary results for the $e^+e^- \to \eta\gamma$ cross section measurement at ϕ are shown in Fig. 19. The systematic uncertainties are now being studied, and the major uncertainty is the difference between the photon detection efficiency in the Monte Carlo and experiment.

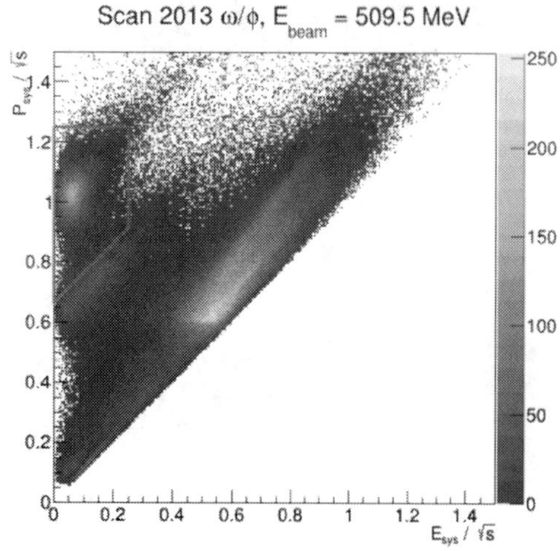

Fig. 16. The total energy vs. total momentum of all the combinations of three photons in the event (data from all $E_{c.m.}$ are used). Combination inside the frame is accepted.

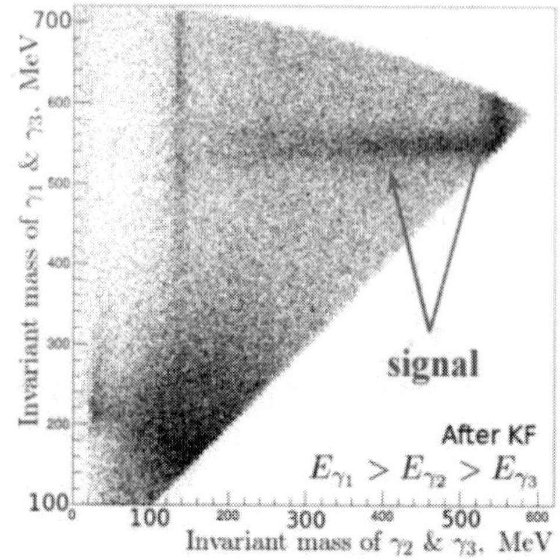

Fig. 17. The $m_{inv}(\gamma_1, \gamma_3)$ vs. $m_{inv}(\gamma_2, \gamma_3)$ distribution for the photons after kinematic fit (data from all $E_{c.m.}$ are used).

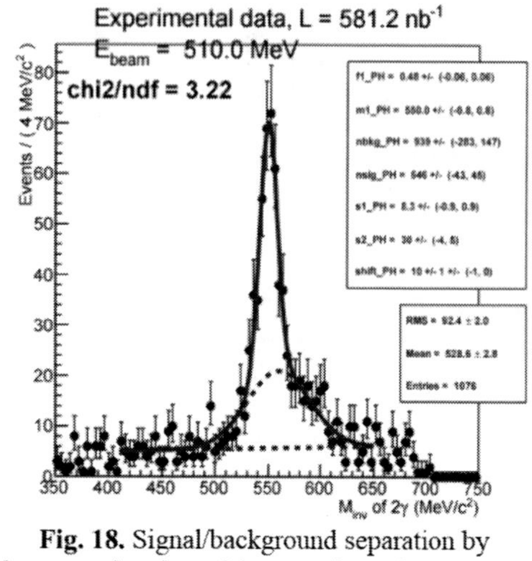

Fig. 18. Signal/background separation by the approximation of the $m_{inv}(\gamma_1, \gamma_3)$ distribution in the experiment ($E_{c.m.} = 1.02$ GeV).

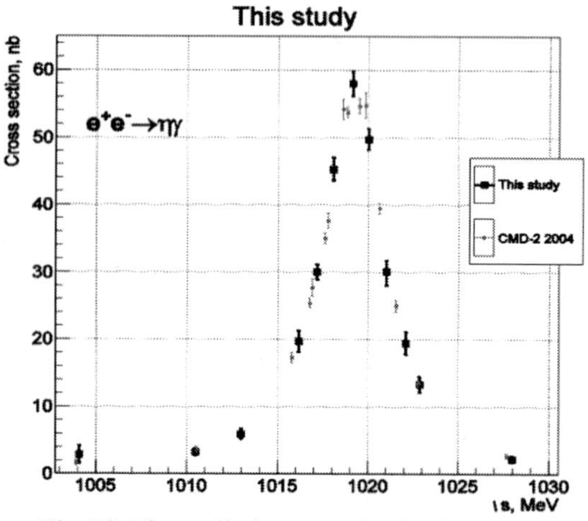

Fig. 19. The preliminary results for the $e^+e^- \to \eta\gamma$ cross section measurement in the ϕ-meson region with the CMD-3 detector.

6 Conclusion

The recent results of the study of the processes $e^+e^- \to K_S K_L$, K^+K^-, $\pi^+\pi^-2\pi^0$, $2\pi^+2\pi^-$, $\eta\gamma$ at the ϕ-meson region with the CMD-3 detector at the VEPP-2000 collider were described. From the combined analysis of $K_S K_L$ and K^+K^- final states the ϕ-meson parameters were measured, and the obtained ratio of $g_{\phi K^+K^-}$ and $g_{\phi K_S K_L}$ couplings was found to be compatible with the isospin symmetry. From the analyzes of the $\pi^+\pi^-2\pi^0$ and

$2\pi^+2\pi^-$ final states the branching ratios $\mathcal{B}(\phi \to \pi^+\pi^-2\pi^0)$ and $\mathcal{B}(\phi \to 2\pi^+2\pi^-)$ were determined. We still are working on the reduction of the systematic uncertainties in the study of processes $e^+e^- \to K^+K^-, \pi^+\pi^-2\pi^0, \eta\gamma$.

We thank the VEPP-2000 personnel for the excellent machine operation. Part of this work is supported by the Russian Foundation for Basic Research grants RFBR 15-02-05674-a and RFBR 16-02-00160-a.

References

1. K. Hagiwara, R. Liao, A.D. Martin, D. Nomura, T. Teubner, J. Phys. G **38**, 085003 (2011)
2. G.W. Bennett et al. (Muon g-2 Collaboration), Phys. Rev. D **73**, 072003, (2006)
3. D.E. Berkaev et al., JETP **113**, 213 (2011)
4. B.I. Khazin et al. (CMD-3 collaboration), Nucl. Phys. B (Proc. Suppl.) **181-182**, 376 (2008)
5. V.M. Aulchenko et al., JINST **10**, P10006 (2015)
6. R.R. Akhmetshin et al. (CMD-3 collaboration), Nucl. Phys. B (Proc. Suppl.) **225-227**, 69 (2012)
7. E.V. Abakumova et al., Phys. Rev. Lett. **110**, 140402 (2013)
8. E.A. Kozyrev et al., Phys. Lett. B **760**, 314-319 (2016)
9. K.A. Olive et al., Particle Data Group, Chin. Phys. C **38**, 090001, (2014)
10. R. R. Akhmetshin et al. (CMD-2 Collaboration), Phys. Lett. B **491**, 81 (2000)

Low-energy antikaon-nuclei interactions studies by AMADEUS: from QCD with strangeness to neutron stars

K. Piscicchia[1,2,⋆], C. Curceanu[1], M. Cargnelli[3], R. Del Grande[1,4], L. Fabbietti[5,6], J. Marton[3], A. Scordo[1], D. Sirghi[1], I. Tucakovic[7], O. Vazquez Doce[5,6], S. Wycech[8], J. Zmeskal[3], G. Mandaglio[9,10], M. Martini[1,11], and P. Moskal[12]

[1] *INFN, Laboratori Nazionali di Frascati, 00044 Frascati, Italy*
[2] *Museo Storico della Fisica e Centro Studi e Ricerche Enrico Fermi, Roma, Italy*
[3] *Stefan-Meyer-Institut für Subatomare Physik, 1090 Wien, Austria*
[4] *Università degli Studi di Roma Tor Vergata, Rome, Italy*
[5] *Excellence Cluster 'Origin and Structure of the Universe', 85748 Garching, Germany*
[6] *Physik Department E12, Technische Universität München, 85748 Garching, Germany*
[7] *Ruđer Bošković Institute, Zagreb, Croatia*
[8] *National Centre for Nuclear Research, 00681 Warsaw, Poland*
[9] *Dipartimento M.I.F.T. dell'Università di Messina, 98166 Messina, Italy*
[10] *INFN Gruppo collegato di Messina, 98166 Messina, Italy*
[11] *Dipartimento di Scienze e Tecnologie applicate, Università 'Guglielmo Marconi', 00193 Roma, Italy*
[12] *Institute of Physics, Jagiellonian University, 30-059 Krakow, Poland*

Abstract. The AMADEUS collaboration aims to provide unique quality results from K^- hadronic interactions in light nuclear targets, in order to solve fundamental open questions in the non-perturbative strangeness QCD sector, like the controversial nature of the $\Lambda(1405)$ state, the yield of hyperon formation below threshold, the yield and shape of multi-nucleon K^- absorption, processes which are intimately connected to the possible existence of exotic antikaon multi-nucleon clusters and to the role of strangeness in neutron stars. AMADEUS takes advantage of the DAΦNE collider, which provides a unique source of monochromatic low-momentum kaons and exploits the KLOE detector as an active target, in order to obtain excellent acceptance and resolution data for K^- nuclear capture on H, ^{4}He, ^{9}Be and ^{12}C, both at-rest and in-flight.

1 Introduction

The AMADEUS (Anti-kaonic Matter At DAΦNE: An Experiment with Unraveling Spectroscopy) [1] experiment investigates the low-energy K^- hadronic interaction in light nuclei (e.g. H, ^{4}He, ^{9}Be and ^{12}C) in order to provide experimental constraints on the non-perturbative QCD in the strangeness sector. It uses the low momentum (about 127 MeV/c), almost monochromatic, charged kaons provided by the decay of ϕ mesons at-rest at the DAΦNE factory [2]. The analyses presented here refers to the data acquired by the KLOE [3] collaboration during the 2004/2005 data taking campaign and the data collected by KLOE in collaboration with AMADEUS in the period November-December 2012. The

⋆e-mail: kristian.piscicchia@lnf.infn.it

latter data taking was performed by implementing in the central region of the KLOE setup a dedicated pure carbon (graphite) solid target.

In the study of the low-energy QCD with u, d and s quarks the chiral perturbation theory is not applicable, due to the presence of the broad $\Lambda(1405)$ state just few MeV below the $\bar{K}N$ threshold. The $\Lambda(1405)$ is a $J^P = 1/2^-$ isospin $I = 0$ strange baryon resonance, assigned to the lowest $L = 1$ supermultiplet of the three-quark system, which decays into $(\Sigma\pi)^0$ through the strong interaction. The $\Sigma^0\pi^0$ decay channel, which is free from the $I = 1$ contribution and from the isospin interference term, represents the cleanest signature of the $\Lambda(1405)$ resonance, because not affected by the neighbour $\Sigma(1385)$ resonance. On the other hand, better invariant mass and momentum resolutions are achievable with the reconstruction of the $\Sigma^+\pi^-$ decay channel, which is more sensitive to the resonance shape, once the background is under control. Despite the fact that $\Lambda(1405)$ is a four-stars resonance in Particle Data Group (PDG) [4], its nature still remains an open issue. The three quark picture (uds) fails to reproduce the observed properties of this state. A review of the theoretical works, and references to the experimental literature can be found in [5]. According to the chiral unitary predictions [6] a high mass pole, coupled to the $\bar{K}N$ production channel and located around 1420 MeV, might contribute to the measured $\Lambda(1405)$ shape. Since the accessible invariant mass, in K^-p absorption processes, is influenced by the binding energy of the proton in the hosting nucleus, our strategy is to unveil the presence of the high mass pole by exploiting K^- captures in-flight [7, 8]. In this case the kinetic energy of the kaon sets the energy threshold just below the $\bar{K}N$ threshold. The shape of the $(\Sigma\pi)^0$ spectra is also distorted by the non-resonant production below threshold. A key issue, which is addressed in the analyses described below, is the investigation of the non-resonant hyperon-pion transition amplitude below threshold.

The position of the $\Lambda(1405)$ clearly reflects the strength of the $\bar{K}N$ interaction, thus influencing the possible formation of $\bar{K}$ multi-nucleon bound states. For the di-baryonic kaonic bound state ppK^- theoretical predictions deliver a wide range of binding energies and widths [9], while the experimental results are contradictory [10–16]. Moreover, the extraction of ppK^- signal in K^- absorption experiments is strongly affected by the yield and the shape of the competing K^- double-nucleon absorption (2NA) process.

In Section 2 the features of the DAΦNE accelerator and the KLOE detector are summarized. In Section 3 the two collected data samples which were analysed are described. The particle identification procedure is summarised in Section 4. Sections 5 and 6 are dedicated to the obtained results and ongoing analyses, regarding K^- multi-nucleon absorption processes, the search for a ppK^- signal and the investigation of resonant and non-resonant Hyperon-pion production in light nuclei. The paper ends with conclusions and perspectives.

2 The KLOE detector at DAΦNE

DAΦNE (Double Anular Φ-factory for Nice Experiments) is a double ring e^+e^- collider, designed to work at the center of mass energy of the ϕ particle; the ϕ meson decay produces charged kaons with low momentum ($\simeq$ 127 MeV/c) which allows to either stop them, or to explore the products of their low-energy nuclear absorptions.

The KLOE detector is centered around the interaction region of DAΦNE and is characterised by an acceptance of $\simeq 98\%$; it consists of a large cylindrical Drift Chamber (DC) [17] and a fine sampling lead-scintillating fibers calorimeter [18], all immersed in the axially symmetrical magnetic field with intensity of 0.52 T, provided by a superconducting solenoid. The chamber is characterized by excellent position and momentum resolutions. Tracks are reconstructed with a resolution in the transverse $R-\phi$ plane $\sigma_{R\phi} \sim 200\,\mu$m and a resolution along the z-axis $\sigma_z \sim 2$ mm. The transverse momentum resolution for low momentum tracks ((50 $< p <$ 300)MeV/c) is $\frac{\sigma_{p_T}}{p_T} \sim 0.4\%$. The calorimeter is composed of a cylindrical barrel and two endcaps, providing a solid angle coverage of 98%. The volume ratio (lead/fibers/glue=42:48:10) is optimized for a high light yield and a high efficiency for photons in the range (20-300) MeV/c. The photon detection efficiency is 99% for energies larger than 80 MeV and it falls to 80% at 20 MeV due to the cutoff introduced by the ADC and TDC thresholds. The position of the clusters along the fibers can be obtained with a resolution $\sigma_{\parallel} \sim 1.4\,\mathrm{cm}/\sqrt{E(\mathrm{GeV})}$. The resolution in the orthogonal direction is $\sigma_{\perp} \sim 1.3\,\mathrm{cm}$. The energy and time resolutions for photon clusters are given by $\frac{\sigma_E}{E_\gamma} = \frac{0.057}{\sqrt{E_\gamma(\mathrm{GeV})}}$ and $\sigma_t = \frac{57\,\mathrm{ps}}{\sqrt{E_\gamma(\mathrm{GeV})}} \oplus 100$ ps.

The DC entrance wall composition is 750 μm of carbon fiber and 150 μm of aluminum foil. Dedicated GEANT Monte Carlo simulations of the KLOE apparatus show that out of the total number of kaons interacting in the DC entrance wall, about 81% are absorbed in the carbon fiber component and the remaining 19% in the aluminum foil. The KLOE DC is filled with a mixture of helium and isobutane (90% in volume ^{4}He and 10% in volume C_4H_{10}).

3 Data samples

The AMADEUS experiment is conceived to integrate the high acceptance and momentum resolution KLOE detector with the low momentum K^- beam of the DAΦNE collider in a future dedicated setup. As a first step, the data collected by the KLOE collaboration during the 2004/2005 data taking, for which the dE/dx information of the reconstructed tracks is available (dE/dx represents the truncated mean of the ADC collected counts due to the ionization in the DC gas), corresponding to ~ 1.74 fb^{-1}, were analysed. The KLOE detector was used as an active target, the hadronic interaction of negative kaons with the materials of the apparatus being investigated. The topology of these data is shown in figure 1, representing the radial position (ρ_Λ) of the $\Lambda(1116)$ decay vertex (see Section 4). Four components are recognizable, from inside to outside we observe K^- absorptions in the DAΦNE beryllium sphere (~ 5 cm), the DAΦNE aluminated beryllium pipe (~ 10 cm), the KLOE DC entrance wall (aluminated carbon fiber ~ 25 cm) and the long tail originating from K^- interactions in the gas filling the KLOE DC (25-200 cm). Extremely rich experimental information is contained in this sample, with K^- hadronic interactions, both at-rest and in-flight [7], in a variety of light nuclear targets (H, ^{4}He, ^{9}Be and ^{12}C), which, of course, turned into a challenging analysis and interpretation efforts.

In order to increase the statistics and as an essential interpretation tool, a high purity carbon target (graphite) was realized in summer 2012 and installed inside the KLOE DC, between the beam pipe and the DC entrance wall.

The geometry of the target was optimized by means of a GEANT3 MC simulation. The final configuration was a three-sector half-cylinder ($\rho = 0.95\mathrm{g/cm^3}$) supported by an aluminium frame, with a length of 600 mm and a mean radius of 24,7 mm. The thickness varied from 4 to 6 mm, in order to optimize the percentage of stopped kaons, taking into account the ϕ boost, derived from the crossing angle of the e^+ e^- beams. The half-cylindrical configuration was chosen in order to take advantage of the K^+ tagging in the opposite direction. The design of the realized target is shown in figure 2. We took data from 6 November to 14 December 2012, for a total integrated luminosity of ~90 pb^{-1}. Up to now we analysed a sample of 37 pb^{-1} reconstructed data. The topology of the reconstructed Λ decay vertices radial position is shown in figure 3, and as expected, the majority of kaons are stopped in the target.

4 Particle Identification

The investigation of the negatively charged kaons interactions in nuclear matter is performed through the reconstruction of hyperon-pion and hyperon-nucleon/nucleus correlated pairs productions, fol-

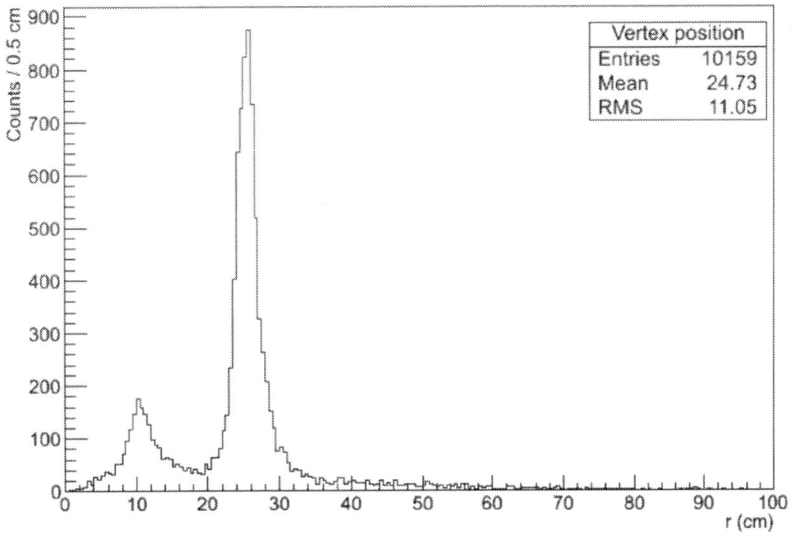

Figure 1. *Radial position distribution ρ_Λ, of the Λ decay vertex, for 2004-2005 KLOE collected data.*

lowing the K^- absorptions in H, ^{4}He, ^{9}Be and ^{12}C. The $\Lambda(1116)$ identification proceeds through the reconstruction of the $\Lambda \to p + \pi^-$ (BR = 63.9 ±0.5%) decay vertex. A spatial resolution below 1 mm is achieved for vertices found inside the DC volume (evaluated with Monte Carlo simulations). The obtained $M_{p\pi^-}$ invariant mass mean value is 1115.753±0.002 MeV/c^2 (only statistical error is given, the systematics being under investigation), with a resolution of σ=0.5 MeV/c^2. The particle identification takes advantage of both dE/dx information from the DC wires and the measurement of the energy released in the Calorimeter, as described in [1]. Σ particles are identified through their decay into $\Lambda\gamma$ or $p\pi$ as reported in [7, 8]. The K^- absorption vertex position, obtained exploiting the correlated production of the hyperon together with an extra-particle (pion, proton etc.) is then used to select the target. As an example, the obtained resolution on the radial coordinate ($\rho_{\Lambda p}$) for the Λp vertex is 1.2 mm. Cuts on the absorption vertex radial position were optimised, based on MC simulations and a study of the Λ decay path, to select the targets with minimal contamination from other components. More details on the particle identification procedure can be found in [19]

5 K^- absorption on two nucleons in the Σ^0p final state

In [19], a high purity sample of Σ^0p events, from K^- captures in the ^{12}C, was reconstructed. Σ^0p is (together with Λp) an expected decay channel of the ppK$^-$ cluster, with the advantage to be free from the Σ N $\to$ Λ N' conversion processes. The conversions strongly affect the uncorrelated Λp production, thus distorting the observed spectra.

A simultaneus fit of the: Σ^0p invariant mass, the relative angle of the Σ^0 and proton in the laboratory system $\cos(\theta_{\Sigma^0 p})$, the Σ^0 and the proton momenta was performed to the following simulated processes:

- $K^-A \to \Sigma^0 - (\pi)p_{spec}(A')$ (1NA),
- $K^-pp \to \Sigma^0 p$ (2NA),

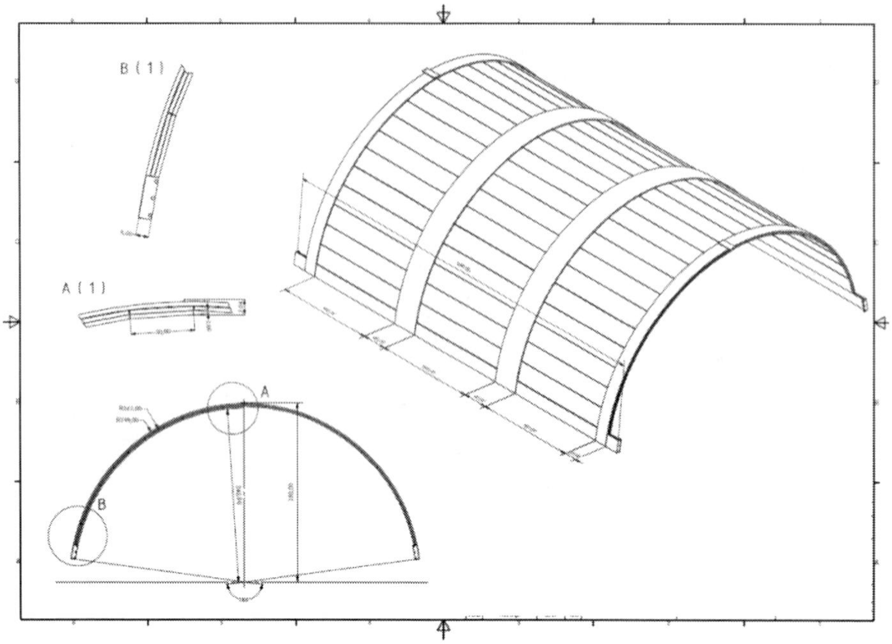

Figure 2. *Design of the graphite target inserted between the beam pipe and the KLOE DC entrance wall.*

- $K^-ppn \rightarrow \Sigma^0 pn$ (3NA),
- $K^-ppnn \rightarrow \Sigma^0 pnn$ (4NA).

Also Final State Interactions (FSI) of the Σ^0 and p emerging from a K^-pp capture were taken into account.

The yield of the 2NA, when the produced Σ^0 and p particles are free form any FSI process, was measured for the first time, with good precision. The obtained results are summarized in Table 1. A second fit was carried out including a ppK^- component, decaying into $\Sigma^0 p$. A systematic scan of possible binding energies and widths, varying within 15-75 MeV and 30-70 MeV respectively, was performed. The best fit resulted in a binding energy of 45 MeV and a width of 30 MeV. The resulting yield, normalised to the number of stopped K^-, is $ppK^-/K^-_{stop} = (0.044 \pm 0.009 stat^{+0.004}_{-0.005} syst) \times 10^{-2}$. The significance of the bound state with respect to a statistical fluctuation was checked by means of an F-test and was found to be significant at the level of 1σ only. Although the measured spectra are compatible with the hypothesis of a ppK^- contribution, the significance of the result is not sufficient to claim the discovery of the state. We refer to [19] for the details of the analysis.

6 $Y\pi$ resonant and non resonant production and the shape of the $\Lambda(1405)$

The position of the $\Lambda(1405)$ state is determined by the strength of the $\bar{K}N$ attractive interaction, thus influencing the possible formation of $\bar{K}$ multi-N states formation. When extracting the $\Lambda(1405)$ shape from K^- induced reactions in light nuclear targets (see for example [23]) the hyperon-pion spectroscopy is influenced by the energy threshold, imposed by the last nucleon binding energy. The $m_{\Sigma\pi}$ invariant mass threshold is about 1412 MeV and 1416 MeV, for K^- capture at-rest in ^4He and ^{12}C respectively, thus the K^- absorption at-rest is not sensitive to the high mass pole predicted by chiral unitary models. The $\bar{K}N$ sub-threshold region is accessible by exploiting K^-N absorptions in-flight.

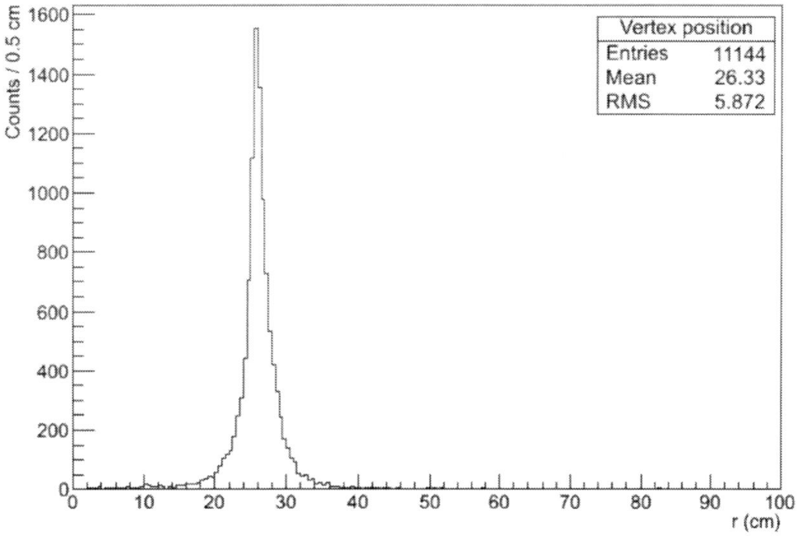

Figure 3. *Radial position distribution of ρ_Λ for carbon target events.*

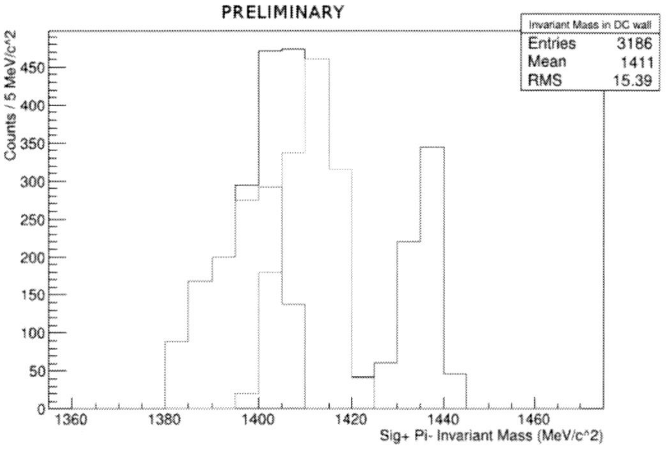

Figure 4. *(Colour online.) $m_{\Sigma\pi}$ invariant mass distributions in-flight (green) and at-rest (violet) in ^{12}C. Blue is the sum of green and violet curves. The red curve refers to K^- absorptions on Hydrogen*

For a mean kaon momentum of 100 MeV/c the $m_{\Sigma\pi}$ threshold is shifted upwards by about 10 MeV. A second bias is represented by the non-resonant $K^-N \rightarrow Y\pi$ formation, which gives rise to highly correlated hyperon-pion pairs production. The corresponding $m_{Y\pi}$ invariant masses spectra is narrow (of the order of 10 MeV) and peaked below the $\bar{K}N$ threshold. The $\Lambda\pi$ and $\Sigma\pi$ non resonant transition amplitudes, for K^- capture in light nuclear targets, was never measured. The Λ and π^- kinematic dis-

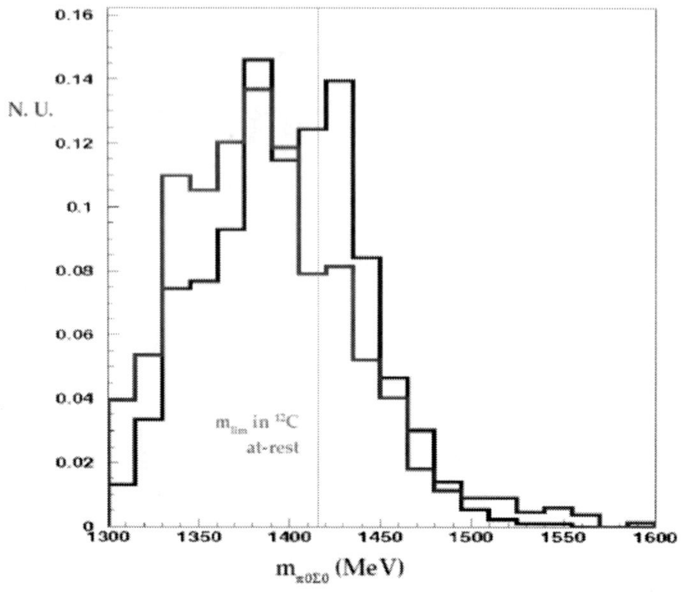

Figure 5. *(Colour online.) $m_{\Sigma^0\pi^0}$ invariant mass distribution from K^- captures in the KLOE DC wall (black curve) and pure carbon graphite target (blue curve)*

Process	yield / $K^-_{stop}\times 10^{-2}$	$\sigma_{stat}\times 10^{-2}$	$\sigma_{syst}\times 10^{-2}$
2NA-QF	0.127	±0.019	$^{+0.004}_{-0.008}$
2NA-FSI	0.272	±0.028	$^{+0.022}_{-0.023}$
Tot 2NA	0.399	±0.033	$^{+0.023}_{-0.032}$
3NA	0.274	±0.069	$^{+0.044}_{-0.021}$
Tot 3 body	0.546	±0.074	$^{+0.048}_{-0.033}$
4NA + bkg.	0.773	±0.053	$^{+0.025}_{-0.076}$

Table 1. Production probability of the Σ^0p final state for different intermediate processes normalised to the number of stopped K^- in the DC wall. The statistical and systematic errors are shown as well [19].

tributions for K^- captures in ^{4}He, both at-rest and in-flight, were calculated in [22]. The momentum probability distribution functions of the emerging hyperon pion pairs, following K^-n absorptions, are expressed in terms of the K^-n transition amplitudes: the isospin $I = 1$ S-wave non- resonant amplitude ($|f^{nr}|$) and the resonant $I = 1$ P-wave amplitude, dominated by the $\Sigma^-(1385)$. Since the resonant amplitude is well known from direct experiments, the measured total momentum distributions can be used to extract the non-resonant $|f^{nr}|$ amplitude module below the $\bar{K}N$ threshold. The goal of the ongoing analyses is to measure the contributions and the shapes of the non resonant $\Lambda\pi$ and $\Sigma\pi$ productions. The knowledge of the $(\Sigma\pi)^0$ isospin $I = 0$ non-resonant transition amplitude will allow to disentangle the resonant $\Lambda(1405)$ shape. Preliminary $\Sigma^+\pi^-$ invariant mass spectra, from K^- captures in the wall of the KLOE DC, not background subtracted nor acceptance corrected, are shown in Fig. 4. The red curve refers to K^- absorptions on Hydrogen, green and violet distributions refer to K^-

captures in-flight and at-rest in ^{12}C respectively, the blue line is the sum of green and violet distributions. The red distribution reflects the non-resonant K^-H absorption in-flight, which corresponds to a narrow invariant mass shape peaked below $m_K + m_p + \langle p_K^2 \rangle / 2m_K$, the third term represents the mean kinetic energy of the non-relativistic kaons. In figure 5 (black distribution) the $\Sigma^0\pi^0$ invariant mass spectrum from K^- captures in the KLOE DC wall is shown [25]. The reduced resolution, due to the reconstruction of the three photons clusters, does not allow in this case to separate the two processes of at-rest and in-flight capture, moreover the absorbing target (H or ^{12}C) can not be distinguished in this case. The $m_{\Sigma^0\pi^0}$ spectrum can be compared with the corresponding distribution from K^- absorptions in a pure carbon target (fig. 5 blue distribution), sample which consists of K^- captures at-rest. In fig. 5 a red line indicates the energy threshold corresponding to K^- absorption in ^{12}C at-rest. In order to compare the two distributions plotted in fig. 5 the two curves are normalised to unity. A rich sample of in-flight $K^-{}^{12}$C captures can be easily identified. A spectroscopic study of the kinematic region ranging between the at-rest energy threshold, and the $\bar{K}N$ threshold, opened by the low momentum in-flight capture process, will allow to clarify the nature of the high mass $\Lambda(1405)$ pole.

7 Conclusions and perspectives

In this work a broad research program of low-energy K^- induced reactions on light nuclear targets is presented. The $\bar{K}N$ interaction in nuclear matter is investigated through the study of the hyperon resonances properties below the $\bar{K}N$ threshold, and the characterization of K^--multi-nucleon captures processes. The latter is found to strongly impact on the K^--multi-nucleon bound state search, as the K^-pp (2NA) absorption overlaps with the bound state in the phase space region where it is expected.

In the search for the K^-pp bound state, and to extract the K^--multi-nucleon absorption yields, we faced the problem of disentangling the elastic and inelastic final state interactions of the hyperons with the residual nucleons. The former process is of particular interest for the measurement of the hyperon-nucleon (multi-nucleon) cross sections, for which the available experimental information is extremely scarce. Moreover, the hyperon-nucleon (multi-nucleon) interaction potentials are fundamental inputs in the determination of the equation of state for the neutron stars, whose structure is strongly debated, following the measurement of two neutron stars exceeding 2M⊙ [26, 27]. Experimental constrains from the hyperon-nucleon scattering processes are mandatory to guide the theory.

Presently a feasibility study [28, 29] is ongoing for the realization of a dedicated AMADEUS experimental setup, in order to deepen and extend the low energy anti-kaon nuclei interaction studies and obtain fundamental input for the study of QCD with strangeness and of neutron stars.

Acknowledgement

We acknowledge the KLOE Collaboration for their support and for having provided us the data and the tools to perform the analysis presented in this paper.

References

[1] C. Curceanu, K. Piscicchia et al., Acta Phys. Polon. B46 1, 203-215, (2015).
[2] R. Baldini et al., Proposal for a Phi-Factory, report LNF-90/031(R) (1990).
[3] F. Bossi, E. De Lucia, J. Lee-Franzini, S. Miscetti, M. Palutan and KLOE coll., Riv. Nuovo Cim. 31 (2008) 531-623
[4] C. Patrignani et al. (Particle Data Group), Chin. Phys. C, 40, 100001 (2016).
[5] T. Hyodo, D. Jido, Prog. Part. Nucl. Phys. 67 (2012) 55.

[6] Phys. Lett. B 500 (2001),
Phys. Rev. C 66 (2002),
Nucl. Phys. A 725 (2003),
Nucl. Phys. A881, 98 (2012)

[7] K. Piscicchia et al., PoS Bormio2013 034 (2013)

[8] A. Scordo et al., PoS Bormio2014 039 (2014)

[9] T. Yamazaki, *et al.*, Phys. Rev. C **76** 045201 (2007)
A. Doté, *et al.*, Phys. Rev. C **79** 014003 (2009)
S. Wycech, *et al.*, Phys. Rev. C **79** 014001 (2009)
N. Barnea, *et al.*, Phys. Lett. B **712** 132 (2012)
N.V. Shevchenko, *et al.*, Phys. Rev. Lett. **98** 082301 (2007)
Y. Ikeda, *et al.*, Phys. Rev. C **79** 035201 (2009)
E. Oset, *et al.*, Nucl. Phys. A **881** 127 (2012)

[10] G. Agakishiev, *et al.*, HADES Coll., Phys. Lett. B **742** 242 (2015)

[11] M. Agnello, *et al. (FINUDA Coll.)*, Phys. Rev. Lett. **94** 212303 (2005)

[12] T. Yamazaki, *et al.*, Phys. Rev. Lett. **104** 132502 (2010)

[13] Y. Ichikawa, *et al.*, Prog. Theor. Exp. Phys. **021D01** (2015)

[14] A. O. Tokiyasu *et al.*, Phys.Lett. B **728**, 616 (2014)

[15] L. Fabbietti *et al.*, Nucl. Phys. A **914**, 60 (2013)

[16] T. Hashimoto *et al.*, Prog. Theor. Exp. Phys. 061D01 (2015)

[17] M. Adinolfi et al., [KLOE Collaboration], Nucl. Inst. Meth. A 488, (2002) 51.

[18] M. Adinolfi et al. [KLOE Collaboration], Nucl. Inst. Meth. A 482, (2002) 368.

[19] O. Vazquez Doce et al., Phys. Lett. B 758, 134-139 (2016)

[20] R. Roosen, J. H. Wickens, Il Nuovo Cim. A Series 11, **66** 101 (1981)

[21] FINUDA collaboration, Phys. Lett. B, **669** 229 (2008)

[22] K. Piscicchia, S. Wycech, C. Curceanu, Nucl. Phys. A **954**, 75-93 (2016)

[23] J. Esmaili et al., Phys. Lett. B 686 (2010) 23-28

[24] FINUDA collaboration, Phys. Lett. B 704, 474–480 (2011)

[25] K. Piscicchia, PhD theis (2013), Lambda(1405) measurement through the decay to $\Sigma^0\pi^0$, resulting from K^- meson absorption on ^{4}He and ^{12}C, with the KLOE detector, http://www.infn.it/thesis/thesis_dettaglio.php?tid=7097

[26] P. B. Demorest, T. Pennucci, S. M. Ransom, M. S. E. Roberts & J. W. T. Hessels, Nature (London) 467, (2011)

[27] D. J. Nice, E. M. Splaver, I. H. Stairs, O. Loehmer, and A. Jessner, Astrophys. J. 634, 1242 (2005).

[28] M. Bazzi, et al., Nuclear Instruments and Methods in Physics Research A 671 (2012) 125–128

[29] M. Bazzi et al., Journal of Instrumentation 8(05) January 2013

From Hadronic Cross Section to the measurement of the Vacuum Polarization at KLOE: a fascinating endeavour

Graziano Venanzoni[1,⋆]

[1] *INFN Sezione di Pisa, Pisa, Italy*

Abstract. The KLOE experiment at the $\phi-factory$ DAΦNE in Frascati is the first to have employed Initial State Radiation (ISR) to precisely determine the $e^+e^- \to \pi^+\pi^-(\gamma)$ cross section below 1 GeV. Such a measurement is particularly important to test the Standard Model (SM) calculation for the $(g-2)$ of the muon, where a long standing 3σ discrepancy is observed. I will review the ISR activity in KLOE in the last 20 years from the measurement of the hadronic cross section to the first direct determination of the time-like complex running $\alpha(s)$ in the region below 1 GeV.

1 Introduction: the Endurance expedition

In August 1914, polar explorer Sir Ernest Shackleton boarded the Endurance and set sail for Antarctica, where he planned to cross the last uncharted continent on foot. In January 1915, after battling its way through a thousand miles of pack ice and only a day's sail short of its destination, the Endurance became locked in an island of ice (see Fig 1, left). Thus began the legendary ordeal of Shackleton and his crew of twenty-seven men. For ten months the ice-moored Endurance drifted northwest before it was finally crushed between two ice floes. With no options left, Shackleton and a 5 members crew left on 24 April 1916 on a 7 m boat, the *James Caird*, from Elephant Island in the South Shetland Islands for a near-impossible journey over 850 miles of the South Atlantic's heaviest seas to the island of South Georgia (see Fig. 1, right). Shackleton indeed succeeded to reach the island, and on 30 August, after four attempts, Shackleton was able to return to Elephant Island to rescue the party stranded there. Every single man survived.

With due proportion, this epic feat of endurance had many distinctive characters of the modern scientific research: experience the raw beauty and test the human spirit in an unknown (and often hostile) territory with perseverance and dedication; draw conclusions from observation, (incomplete) information, and inspiration; learn from errors. Such a mental attitude especially applies when the object of a research is a precision measurement, as was the trip of the *James Caird* boat to South Georgia, a tiny dot of land in a vast and hostile ocean.

2 1999-2004: The first KLOE measurement of $\sigma(e^+e^- \to \pi^+\pi^-)$ with ISR

After more than twenty years from the first calculations [1], the late nineties saw a vigorous and renovated interest towards the use of Initial State Radiation process (ISR in the following) for the precise

⋆e-mail: graziano.venanzoni@lnf.infn.it

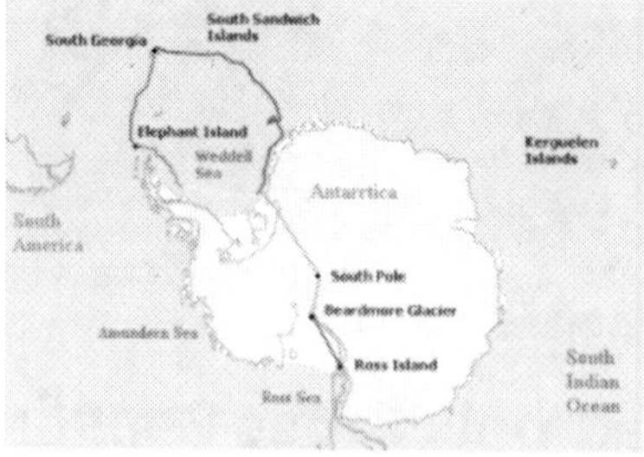

Figure 1. Left: The Endurance trapped in ice; *—Photo by Frank Hurley, courtesy of the Macklin Collection*; Right: Map of the sea routes of Endurance and James Caird: in red the voyage of Endurance; in yellow the drift in pack ice; in green the sea-ice drift after Endurance sinks; in blue the voyage of the boat James Caird.

measurement of the hadronic cross section [2]. The high luminosity reached at flavor factories made it possible to compensate the smaller cross section of the radiative process, making ISR an alternative and competitive method with respect to the traditional energy scan [3]. From the theoretical side the development of the PHOKHARA Monte Carlo event generator [4], based on exact QED NLO predictions, together with improved Monte Carlo codes for luminosity measurement, like BABAYAGA [5] or BHWIDE [6] and MCGPJ [7], made it possible to reach the sub per cent accuracy on the measurement of ISR cross sections [8].

From the experimental side ISR was first performed at KLOE [9, 10], to determine the cross section $e^+e^- \to \pi^+\pi^-\gamma$, soon followed by BaBar [11]. In the recent years this method was used to study a multitude of final states at BaBar [12], Belle [13, 14] and BES-III [15].

The first KLOE analysis of $e^+e^- \to \pi^+\pi^-\gamma$ cross section (called KLOE05 in the following) was performed during the years from 1999 to 2004 [16]. This analysis was mainly motivated by the discrepancy between the Standard Model prediction (SM in the following) and the measured value of the muon g-2 [17]. A large part of the uncertainty on the theoretical estimates comes from the leading order hadronic contribution $a_\mu^{\rm HLO}$, which at low energies is not calculable by perturbative QCD, but has to be evaluated with a dispersion integral using measured hadronic cross sections.

The process $e^+e^-\to\pi^+\pi^-$ below 1 GeV accounts for $\sim$ 70% of $a_\mu^{\rm HLO}$ [18], whose uncertainty dominates the SM evaluation of a_μ.

At the time of KLOE05 there was a rather strong disagreement between $a_\mu^{\rm HLO}$ value obtained using data from τ decays into two- and four-pion final states after isospin-breaking corrections and e^+e^- based data. In particular while the SM prediction using τ data showed a difference less than 1σ with the BNL experimental result, the e^+e^- based result disagreed by $\sim 3\sigma$ [20] (see Fig. 2, left). In addition the $\pi^+\pi^-$ spectral function from e^+e^- and isospin-corrected τ data were showing a systematic deviation in shape and normalisation [20].

KLOE05 and the following KLOE08 analyses used the so called *Small Angle (SA)* selection cuts: photons remaining undetected are restricted to a cone of $\theta_\gamma < 15°$ around the beamline (narrow cones in Fig. 2, right) and the two charged pion tracks are detected within $50° < \theta_\pi < 130°$ (wide cones in Fig. 2, right). In this configuration, the direction of the photons is reconstructed from the tracks' momenta by closing kinematics: $\vec{p}_\gamma \simeq \vec{p}_{\rm miss} = -(\vec{p}_{\pi^+} + \vec{p}_{\pi^-})$. While these cuts guarantee a high statistics for ISR signal events and a reduced contamination from the resonant process $e^+e^- \to \phi \to \pi^+\pi^-\pi^0$, in which the π^0 mimics the missing momentum of the photon(s), and from the final state radiation process $e^+e^- \to \pi^+\pi^-\gamma_{\rm FSR}$, a highly energetic photon emitted at small angle forces

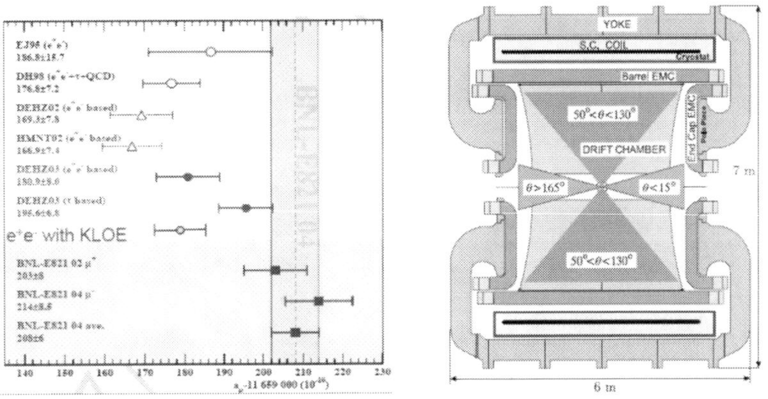

Figure 2. Left: Comparison between different theoretical predictions of the anomalous magnetic moment of the muon and the experimental measurement from E821 around 2004 when KLOE data were available. From F. Teubert presentation at ICHEP04 [19]. Right: Schematic view of the KLOE detector with selection regions.

the pions also to be at small angles (and thus outside the selection cuts), resulting in a kinematic suppression of events with $M^2_{\pi\pi} < 0.35$ GeV^2. The measurement of $\pi\pi\gamma$ cross section was normalized to the DAΦNE luminosity using large angle Bhabha scattering with 0.3% total systematic uncertainty (using BABAYAGA), and the pion form factor was derived in the energy region between 0.35 and 0.95 GeV^2 by dividing the $\pi\pi\gamma$ cross section by the so called radiator function obtained with NLO PHOKHARA MC generator. Although with a limited accuracy of 1.3% equally divided between theory and experiment, the importance of KLOE05 was twofold:

- It established the use of ISR as a working technique to precisely measure the hadronic cross section with a competitive accuracy of the energy scan;
- It confirmed the 3σ discrepancy of the muon g-2 between the direct measurement and the SM prediction based on e^+e^- data, as shown in Fig 2, left.

In addition KLOE05 allowed the KLOE Collaboration to get confidence into a novel technique which would have been refined in the following years.

3 2005-2010: KLOE08 and KLOE10 analyses

As most of the adventures in a new territory, the KLOE05 analysis was not free from errors. Particularly a bias in the evaluation of the trigger correction was found, which together with a reevaluation of the luminosity (due to an improved version of Bhabha generator BABAYAGA [5]), resulted in a few % discrepancy in the two spectra below the ρ peak, and a slightly smaller value of $a^{\pi\pi}_{\mu}$ (below one standard deviation) [21]. In the meantime a new analysis called (KLOE08 in the following) was carried out with the same SA selection criteria for pion tracks and photon angles as of KLOE05 on 240 pb^{-1} of data taken in 2002 [21]. With respect to KLOE05, KLOE08 profits from lower machine background and more stable DAΦNE operation in 2002. Data filters and analysis technique were improved, together with a refined knowledge of the detector response and of the KLOE simulation and analysis, including a new version of generator PHOKHARA, with next-to-leading-order ISR and FSR corrections, as well as simultaneous emission of one ISR and one FSR photon [4].

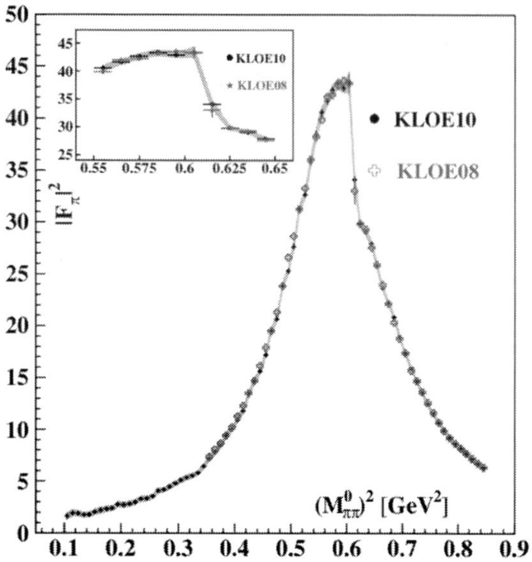

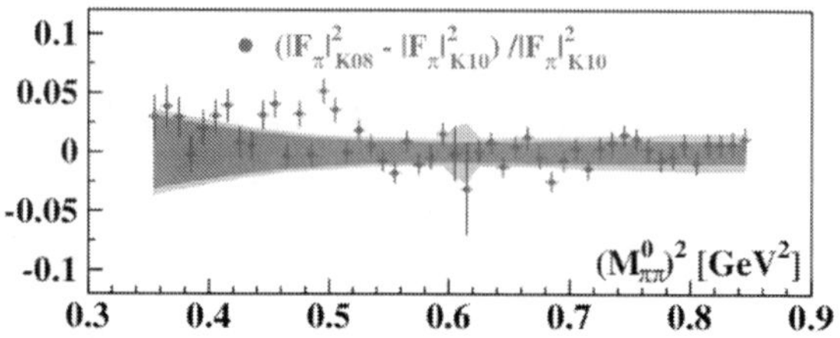

Figure 3. Comparison of KLOE10 with the previous KLOE result, KLOE08. Left: Pion form factor $|F_\pi|^2$. Right: Fractional difference between KLOE08 and KLOE10 results. The dark (light) gray band gives the statistical (total) error for the present result. Errors on KLOE08 points contain the combined statistical and systematic uncertainty.

All these refinement resulted in an improved 0.9% total systematic uncertainty consistent with KLOE05, with a total error smaller by 30%, confirming the 3σ discrepancy on the muon g-2. In view of the improvements of the analysis and the superior quality of the 2002 data, KLOE08 superseded KLOE05 [21].

Although the KLOE08 measurement had reached a remarkable 0.9% systematic error, the measured region was limited to 600 MeV, due to the SA selection requirement. To reach the dipion threshold, a new measurement (called KLOE10 in the following) was performed where a photon was detected in the calorimeter at large polar angles, $50^o < \theta_\gamma < 130^o$, in the same region where also pion tracks are detected (*so called* Large Angle (LA) selection) [22]. However, compared to the measurements with photons at small angles, these conditions imply a reduction in statistics of about a factor of 5, and an increase of the background from the process $\phi \to \pi^+\pi^-\pi^o$, as well as the irreducible background from events with final state radiation and from ϕ radiative decays. To reduce the background contamination and the related uncertainties, the analysis was performed on 232.6 pb^{-1} of data taken in 2006 at 1 GeV (*i.e.* 20 MeV below the ϕ mass). The KLOE10 analysis measured $a_\mu^{\pi\pi}$ in the region 0.1–0.85 GeV^2 with a 1% experimental systematic error and 0.9% theoretical one dominated by uncertainty on final state radiation (FSR) model. KLOE10 showed a good agreement with the previous KLOE measurements (see Fig. 3), allowing to cover 70% of the leading order hadronic contribution to the muon anomaly with 1% total error. It showed also a reasonable agreement with the results from the Novosibirsk experiments CMD-2 and SND (see Fig. 4, left); while comparing with the BaBar result it showed agreement below 0.4 GeV^2 (corresponding to a value of $M_{\pi\pi}$ of 630 MeV), while above 0.4 GeV^2 the BaBar result is higher by 2-3% (Fig. 4, right).

The KLOE10 data allowed also to test the validity of different models for the $\pi^+\pi^-$ final state photon emission in the forward-backward asymmetry, showing good agreement with the point-like pions (*so called* scalar QED) assumption (see Fig. 5) [8, 23–25].

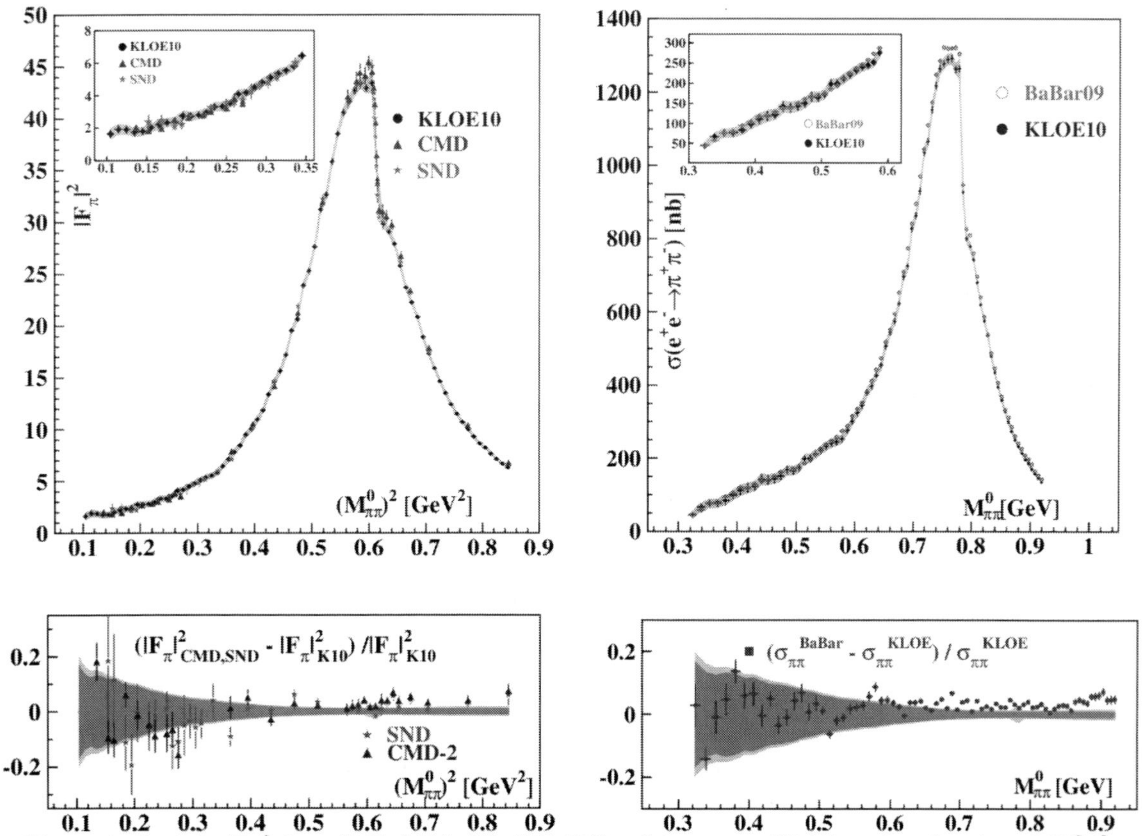

Figure 4. Top left: $|F_\pi|^2$ from CMD-2 [26, 27], SND [28] and the present KLOE result as function of $(M^0_{\pi\pi})^2$. Bottom left: Fractional difference between CMD-2 or SND and KLOE. Top right: $\sigma^{\text{bare}}_{\pi\pi}$ from BaBar [29] and the new KLOE result as function of $M^0_{\pi\pi}$. Bottom right: Fractional difference between BaBar and KLOE. CMD-2, SND and BaBar data points have the total uncertainty attached. The dark (light) band in the lower plots shows statistical (total) error of the KLOE result.

4 2010-2012: KLOE12 analysis

While KLOE08 and KLOE10 were independent measurements with respect to the data set and angular selection, they were both normalized to the DAΦNE luminosity, and used the radiation function to obtain the pion form factor and a_μ. As shown in Ref. [29] it's possible to extract the pion form factor by normalizing to the muon ISR differential cross section. In this approach the integrated luminosity as well as the radiation function H cancel in the ratio, as does the vacuum polarisation. In addition using the same fiducial volume, acceptance corrections to $\pi\pi\gamma$ and $\mu\mu\gamma$ spectra almost cancel resulting in a per mille systematic uncertainty. However the price to pay is to perform an additional analysis on muon events at subpercent level. The same sample of 239.2 pb^{-1} of KLOE08 was analyzed with the small angle photon selection (called KLOE12 in the following) [30]. While the analysis for $\pi\pi\gamma$ is essentially the same as for KLOE08, the analysis for $\mu\mu\gamma$ is new and is based on the following main features: 1) separation between $\mu\mu\gamma$ and $\pi\pi\gamma$ events obtained assuming the final state of two charged particles with equal track mass M_{TRK} and one photon: the $M_{TRK} < 115$ MeV ($M_{TRK} > 130$ MeV) selection leads to 9×10^5 (3.1×10^6) candidate $\mu\mu\gamma$ ($\pi\pi\gamma$) events, this selection is checked against other

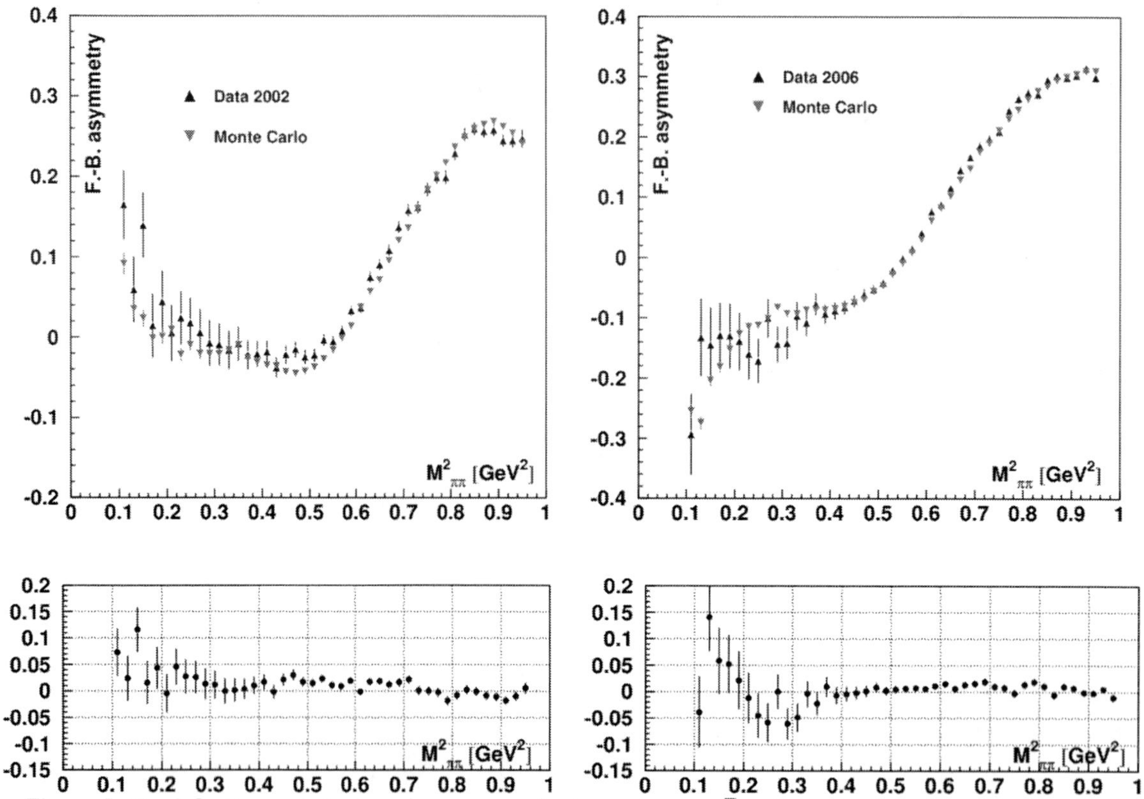

Figure 5. Top left: Forward–Backward asymmetry for data taken at $\sqrt{s} = M_\phi$ in 2002, and the corresponding Monte Carlo prediction using the PHOKHARA v6.1 generator. Bottom left: Absolute difference between the asymmetries from data and Monte Carlo prediction. Top right: Forward–Backward asymmetry for data taken at $\sqrt{s} \simeq 1000$ MeV in 2006, and the corresponding Monte Carlo prediction using the PHOKHARA v6.1 generator. Bottom right: Absolute difference between the asymmetries from data and Monte Carlo prediction. From Ref. [8]

techniques, such as a kinematic fit or tighter cuts on the quality of the charged tracks, all providing consistent results; 2) trigger, particle identification and tracking efficiencies checked from data control samples. The $\mu\mu\gamma$ cross section measurement was compared with the one obtained by PHOKHARA MC [4], and a good agreement is found [30]. Then the pion form factor has been extracted and compared with the one from KLOE10, showing good agreement (see Fig. 6, left). The value for $a_\mu^{\pi\pi}$ computed with KLOE12 was in good agreement with the previous KLOE results, providing therefore a strong cross check of the systematics and confirming the 3σ discrepancy between the experimental value and the Standard Model prediction of the muon g-2 (see Fig. 6 right).

5 2013-2016: Measurement of the running of the fine structure constant below 1 GeV

With about 2 fb^{-1} of luminosity collected in 2004-2005, the KLOE-2 Collaboration - the successor of KLOE - embarked in a full statistics analysis of the $\mu\mu\gamma$ cross section with a twofold motivation:

1. to search for a light vector boson [38];

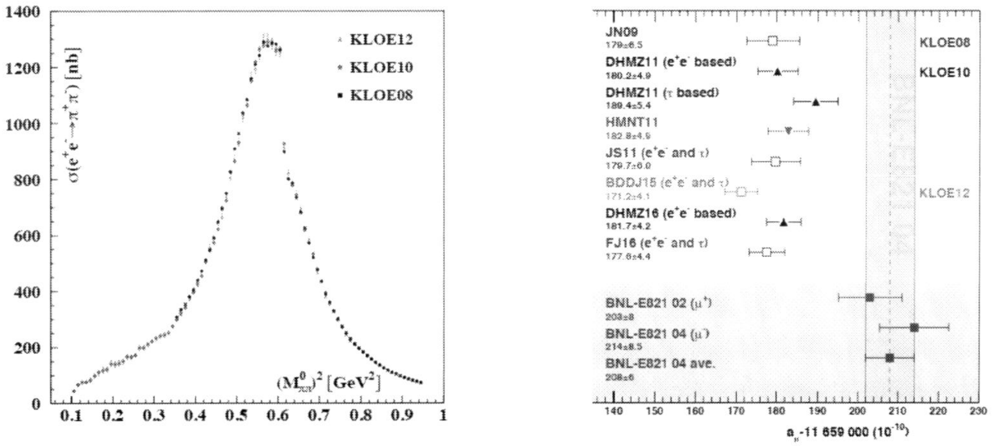

Figure 6. Left: pion form factor $|F_\pi|^2$ obtained by $\pi\pi\gamma/\mu\mu\gamma$ ratio (KLOE12) compared with previous KLOE measurements. Right: compilation of recent results for a_μ compared with the experimental value. The SM predictions refer to: JN09 [31], DHMZ11 [32], HLMNT11 [33], JS11 [34], BDDJ15 [35], DHMZ16 [36], FJ16 [37]. KLOE08 [21], KLOE10 [22], and KLOE12 [30] refer to the compilations where KLOE data are included.

2. to measure the running of the QED coupling constant $\alpha(s)$ below 1 GeV.

The analysis of the so called u–boson is in progress. In the following we will discuss the measurement of the running of the fine structure constant α in the time-like region 0.6< $\sqrt{s}$ <0.975 GeV [39].

The strength of the coupling constant is measured as a function of the momentum transfer of the exchanged photon $\sqrt{s} = M_{\mu\mu}$ where $M_{\mu\mu}$ is the $\mu^+\mu^-$ invariant mass. The value of $\alpha(s)$ is extracted from the ratio of the differential cross section for the process $e^+e^- \to \mu^+\mu^-\gamma(\gamma)$ with the photon emitted in the Initial State (ISR) to the corresponding cross section obtained from Monte Carlo (MC) simulation with the coupling set to the constant value $\alpha(s) = \alpha(0)$:

$$|\frac{\alpha(s)}{\alpha(0)}|^2 = \frac{d\sigma_{data}(e^+e^- \to \mu^+\mu^-\gamma(\gamma))|_{ISR}/d\sqrt{s}}{d\sigma^0_{MC}(e^+e^- \to \mu^+\mu^-\gamma(\gamma))|_{ISR}/d\sqrt{s}} \tag{1}$$

To obtain the ISR cross section, the observed cross section must be corrected for events with one or more photons in the final state (FSR). This has been done by using the PHOKHARA MC event generator, which includes next-to-leading-order ISR and FSR contributions [4]. Figure 7, left, shows the ratio of the $\mu^+\mu^-\gamma$ cross-section from data with the corresponding NLO QED calculation from PHOKHARA generator including the Vacuum Polarization effects.

The agreement between the two cross sections is excellent.

We use Eq. (1) to extract the running of the effective QED coupling constant $\alpha(s)$. By setting in the MC the electromagnetic coupling to the constant value $\alpha(s) = \alpha(0)$, the hadronic contribution to the photon propagator, with its characteristic $\rho - \omega$ interference structure, is clearly visible, see Fig. 7, right. The prediction from Ref.[40] is also shown. While the leptonic part is obtained by perturbation theory, the hadronic contribution to $\alpha(s)$ is obtained via an evaluation in terms of a weighted average

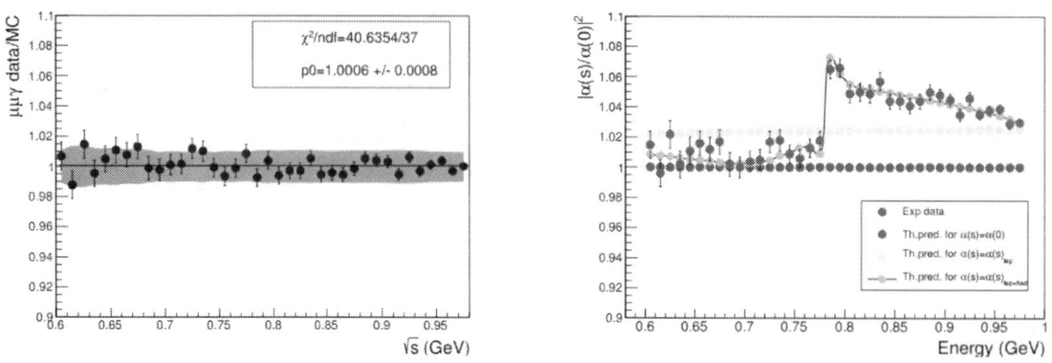

Figure 7. Left: Ratio between the measured differential $\mu^+\mu^-\gamma$ cross section and the MC prediction from PHOKHARA. The green band shows the systematic error. Right: The square of the modulus of the running $\alpha(s)$ in units of $\alpha(0)$ compared with the prediction (provided by the `alphaQED` package [40]) as a function of the dimuon invariant mass. The red points are the KLOE data with statistical errors; the violet points are the theoretical prediction for a fixed coupling ($\alpha(s) = \alpha(0)$); the yellow points are the prediction with only virtual lepton pairs contributing to the shift $\Delta\alpha(s) = \Delta\alpha(s)_{lep}$, and finally the points with the solid line are the full QED prediction with both lepton and quark pairs contributing to the shift $\Delta\alpha(s) = \Delta\alpha(s)_{lep+had}$.

compilation of $R_{had}(s)$, based on the available experimental $e^+e^- \to$ hadrons annihilation data (for an up to date compilation see [41] and references therein).

For comparison, the prediction with constant coupling (*no running*) and with only lepton pairs contributing to the running of $\alpha(s)$ is given.
By including statistical and systematics errors, we exclude the only-leptonic hypothesis at $6\,\sigma$ which is the strongest direct evidence ever achieved by a collider experiment [1].

5.1 Extraction of Real and Imaginary part of $\Delta\alpha(s)$

By using the definition of the running of α the real part of the shift $\Delta\alpha(s)$ can be expressed in terms of its imaginary part and $|\alpha(s)/\alpha(0)|^2$:

$$\mathrm{Re}\,\Delta\alpha = 1 - \sqrt{|\alpha(0)/\alpha(s)|^2 - (\mathrm{Im}\,\Delta\alpha)^2}. \tag{2}$$

The imaginary part of $\Delta\alpha(s)$ can be related to the total cross section $\sigma(e^+e^- \to \gamma^* \to \textit{anything})$, where the precise relation reads [18, 31, 45]: $\mathrm{Im}\,\Delta\alpha = -\frac{\alpha}{3} R(s)$, with $R(s) = \sigma_{tot}/\frac{4\pi|\alpha(s)|^2}{3s}$. $R(s)$ takes

into account leptonic and hadronic contribution $R(s) = R_{lep}(s) + R_{had}(s)$, where the leptonic part corresponds to the production of a lepton pair at lowest order taking into account mass effects:

$$R_{lep}(s) = \sqrt{1 - \frac{4m_l^2}{s}}\left(1 + \frac{2m_l^2}{s}\right), \quad (l = e, \mu, \tau). \tag{3}$$

[1]The first evidence for the hadronic VP came from the ACO experiment which found an evidence at 3σ of the ϕ contribution to the process $e^+e^- \to \mu^+\mu^-$ in the region ± 6 MeV around the ϕ peak (at 1019.4 MeV) [42]. The strongest evidence for hadronic VP effect comes from the muon g-2 experiment; at CERN it was determined at more than 7σ level [43, 44].

In the energy region around the ρ-meson we can approximate the hadronic cross section by the 2π dominant contribution:

$$R_{had}(s) = \frac{1}{4}\left(1 - \frac{4m_\pi^2}{s}\right)^{\frac{3}{2}} |F_\pi^0(s)|^2, \tag{4}$$

where F_π^0 is the pion form factor deconvolved: $|F_\pi^0(s)|^2 = |F_\pi(s)|^2 \left|\frac{\alpha(0)}{\alpha(s)}\right|^2$.

The results obtained for the 2π contribution to the imaginary part of $\Delta\alpha(s)$ by using the KLOE pion form factor measurement[30], are shown in Fig. 8 and compared with the values given by the $R_{had}(s)$ compilation of Ref. [40] using only the 2π channel, with the KLOE data removed (to avoid correlations).

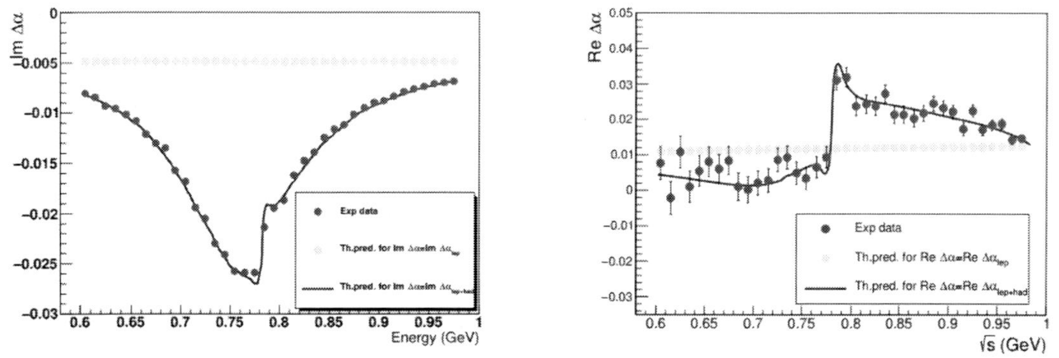

Figure 8. Left: $\mathrm{Im}\,\Delta\alpha$ extracted from the KLOE data compared with the values provided by `alphaQED` routine (without the KLOE data) for $\mathrm{Im}\,\Delta\alpha = \mathrm{Im}\,\Delta\alpha_{\mathrm{lep}}$ (yellow points) and $\mathrm{Im}\,\Delta\alpha = \mathrm{Im}\,\Delta\alpha_{\mathrm{lep+had}}$ only for $\pi\pi$ channels (blue solid line). Right: $\mathrm{Re}\,\Delta\alpha$ extracted from the experimental data with only the statistical error included compared with the `alphaQED` prediction (without the KLOE data) when $\mathrm{Re}\,\Delta\alpha = \mathrm{Re}\,\Delta\alpha_{\mathrm{lep}}$ (yellow points) and $\mathrm{Re}\,\Delta\alpha = \mathrm{Re}\,\Delta\alpha_{\mathrm{lep+had}}$ (blue solid line).

The extraction of the $\mathrm{Re}\,\Delta\alpha$ has been performed using the Eq. (2) and it is shown in Fig. 8, right. The experimental data with only the statistical error included have been compared with the `alphaQED` prediction when $\mathrm{Re}\,\Delta\alpha = \mathrm{Re}\,\Delta\alpha_{\mathrm{lep}}$ (yellow points in the colour Figure) and $\mathrm{Re}\,\Delta\alpha = \mathrm{Re}\,\Delta\alpha_{\mathrm{lep+had}}$ (dots with solid line). As can be seen, an excellent agreement for $\mathrm{Re}\,\Delta\alpha(s)$ has been obtained with the data-based compilation.

Finally Re $\Delta\alpha$ has been fitted by a sum of the leptonic and hadronic contributions, where the hadronic contribution is parametrized as a sum of the $\rho(770)$, $\omega(782)$ and $\phi(1020)$ resonance components and a non-resonant term. The product of the branching fractions has been extracted:

$$BR(\omega \to \mu^+\mu^-)BR(\omega \to e^+e^-) = (4.3 \pm 1.8_{stat} \pm 2.2_{syst}) \cdot 10^{-9}, \tag{5}$$

where the first error is statistical and the second systematic. By multiplying by the phase space factor $\xi = \left(1 + 2\frac{m_\mu^2}{m_\omega^2}\right)\left(1 - 4\frac{m_\mu^2}{m_\omega^2}\right)^{1/2}$ and assuming lepton universality, $BR(\omega \to \mu^+\mu^-)$ can be extracted:

$$BR(\omega \to \mu^+\mu^-) = (6.6 \pm 1.4_{stat} \pm 1.7_{syst}) \cdot 10^{-5} \tag{6}$$

compared to $BR(\omega \to \mu^+\mu^-) = (9.0 \pm 3.1) \cdot 10^{-5}$ from PDG [46].

6 Conclusions and Outlook

The last 20 years have seen a flourishing use of the ISR to measure the hadronic cross sections at flavor factories. Being more than a poor man's alternative to scan energies the ISR represents an alternate, independent and complementary way to determine those cross sections with different systematic errors. Such a success story was possible thanks to dedicated work both from the theoreticians developing Monte Carlo generators like PHOKHARA and BABAYAGA@NLO which were fundamental to achieve sub percent precision, as well as from experimental groups using ISR not only to determine hadronic cross sections but also for spectroscopy studies. In that respect the KLOE measurements of hadronic cross section was a significant contribution to this field [2]:

- It proved the ISR as a working tool to precisely measure the hadronic cross sections;
- It confirmed the 3σ discrepancy between the Standard Model prediction for the g-2 and the measured value, and the disagreement with the SM prediction based on τ data. These two sets of data are now in agreement with each other, once the vector meson and $\rho - \gamma$ mixing effects are properly accounted for [47];
- It allowed a first direct measurement of the time-like complex running $\alpha(s)$. That procedure would be helpful to solve a problem in the treatment of the narrow resonances to a_μ and to the running of α [41];
- By the close contact and interactions with theorists it contributed to the inter-comparisons and refinements of the Monte Carlo codes [8].

Thanks to the worldwide (experimental and theoretical) efforts the accuracy of leading-order hadronic vacuum polarization contribution, a_μ^{HLO}, has improved to $\sim 4\times10^{-10}$. Although corresponding to a remarkable fractional uncertainty of 0.6%, it still constitutes the main uncertainty of the SM prediction to a_μ. New experiments at Fermilab and J-PARC aiming at a fourfold improved precision are underway [48, 49]. Together with a fourfold improved precision on the experimental side, an improvement of the LO hadronic contribution is highly desirable. An interesting possibility in this respect is a novel approach aiming to determine a_μ^{HLO} from a direct space-like measurement of $\alpha(t)$ as proposed in Refs. [50, 51] (see also [52]) recently.

[2] As underlined also by more than 600 citations of the corresponding articles [16, 21, 22, 30].

7 Acknoweldgements

The ISR measurement at KLOE was first of all a story of passion, friendship, committement and dedication. I would like to express my gratitude to the members of the KLOE/KLOE-2 Collaboration who made this possible and with whom I had the privilige to collaborate, particularly: P. Beltrame, F. Curciarello, V. De Leo, A. Denig, S. Di Falco, M. Incagli, W. Kluge, J. Lee-Franzini, P. Lukin, G. Mandaglio, S. Müller, F. Nguyen, A. Palladino, and B. Valeriani. Many friends and colleagues gave useful suggestions, help, criticism and support during the development of the analyses: A. Arbuzov, M. Benayoun, C.M. Carloni Calame, H. Czyż, S. Eidelman, G. Fedotovich, F. Ignatov, F. Jegerlehner, A. Keshavarzi, J.H. Kühn, G. Montagna, O. Nicrosini, L. Pancheri, M. Passera, F. Piccinini, G. Rodrigo, O. Shekhtostova, T. Teubner, and L. Trentadue. To all of them, and to the whole Radio Monte Carlo Working Group, goes my sincere gratitude and appreciation. Finally I would like to thank Rita Bertelli for her precious work and assistance during all these years.

References

[1] V. N. Baier, V. A. Khoze, ZHETF **48**(1965) 946, Yad. Fiz. **2** (1965) 287; V. N. Baier, V. A. Khoze, Sov. Phys. JETP **21** (1965) 629, 1145
[Zh. Eksp. Theor. Fiz. **48** (1965) 946; V. N. Baier and V. S. Fadin Phys. Lett. **27B** (1968) 223; G. Pancheri, Nuovo Cim. **A 60** (1969) 321; M. Greco, G. Pancheri, Y. N. Srivastava, Nucl. Phys. **B 101** (1975) 234; M. S. Chen, P. M. Zerwas,
Phys. Rev. **D 11** (1975) 58.

[2] A. B. Arbuzov et al., JHEP **9710**:006, 1997; A. B. Arbuzov, E. A. Kuraev, N. P. Merenkov, L. Trentadue, JHEP **9812**:009, 1998; M. Benayoun, S. I. Eidelman, V. N. Ivanchenko, Z. K. Silagadze, Mod. Phys. Lett. **A 14** (1999) 2605; S. Binner, J. H. Kühn, K. Melnikov, Phys. Lett. **B 459** (1999) 279.

[3] See for example W. Kluge, Nucl. Phys. Proc. Suppl. **181-182** (2008) 280; V. P. Druzhinin, S. I. Eidelman, S. I. Serednyakov and E. P. Solodov, Rev. Mod. Phys. **83** (2011) 1545.

[4] F. Campanario, H. Czyż, J. Gluza, M. Gunia, T. Riemann, G. Rodrigo and V. Yundin, JHEP **1402** (2014) 114; H. Czyż, M. Gunia and J. H. Kühn, JHEP **1308** (2013) 110; H. Czyż, A. Grzelinska, J. H. Kühn and G. Rodrigo, Eur. Phys. J. C **39** (2005) 411; H. Czyż, A. Grzelinska, J. H. Kühn and G. Rodrigo, Eur. Phys. J. C **33** (2004) 333; H. Czyż, A. Grzelinska, J. H. Kühn and G. Rodrigo, Eur. Phys. J. C **27** (2003) 563; J. H. Kühn and G. Rodrigo, Eur. Phys. J. C **25** (2002) 215; G. Rodrigo, H. Czyż, J. H. Kühn and M. Szopa, Eur. Phys. J. C **24** (2002) 71; G. Rodrigo, A. Gehrmann-De Ridder, M. Guilleaume and J. H. Kühn, Eur. Phys. J. C **22** (2001) 81.

[5] G. Balossini, C. Bignamini, C. M. C. Calame, G. Montagna, O. Nicrosini and F. Piccinini, Phys. Lett. B **663** (2008) 209; G. Balossini, C. M. Carloni Calame, G. Montagna, O. Nicrosini and F. Piccinini, Nucl. Phys. B **758** (2006) 227; C. M. Carloni Calame, Phys. Lett. B **520** (2001) 16; C. M. Carloni Calame, C. Lunardini, G. Montagna, O. Nicrosini and F. Piccinini, Nucl. Phys. B **584** (2000) 459.

[6] S. Jadach, W. Placzek and B. F. L. Ward, Phys. Lett. B **390** (1997) 298.

[7] A. B. Arbuzov, G. V. Fedotovich, F. V. Ignatov, E. A. Kuraev and A. L. Sibidanov, Eur. Phys. J. C **46** (2006) 689.

[8] S. Actis *et al.*, Eur. Phys. J. C **66** (2010) 585.

[9] S. Spagnolo, Eur. Phys. J. **C 6** (1999) 637.

[10] G. Cataldi, A. Denig, W. Kluge, G. Venanzoni, KLOE memo 195 (Aug. 1999);
Frascati Physics Series (2000), 569.

[11] BaBar, E. P. Solodov, SLAC, April 30 - May 2, 2001, hep-ex/0107027.
[12] Physics of the B factory book, Eur. Phys. J. C74 (2014) 3026.
[13] H. Nakazawa [Belle Collaboration], Nucl. Part. Phys. Proc. **260** (2015) 98.
[14] G. Pakhlova, Nucl. Phys. Proc. Suppl. **181-182** (2008) 117.
[15] M. Ablikim *et al.* [BESIII Collaboration], Phys. Lett. B **753** (2016) 629.
[16] A. Aloisio *et al.* [KLOE Collaboration], Phys. Lett. B **606** (2005) 12.
[17] G. W. Bennett *et al.* [Muon G-2 Collaboration], Phys. Rev. D **73** (2006) 072003. For a review see J. P. Miller, E. de Rafael, B. Lee Roberts, Rep. Prog. Phys. **70** (2007) 795.
[18] S. Eidelman and F. Jegerlehner, Z. Phys. C **67** (1995) 585
[19] F. Teubert, talk at ICHEP04 Conference, http://www.ihep.ac.cn/data/ichep04/ppt/plenary/p21-teubert-f.ppt
[20] M. Davier, S. Eidelman, A. Hocker and Z. Zhang, Eur. Phys. J. C **27** (2003) 497.
[21] F. Ambrosino *et al.* [KLOE Collaboration], Phys. Lett. B **670** (2009) 285
[22] F. Ambrosino *et al.* [KLOE Collaboration], Phys. Lett. B **700** (2011) 102.
[23] H. Czyż, A. Grzelinska and J. H. Kühn, Phys. Lett. B **611** (2005) 116.
[24] F. Ambrosino *et al.* [KLOE Collaboration], Phys. Lett. B **634** (2006) 148.
[25] G. Pancheri, O. Shekhovtsova and G. Venanzoni, J. Exp. Theor. Phys. **106** (2008) 470; G. Pancheri, O. Shekhovtsova and G. Venanzoni, Phys. Lett. B **642** (2006) 342; K. Melnikov, F. Nguyen, B. Valeriani and G. Venanzoni, Phys. Lett. B **477** (2000) 114.
[26] R. R. Akhmetshin et al. [CMD-2 Coll.], *Phys. Lett. B* **648** (2007) 28.
[27] R. R. Akhmetshin et al. [CMD-2 Coll.], *JETP Lett.* **84** (2006) 413.
[28] M. N. Achasov et al. [SND Coll.], *J. Exp. Theor. Phys.* **103** (2006) 380.
[29] B. Aubert *et al.* [BABAR Collaboration], Phys. Rev. Lett. **103** (2009) 231801.
[30] D. Babusci *et al.* [KLOE Collaboration], Phys. Lett. B **720** (2013) 336.
[31] F. Jegerlehner and A. Nyffeler, Phys. Rept. **477** (2009) 1.
[32] M. Davier, A. Hoecker, B. Malaescu and Z. Zhang, Eur. Phys. J. C **71** (2011) 1515 Erratum: [Eur. Phys. J. C **72** (2012) 1874].
[33] K. Hagiwara, R. Liao, A. D. Martin, D. Nomura and T. Teubner, J. Phys. G **38** (2011) 085003.
[34] F. Jegerlehner and R. Szafron, Eur. Phys. J. C **71** (2011) 1632.
[35] M. Benayoun, P. David, L. DelBuono and F. Jegerlehner, Eur. Phys. J. C **75** (2015) no.12, 613 doi:10.1140/epjc/s10052-015-3830-x.
[36] M. Davier, arXiv:1612.02743 [hep-ph].
[37] F. Jegerlehner, arXiv:1705.00263 [hep-ph].
[38] F. Curciarello, EPJ Web Conf. **118** (2016) 01008.
[39] A. Anastasi *et al.* [KLOE-2 Collaboration], Phys. Lett. B **767** (2017) 485.
[40] F. Jegerlehner, `alphaQED` package [version April 2012] http://www-com.physik.hu-berlin.de/~fjeger/alphaQED.tar.gz; see also F. Jegerlehner, Nuovo Cim. C **034S1** (2011) 31; Nucl. Phys. Proc. Suppl. **162** (2006) 22.
[41] F. Jegerlehner, EPJ Web Conf. **118** (2016) 01016.
[42] J. E. Augustin *et al.*, Phys. Rev. Lett. **30** (1973) 462.
[43] J. Bailey *et al.* [CERN Muon Storage Ring Collaboration], Phys. Lett. **67B** (1977) 225 [Phys. Lett. **68B** (1977) 191].
[44] J. Bailey *et al.* [CERN-Mainz-Daresbury Collaboration], Nucl. Phys. B **150** (1979) 1.
[45] F. Jegerlehner, "The anomalous magnetic moment of the muon," Springer Tracts Mod. Phys. **226** (2008) 1.
[46] K. A. Olive *et al.* [Particle Data Group], Chin. Phys. C **38** (2014) 090001.

[47] M. Benayoun, EPJ Web Conf. **118** (2016) 01001; M. Benayoun, P. David, L. DelBuono and F. Jegerlehner, Eur. Phys. J. C **73** (2013) 2453; M. Benayoun, P. David, L. DelBuono and F. Jegerlehner, Eur. Phys. J. C **72** (2012) 1848; F. Jegerlehner and R. Szafron, Eur. Phys. J. C **71** (2011) 1632; M. Benayoun, P. David, L. DelBuono and O. Leitner, Eur. Phys. J. C **68** (2010) 355; M. Benayoun, P. David, L. DelBuono and O. Leitner, Eur. Phys. J. C **65** (2010) 211; M. Benayoun, P. David, L. DelBuono, O. Leitner and H. B. O'Connell, Eur. Phys. J. C **55** (2008) 199.

[48] J. Grange et al. [Muon g-2 Collaboration], arXiv:1501.06858 [physics.ins-det].

[49] N. Saito [J-PARC g-2/EDM Collaboration], AIP Conf. Proc. **1467** (2012) 45.

[50] C. M. Carloni Calame, M. Passera, L. Trentadue and G. Venanzoni, Phys. Lett. B **746** (2015) 325.

[51] G. Abbiendi *et al.*, Eur. Phys. J. C **77** (2017) no.3, 139.

[52] L. Trentadue, these proceedings.

Muon $g-2$ theory: The hadronic part

Fred Jegerlehner[1,2,★]

[1]*Deutsches Elektronen–Synchrotron (DESY), Platanenallee 6, D–15738 Zeuthen, Germany*
[2]*Humboldt–Universität zu Berlin, Institut für Physik, Newtonstrasse 15, D–12489 Berlin, Germany*

Abstract. I present a status report of the hadronic vacuum polarization effects for the muon $g-2$, to be considered as an update of [1]. The update concerns recent new inclusive R measurements from KEDR in the energy range 1.84 to 3.72 GeV. For the leading order contributions I find $a_\mu^{\rm had(1)} = (688.07\pm4.14)[688.77\pm3.38]\times10^{-10}$ based on e^+e^-data [incl. τ data], $a_\mu^{\rm had(2)} = (-9.93\pm0.07)\times10^{-10}$ (NLO) and $a_\mu^{\rm had(3)} = (1.22\pm0.01)\times10^{-10}$ (NNLO). Collecting recent progress in the hadronic light-by-light scattering I adopt π^0, η, η' [95 $\pm$ 12] + axial–vector [8$\pm$ 3] + scalar [$-6\pm$ 1] + π, K loops [-20 ± 5] + quark loops [22$\pm$ 4] + tensor [1 $\pm$ 0] + NLO [3 $\pm$ 2] which yields $a_\mu^{(6)}(\mathrm{lbl, had}) = (103 \pm 29) \times 10^{-11}$. With these updates I find $a_\mu^{\rm exp} - a_\mu^{\rm the} = (31.3 \pm 7.7) \times 10^{-10}$ a 4.1 σ deviation. Recent lattice QCD results and future prospects to improve hadronic contributions are discussed.

1 Overview: hadronic effects in $g-2$.

This review of the hadronic vacuum polarization (HVP) contributions to the muon $g-2$ is to be considered as a complement to the theory reviews by Marc Knecht and Massimiliano Procura which focus on the hadronic light-by-light (HLbL) part and the reviews on hadronic cross sections by Graziano Venanzoni, Simon Eidelman and Achim Denig in these Proceedings.

The present experimental muon $g-2$ result from Brookhaven (BNL) $a_\mu^{\rm exp} = (11\,659\,209.1 \pm 5.4 \pm 3.3[6.3]) \times 10^{-10}$ [2] soon will be improved by the new muon $g-2$ experiments at Fermilab and J-PARC. The Fermilab experiment will be able to reduce the error by a factor 4, the J-PARC experiment will provide an important cross check with a very different technique [3]. It means that the new muon $g-2$ experiments are expected to establish a possible new physics contribution at the level $\Delta a_\mu = a_\mu^{\rm exp} - a_\mu^{\rm the} = 6.7\,\sigma$ provided theory remains as it is today and the central value does not move significantly. If we achieve a reduction of the hadronic uncertainty by factor 2 we would arrive at $\Delta a_\mu = 11.6\,\sigma$. That's what we hope to achieve. Figure 1 illustrates the present status and what has been achieved so far. The present results $a_\mu^{\rm HVP\ LO} = (6888 \pm 34) \times 10^{-11}$ amounts to +59.09 ±0.30 ppm, which poses the major challenge. The subleading results $a_\mu^{\rm HVP\ NLO} = (-99.3 \pm 0.7) \times 10^{-11}$ and $a_\mu^{\rm HVP\ NNLO} = (12.2 \pm 0.1) \times 10^{-11}$ although relevant will be known well enough. These number also compare with the well established weak $a_\mu^{\rm EW} = (154\pm1)\times10^{-11}$ at 1.3 ± 0.0 ppm and the problematic HLbL estimated to contribute $a_\mu^{\rm HLbL} = (103 \pm 29\ [105 \pm 26]) \times 10^{-11}$, which is representing a +0.90 ±0.25 ppm effect.

★e-mail: fjeger@physik.hu-berlin.de

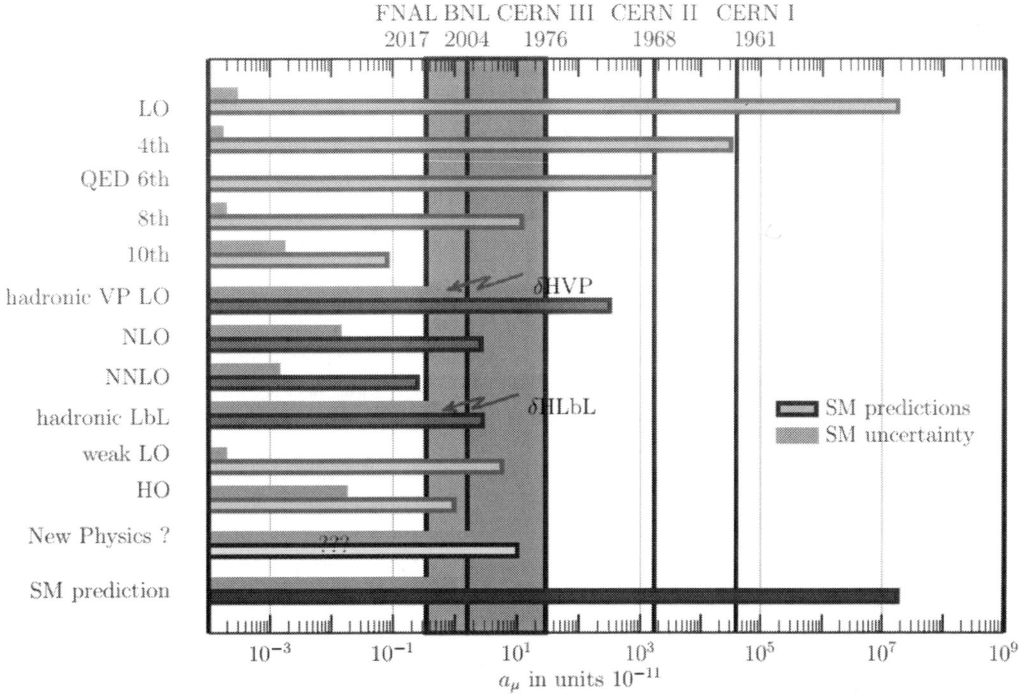

Figure 1. Past and future $g-2$ experiments testing various contributions. As New Physics ? we display the deviation $(a_\mu^{\rm exp} - a_\mu^{\rm the})/a_\mu^{\rm exp}$. Arrows point to what is limiting theory precision presently: the Hadronic Vacuum Polarization (HVP) and Hadronic Light-by-Light (HLbL) contributions.

Virtual effect form low energy hadronic excitations are the standard problem in electroweak precision physics. At a certain level of precision predictions are hampered by non-perturbative effects, which technically are not under desirable control on the theory side. For the muon $g-2$ the leading hadronic effects are related to the diagrams in figure 2 and concern
(a) Hadronic vacuum polarization (HVP) of order $O(\alpha^2)$, $O(\alpha^3)$,
(b) Hadronic light-by-light scattering (HLbL) of order $O(\alpha^3)$,
(c) Hadronic effects in 2-loop hadronic electroweak (HEW) corrections of order sub–$O(\alpha G_F m_\mu^2)$.
Light quark loops appear as non-perturbative hadronic "blobs". The evaluation of the corresponding non-perturbative effects relies on hadron production data in conjunction with Dispersion Relations (DR), or on low energy effective modeling by the Resonance Lagrangian Approach (RLA), specifically by the Hidden Local Symmetry (HLS) model [4], or the Extended Nambu–Jona-Lasinio (ENJL) model [5], large–N_c QCD inspired methods [6] and on lattice QCD. Different strategies apply for the different kinds of contributions:

(a) HVP one evaluates via a dispersion integral over $e^+e^- \to$ hadrons data. Here 1 independent amplitude is to be determined by one specific data set. Global fits based on the RLA (like HLS) allow to improve the data-driven evaluations [7]. Lattice QCD is the ultimate tool to get QCD predictions in future.

(b) HLbL so far has been evaluated by modeling via the Resonance Lagrangian Approach (RLA) (chiral perturbation theory (CHPT) extended by vector meson dominance (VMD) in accord with chi-

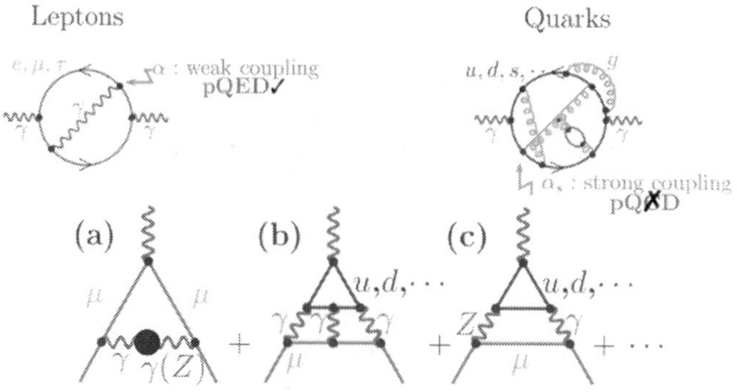

Figure 2. In the upper panel we compare leptonic with hadronic vacuum polarization effects. The lower panel illustrates the three classes of $g-2$ contributions exhibiting substantial hadronic corrections.

ral structure of QCD) or by large–N_c inspired methods and operator product expansions (OPE). A data driven approach based on dispersion relations [8] is attempting to exploit $\gamma\gamma \to$ hadrons – data systematically (here 28 independent amplitudes are to be determined by as many independent data sets, fortunately not all are equally important numerically). Also in this case lattice QCD for me is the ultimate approach, although tough to be achieved with limited computing resources.

(c) HEW corrections due to quark triangle diagrams: since triple vector amplitudes vanish $VVV = 0$ by Furry's theorem only VVA (of $f\bar{f}Z$ -vertex) contributes. Thus it is ruled by the ABJ anomaly, which is perturbative and non-perturbative simultaneously, i.e. the leading effects are calculable. The anomaly cancellation condition intimately relates quark and lepton contributions and the potentially large leading corrections cancel [9–11] such that hadronic corrections are well under control.

2 Evaluation of the leading order $a_\mu^{\rm had}$

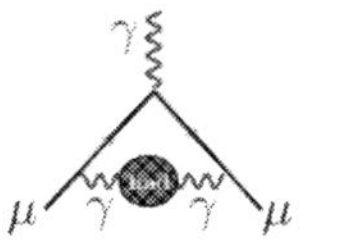

The hadronic contribution to the vacuum polarization can be evaluated, with the help of dispersion relations, from the energy scan of the ratio $R_\gamma(s) \equiv \sigma^{(0)}(e^+e^- \to \gamma^* \to \text{hadrons})/\frac{4\pi\alpha^2}{3s}$ which can be measured up to some energy $E_{\rm cut}$ above which we can safely use perturbative QCD (pQCD) thanks to asymptotic freedom of QCD. We apply pQCD from 5.2 GeV to 9.46 GeV and above 11.5 GeV (see figure 5 below). Note that the DR requires the undressed (bare) cross–section $\sigma^{(0)}(e^+e^- \to \gamma^* \to \text{hadrons}) = \sigma(e^+e^- \to \gamma^* \to \text{hadrons})\,|\alpha(0)/\alpha(s)|^2$. The lowest order (LO) VP contribution is given by

$$a_\mu^{\rm had} = \left(\frac{\alpha m_\mu}{3\pi}\right)^2 \left(\int\limits_{m_{\pi^0}^2}^{E_{\rm cut}^2} ds\, \frac{R_\gamma^{\rm data}(s)\, \hat{K}(s)}{s^2} + \int\limits_{E_{\rm cut}^2}^{\infty} ds\, \frac{R_\gamma^{\rm pQCD}(s)\, \hat{K}(s)}{s^2} \right), \qquad (1)$$

where $\hat{K}(s)$ is a known kernel function growing form $0.63\cdots$ at the $2m_\pi$ threshold to 1 as $s \to \infty$. The integral is dominated by the ρ resonance peak shown in figure 3. The experimental errors imply the dominating theoretical uncertainties. As a result I obtain

$$a_\mu^{\rm had(1)} = (688.07 \pm 4.14)[688.77 \pm 3.38]\ 10^{-10}\ ;\ \ e^+e^- - \text{data based [incl. } \tau] \,. \qquad (2)$$

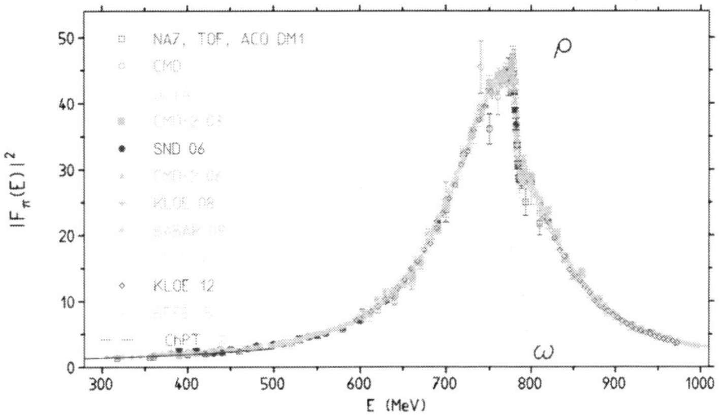

Figure 3. A compilation of the modulus square of the pion form factor in the ρ meson region, which yields about 75% of a_μ^{had}. The corresponding $R(s)$ is $R(s) = \frac{1}{4}\beta_\pi^3\,|F_\pi^{(0)}(s)|^2$, $\beta_\pi = \sqrt{1-4m_\pi^2/s}$ is the pion velocity.

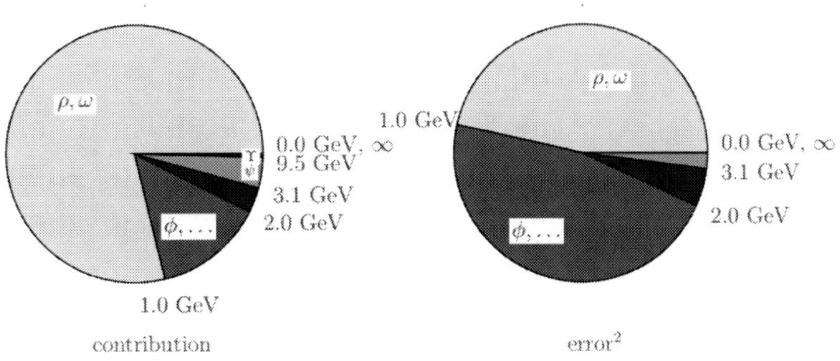

Figure 4. Distribution of contributions and error squares from different energy regions.

Figure 4 shows the distribution of contributions and errors between different energy ranges. One of the main issues is $R_\gamma(s)$ in the region from 1.2 GeV to 2.0 GeV, where more than 30 exclusive channels must be measured and although it contributes about 20% only of the total it contributes about 50% of the uncertainty. In the low energy region, which is particularly important for the dispersive evaluation of the hadronic contribution to the muon $g-2$, data have improved dramatically in the past decade for the dominant $e^+e^- \to \pi^+\pi^-$ channel (CMD-2 [12], SND/Novosibirsk [13], KLOE/Frascati [14–16], BaBar/SLAC [17], BES-III/Beijing [18]) and the statistical errors are a minor problem now. Similarly, the important region between 1.2 GeV to 2.4 GeV has been improved a lot by the BaBar exclusive channel measurements in the ISR mode [19–22]. Recent data sets collected are: $e^+e^- \to 3(\pi^+\pi^-)$, $e^+e^- \to \bar{p}p$ and $e^+e^- \to K_S^0K_L^0, K^+K^-$ from CMD-3 [23, 24], and $e^+e^- \to \bar{n}n$, $e^+e^- \to \eta\pi^+\pi^-$, $e^+e^- \to \pi^0\gamma$, $e^+e^- \to \omega\eta\pi^0$, $e^+e^- \to \omega\eta$, $e^+e^- \to K^+K^-$ and $e^+e^- \to \omega\pi^0 \to \pi^0\pi^0\gamma$ from SND [25–27].

Above 2 GeV fairly accurate BES-II data [28] are available. Recently, a new inclusive determination of $R_\gamma(s)$ in the range 1.84 to 3.72 GeV has been obtained with the KEDR detector at Novosibirsk [29] (see figure 5). A big step in improving low energy cross section measurements has been

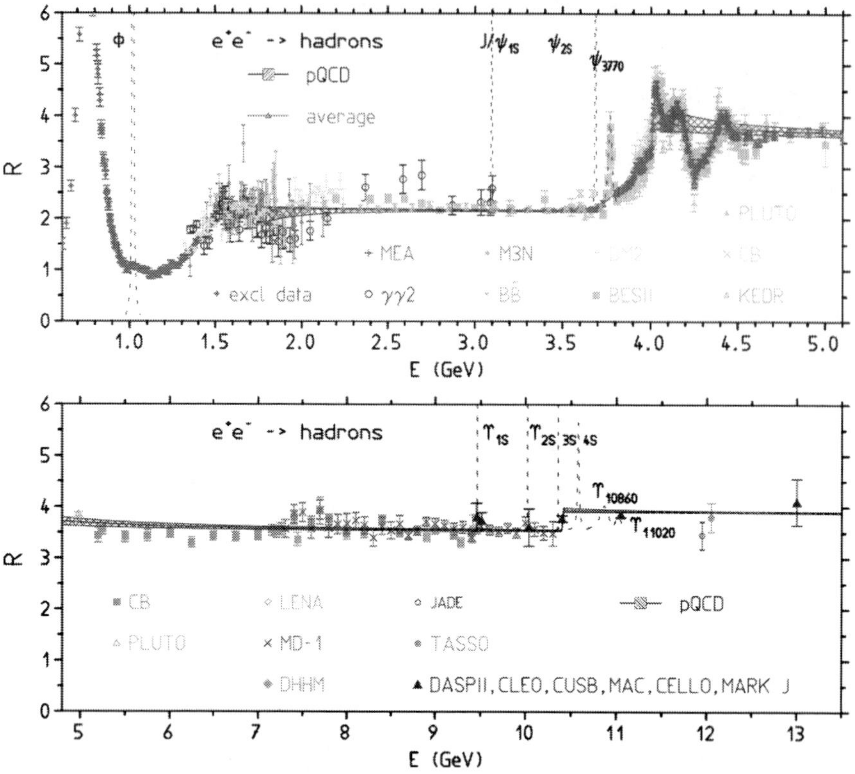

Figure 5. Compilation of the $R(s)$ data. New are the KEDR data between 1.84 and 3.72 GeV. Perturbative QCD predictions are also shown.

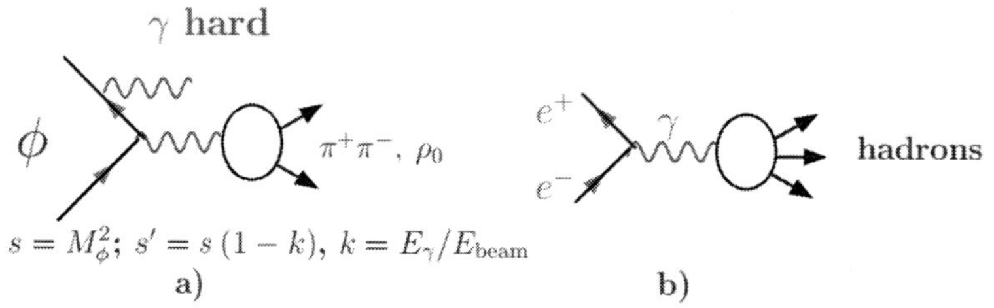

Figure 6. a) Initial state radiation (ISR) on resonance, b) Standard energy scan by tuning the beam energy.

possible with the radiative return or Initial State Radiation (ISR) method figure 6 which has been pioneered by the KLOE Collaboration, followed by BaBar and BES3 experiments. Recent new experimental input for HVP has been obtained by CMD-3 and SND at VEPP-2000 via energy scan and by BESIII at PEPC in the ISR setup (see Contributions by G. Venanzoni, S. Eidelman and A. Denig).

3 NLO and NNLO HVP effects

The next-to-leading order (NLO) HVP is represented by diagrams in figure 7. With kernels from [30],

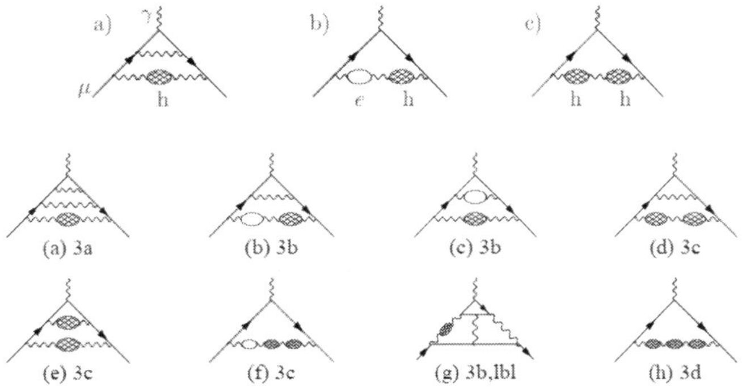

Figure 7. Feynman diagrams with hadronic insertions at NLO (top row) and NNLO.

the results of an updated evaluation are presented in table 1. The next-to-next leading order (NNLO)

Table 1. NLO contributions diagrams a) - c) (in units 10^{-11})

$a_\mu^{(2a)}$	$a_\mu^{(2b)}$	$a_\mu^{(2c)}$	$a_\mu^{\text{had}(2)}$
-206.13(1.30)	103.49(0.63)	3.37(0.05)	-99.27 (0.67)

Table 2. NNLO contributions diagrams (a) - (h) (in units 10^{-11})

$a_\mu^{(3a)}$	$a_\mu^{(3b)}$	$a_\mu^{(3b,\text{lbl})}$	$a_\mu^{(3c)}$	$a_\mu^{(3d)}$	$a_\mu^{\text{had}(3)}$	Ref.
8.0	−4.1	9.1	−0.6	0.005	12.4(1)	[31]
7.834 (61)	−4.033 (28)	9.005 (63)	−0.569 (5)	0.00518 (12)	12.24 (10)	[1]

contributions have been calculated recently [31]. Diagrams are shown in figure 7 and corresponding contributions evaluated with kernels from [31] are listed in table 2.

4 News on VP subtraction

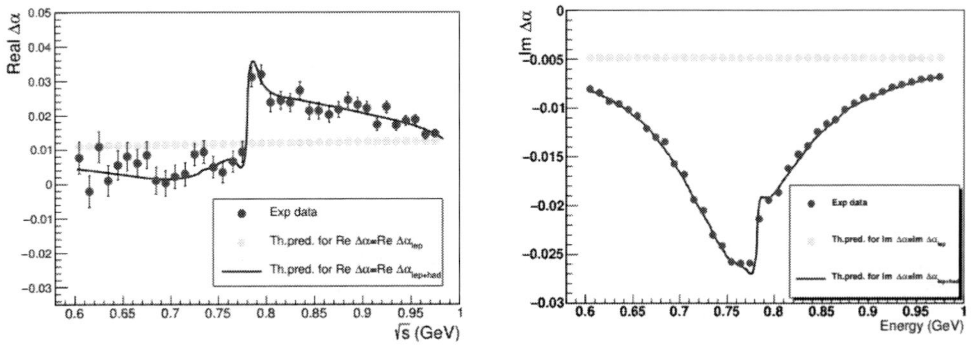

Figure 8. Real and imaginary part of the complex shift $\Delta\alpha$ of the fine structure constant as measured by KLOE. Reprinted from [33] in Phys. Lett. B `doi:10.1016/j.physletb.2016.12.016`

The first direct measurement of the timelike complex VP function in the ρ resonance region by KLOE [33] (see figure 8) nicely confirms dispersion relation calculation and demonstrates the importance of including the imaginary part in vacuum polarization subtraction in obtaining the undressed $\sigma^{(0)}(s)$ version of the physical hadronic cross-sections $\sigma(s)$. The complex running fine structure constant $\alpha(s) = \frac{\alpha(0)}{1-\Delta\alpha(s)}$ is defined in terms of the complex shift $\Delta\alpha(s) = -[\Pi'_\gamma(s) - \Pi'_\gamma(0)]$. Measuring $\left|\frac{\alpha(s)}{\alpha(0)}\right|^2 = \frac{\sigma(e^+e^-\to\mu^+\mu^-)}{\sigma(e^+e^-\to\mu^+\mu^-)_{\rm pt}}$ as well as $R(s) = \frac{\sigma(e^+e^-\to\pi^+\pi^-)}{\sigma(e^+e^-\to\mu^+\mu^-)}$, which determines $\mathrm{Im}\,\alpha(\mathrm{s}) = -\frac{\alpha}{3}\,\mathrm{R}(\mathrm{s})$, and knowing the modulus $|\alpha(s)|$ one can extract $\mathrm{Re}\,\alpha(\mathrm{s})$ as well (see G. Venanzoni's Contribution fort details).

The imaginary parts in the perturbative regions usually are small relative to the leading logarithms which govern the running couplings (renormalization group approach). In the hadronic shift however, resonances are accompanied by imaginary parts which may be huge in particular near resonances which can decay via OZI suppressed channels only (see Sect. 5 of [1]).

5 Low energy effective Lagrangian theory

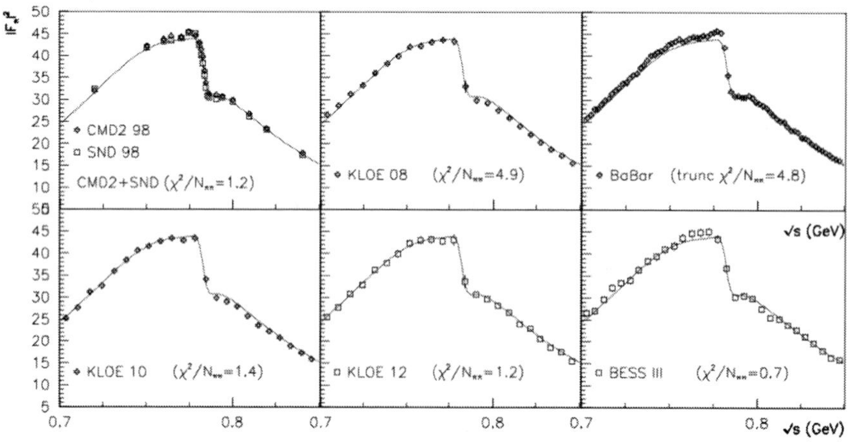

Figure 9. Comparing the τ+PDG prediction (red curve) of the pion form factor in e^+e^- annihilation in the $\rho-\omega$ interference region. Reprinted from [7], with kind permission of The European Physical Journal (EPJ).

A low energy effective field theory description of hadronic data requires an extension of CHPT towards higher energies, which mainly requires to include spin 1 resonances $\rho, \omega, \phi, \cdots$ in accord with the symmetries of QCD. Principles to be included are the chiral structure of QCD, VMD and electromagnetic gauge invariance. The spin 1 resonances are important in the evaluation of HVP as well as of HLbL effects. One possible implementation is the HLS model, which for what concerns HVP can also be seen as a generalized Gounaris-Sakurai model. In the neutral channel $e^+e^- \to$ hadrons $\gamma, \rho^0, \omega, \phi$ mixing makes the channel rather complicated, while in the charged channel of $\tau \to \nu_\tau$hadrons [34–38] is much simpler as the $\rho^\pm$ do not mix with other hadrons. It is thus tempting to start with the isospin rotated $\tau^\pm \to \nu_\tau\pi^\pm\pi^0$ decay spectra and supplement them with appropriate isospin breaking and mixing effects to predict $e^+e^- \to \pi^+\pi^-$, with the result shown in figure 9. It shows that there is no τ vs. e^+e^- conflict and actually simultaneous fits allows one to reduce uncertainties

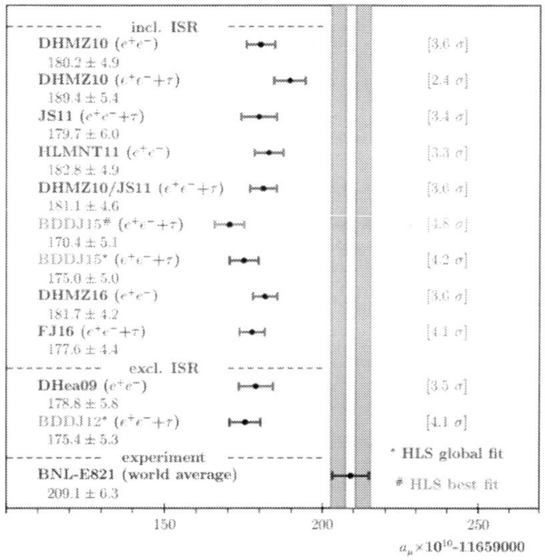

Figure 10. Dependence of a_μ predictions on recent evaluations of $a_\mu^{\rm had,LO}$. The HLS best fit BDDJ15# (NSK+KLOE10+KLOE12) does not include BaBar $\pi\pi$ data [39], while BDDJ15* does. JS11 [40], FJ16 [1] is updated and includes the BES III and KEDR data. Further points are BDDJ12 [7], DHMZ10 [41], DHMZ16 [22, 44], HLMNT11 [43] and DHea09 [42]. The DHMZ10 ($e^+e^-+\tau$) result is not including the $\rho-\gamma$ mixing correction, i.e. it misses important isospin breaking effects. In contrast, DHMZ10/JS11 is obtained by including this correction, which brings the point into much better agreement with standard analyses based on e^+e^- data alone, as for example the DHMZ10 (e^+e^-) result. (see also [21, 45]). Note: results depend on which value has been taken for HLbL. JS11 and BDDJ13 includes $116(39)\times10^{-11}$ [JN] [46], DHea09, DHMZ10, HLMNT11 and BDDJ12 use $105(26)\times10^{-11}$ [PdRV] [47], while FJ16 includes an updated $103(29)\times10^{-11}$.

of HVP by using indirect constraints. The global fit strategy followed in [7] takes into account data below $E_0 = 1.05$GeV (just above the ϕ) to constrain the effective Lagrangian couplings. Used are 45 different data sets, 6 annihilation channels and 10 partial width decays. The effective theory then allows us to predict cross sections for the channels $\pi^+\pi^-, \pi^0\gamma, \eta\gamma, \eta'\gamma, \pi^0\pi^+\pi^-, K^+K^-, K^0\bar{K}^0$, which account for 83.4% of $a_\mu^{\rm had}$. The missing channels $4\pi, 5\pi, 6\pi, \eta\pi\pi, \omega\pi$ and the higher energy tail $E > E_0$ is evaluated using data directly and pQCD for the perturbative region and tail. All mixing effects, as $\gamma\rho$ -mixing, $\rho\omega$ -mixing, $\cdots$, as well as the decay branching fractions are dynamically generated by including self-energy effects of the spin 1 mesons. One thus is taking into account proper phase space, energy dependent widths etc. Such fit strategy is able to shed light on incompatibilities in the data, e.g. KLOE vs BaBar, by comparing the fit qualities, but also reveals the compatibility of τ–decay spectra with e^+e^-–data after accounting for the mixing effects like including $\gamma-\rho^0$ mixing. HLS estimates are included in table 10 together with other recent results.

6 HVP from lattice QCD (following H. Wittig at LATTICE 2016)

The need for ab initio calculation of $a_\mu^{\rm had}$ is well motivated: – the problems to determine non-perturbative contributions to the muon $g-2$ from experimental data at sufficient precision persists and is not easy to improve, – a model–independent extension of CHPT to the relevant energies ranges up to 2 GeV is missing, while the new experiments E989 FNAL and E34 J-PARC require an improvement of the hadronic uncertainties by a factor two to four.

The hope is that LQCD can deliver estimates of accuracy

$$\delta a_\mu^{\mathrm{HVP}}/a_\mu^{\mathrm{HVP}} < 0.5\% \;, \quad \delta a_\mu^{\mathrm{HLbL}}/a_\mu^{\mathrm{HLbL}} \lesssim 10\% \tag{3}$$

in the coming years.

Primary object for getting HVP in LQCD is the e.m. current correlator in configuration space

$$\langle J_\mu(\vec{x},t)\, J_\nu(\vec{0},0)\rangle \;; \quad J_\mu = \frac{2}{3}\bar{u}\gamma_\mu u - \frac{1}{3}\bar{d}\gamma_\mu d - \frac{1}{3}\bar{s}\gamma_\mu s + \cdots \tag{4}$$

In principle, a Fourier transform

$$\Pi_{\mu\nu}(Q) = \int \mathrm{d}^4x e^{\mathrm{i}\,Qx} \langle J_\mu(x)\, J_\nu(0)\rangle = \left(Q_\mu Q_\nu - \delta_{\mu\nu}\, Q^2\right) \Pi(Q^2) \tag{5}$$

yields the vacuum polarization function $\Pi(Q^2)$ needed to calculate

$$a_\mu^{\mathrm{HVP}} = 4\alpha^2 \int_0^\infty \mathrm{d}Q^2\, f(Q^2) \left\{\Pi(Q^2) - \Pi(0)\right\} . \tag{6}$$

The integration kernel in this representation is

$$f(Q^2) = w(Q^2/m_\mu^2)/Q^2 \;; \quad w(r) = \frac{16}{r^2\left(1+\sqrt{1+4/r}\right)^4 \sqrt{1+4/r}} . \tag{7}$$

As LQCD per se has to work on a lattice in a finite box, momenta are quantized $Q_{\min} = 2\pi/L$, where

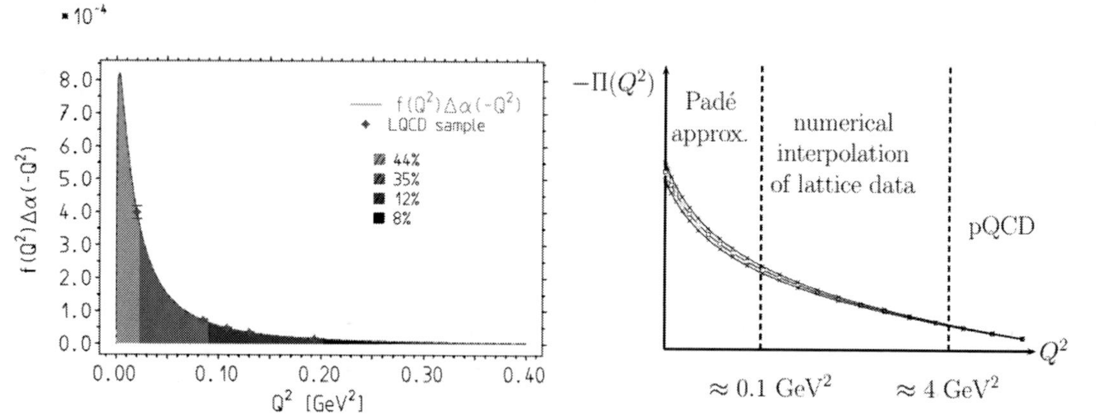

Figure 11. Left: the integrand of (6) as a function of Q^2. Also shown are ranges between Q_i = 0.00, 0.15, 0.30, 0.45 and 1.0GeV and their percent contribution to a_μ^{had} and a possible "LQCD sample". Right: typical ranges for lattice data and their extrapolation to low Q^2.

L is the lattice box length. Therefore, approaching low momenta $Q_{\min} \to 0 \Leftrightarrow L \to \infty$ requires a sufficiently large volume. Present state of the art calculations reach $Q_{\min} = 2\pi/L$ with $m_\pi aL \gtrsim 4$ for $m_\pi \sim$ 200MeV, such that $Q_{\min} \sim$ 314MeV. This means that about 44% of the low Q contribution to a_μ^{had} is not covered by data yet. Typically, lattice data are available for $Q^2 > (2\pi/L)^2$, which one has to extrapolate to $Q^2 = 0$ by VMD type modeling [48] or via Padé's [49] or analytic continuation [51]. The method requires a reliable estimate of the bare $\Pi(0)$ (see e.g. [50]). In order to reach the required accuracy one needs LQCD data down to $Q^2_{\min} \approx 0.1\mathrm{GeV}^2$. [51–55] Some recent results are collected in figure 12.

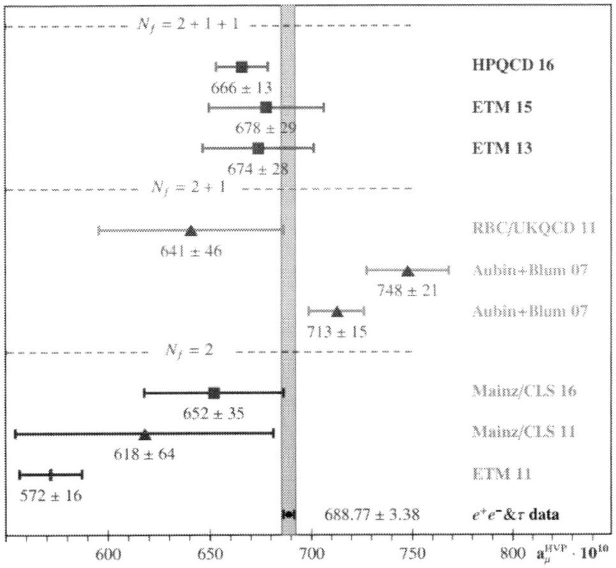

Figure 12. Summary of recent LQCD results for the leading order a_μ^{HVP}, in units 10^{-10}. Labels: ■ marks u, d, s, c, ▲ u, d, s and **I** u, d contributions. Individual flavor contributions from light (u, d) amount to about 90%, strange about 8% and charm about 2%. Results shown are from HPQCD 16 [56], ETM 15 [57], ETM 13 [58], RBC/UKQCD 11 [52], Aubin+Blum 07 [59], Mainz/CLS 16 [60], Mainz/CLS 11 [61] and ETM 11 [48]. The vertical band shows the e^+e^- data driven DR estimate (2).

7 Alternative method to get a_μ^{had}: using $\alpha(t = -Q^2)$ measured via t–channel exchange processes.

A promising alternative method to determine a_μ^{had} is possible by a dedicated measurement of $\alpha(t)$ at spacelike momentum transfer as advocated in [62] and [63]. Given $\alpha(-Q^2)$ and the fact that the leptonic contribution is well under control in perturbation theory one can extract the hadronic shift

$$\Delta\alpha^{had}(-Q^2) = 1 - \frac{\alpha}{\alpha(-Q^2)} - \Delta\alpha^{lep}(-Q^2) \tag{8}$$

and determine a_μ^{had} via the representation

$$a_\mu^{had} = \frac{\alpha}{\pi} \int_0^1 \mathrm{dx}\,(1-\mathrm{x})\,\Delta\alpha^{had}\left(-\mathrm{Q}^2(\mathrm{x})\right) \tag{9}$$

where $Q^2(x) \equiv \frac{x^2}{1-x} m_\mu^2$ is the spacelike square momentum–transfer. In the Euclidean region the integrand is highly peaked around half of the ρ meson mass scale (see figure 13). The method is very different from the standard approach based on (1): radiative corrections are very different (much simpler) as no hadronic final states need to be understood, no VP subtraction is to be performed, no exclusive channel collection etc. So, even a 1% level measurement can provide important independent information. This in view of the problem to get accurate hadronic total cross–section in the range

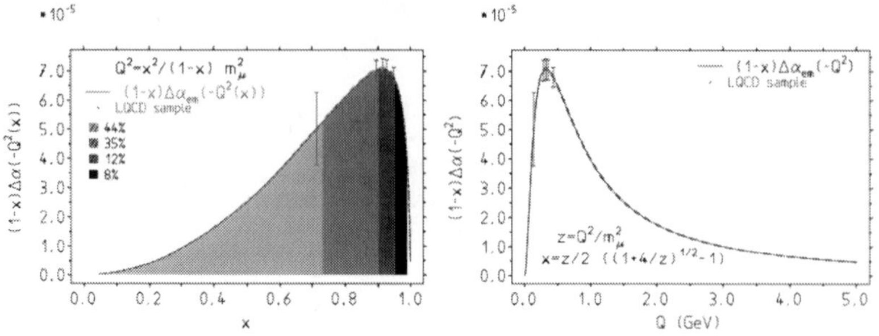

Figure 13. The integrand of the $a_\mu^{\rm had}$ integral (9) as functions of x and Q is strongly peaked at about 330MeV. Legend as in figure 11.

between 1 and 2 GeV and possible unsettled problems (non-convergence of the Dyson resummation near OZI suppressed resonances) in VP subtraction as addressed recently in Sect. 5 of [1].

The possible processes to measure $\alpha(t)$ are Bhabha scattering $e^+(p_+)\ e^-(p_-) \to e^+(p'_+)\ e^-(p'_-)$ or muon electron scattering (see figure 14). The Bhabha process has two tree level diagrams a $t-$ and

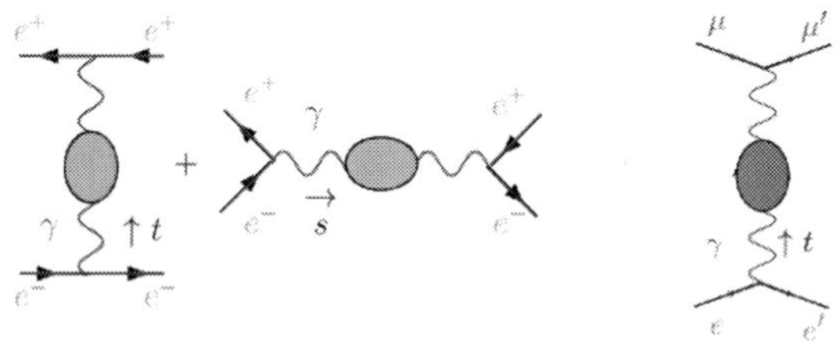

Figure 14. Measuring the running charge in the spacelike regime. Left: VP dressed tree level Bhabha scattering in QED, for small t (small angels) the s channel is suppressed. Right: getting $\alpha(t)$ from the $\mu^- e^- \to \mu^- e^-$ process.

a s–channel one. With the positive c.m. energy square $s = (p^+ + p^-)^2$ and the negative momentum transfer square $t = (p_- - p'_-)^2 = -\frac{1}{2}\,(s - 4m_e^2)\,(1 - \cos\theta)$, θ the e^- scattering angle, there are two very different scales involved which helps to isolate the t channel of interest. The VP dressed lowest order cross–section is

$$\frac{d\sigma}{d\cos\Theta} = \frac{s}{48\pi}\sum\nolimits_{ik}|A_{ik}|^2\,, \tag{10}$$

where A_{ik} are tree level helicity amplitudes, i, k =L,R denote left– and right–handed electrons.

The dressed transition amplitudes in the massless limit ($m_e \approx 0$) read

$$|A_{\rm LL,RR}|^2 = \frac{3}{8}\,(1+\cos\theta)^2\left|\frac{e^2(s)}{s}+\frac{e^2(t)}{t}\right|^2 \;;\; |A_{\rm LR,RL}|^2 = \frac{3}{8}\,(1-\cos\theta)^2\left|\frac{e^2(s)}{s}+\frac{e^2(t)}{t}\right|^2\,. \tag{11}$$

Preferably one uses small angle Bhabha scattering (small $|t|$) as a normalizing process which is dominated by the t–channel $\sim 1/t$, however, detecting electrons and positrons along the beam axis often has its technical limitations. Care also is needed concerning the ISR corrections because cuts for the Bhabha process ($e^+e^- \to e^+e^-$) typically are different from the ones applied to $e^+e^- \to$ hadrons. Usually, experiments have included corresponding uncertainties in their systematic errors, if they have not been explicitly accounted for by applying appropriate radiative corrections. For details I refer to [62] and the Contribution by Luca Trentadue.

While the Bhabha process requires to sort out the s channel from the t channel, the pure t channel reaction of $\mu^- e^-$ scattering $\mu^-(p_-)\ e^-(q_-) \to \mu^-(p'_-)\ e^-(q'_-)$ provides a much simpler setup and could be realized as a fixed target experiment [63] at existing facilities. The leading order cross–section in this case has the simple form

$$\frac{\mathrm{d}\sigma^{\mathrm{unpol.}}_{\mu^- e^- \to \mu^- e^-}}{\mathrm{dt}} = 4\pi\,\alpha(t)^2\,\frac{1}{\lambda(s, m_e^2, m_\mu^2)}\left\{\frac{\left(s - m_\mu^2 - m_e^2\right)^2}{t^2} + \frac{s}{t} + \frac{1}{2}\right\}, \tag{12}$$

exhibiting the effective charge as an overall factor. For details see Luca Trentadue's Contribution. Such an experiment would provide data for the Euclidean electromagnetic current correlator $\Pi'_\gamma(Q^2) - \Pi'_\gamma(0) = -\Delta\alpha^{\mathrm{had}}(-Q^2) = \frac{\alpha}{\alpha(-Q^2)} + \Delta\alpha^{\mathrm{lep}}(-Q^2) - 1$ and would allow for a directly check of lattice QCD data. In addition one could determine $\Delta\alpha^{\mathrm{had}}\left(-Q^2\right)$ at $Q \approx 2.5\mathrm{GeV}$ by this method (one single number!) as the non-perturbative part of $\Delta\alpha^{\mathrm{had}}\left(M_Z^2\right) = \Delta\alpha^{\mathrm{had}}\left(-Q^2\right) + \mathrm{pert.}$ when evaluated in "Adler function" approach advocated in [64].

8 Theory confronts experiment

Table 3. Standard model theory and experiment comparison [in units 10^{-10}].

Contribution	Value	Error	Reference
QED incl. 4-loops+5-loops	11 658 471 . 8851	0 . 036	Remiddi, Kinoshita et al.
Leading hadronic vac. pol.	688 . 77	3 . 38	data-driven $e^+e^- + \tau$
Subleading hadronic vac. pol.	-9 . 927	0 . 072	2016 update
NNLO hadronic vac. pol.	1 . 224	0 . 010	[31]
Hadronic light–by–light	10 . 34	2 . 88	[46, 69]
Weak incl. 2-loops	15 . 36	0 . 11	[11, 70]
Theory	11 659 177 . 6	4 . 4	–
Experiment	11 659 209 . 1	6 . 3	[2] updated
Exp.- The. 4.1 standard deviations	31 . 3	7 . 7	–

Table 3 summarizes the present status of the SM prediction for a_μ in comparison with the experimental value [2]. For a recent update of the weak contribution see [70]. As an estimate based on [4–6, 10, 46, 65, 68, 69] we adopt π^0, η, η' [95 ± 12] + axial–vector [8 ± 3] + scalar [−6 ± 1] + π, K loops [−20 ± 5] + quark loops [22 ± 4] + tensor [1 ± 0] + NLO [3 ± 2] which yields

$$a_\mu^{(6)}(\mathrm{lbl, had}) = (103 \pm 29) \times 10^{-11}\,. \tag{13}$$

The result differs little from the "agreed" value $(105 \pm 26) \times 10^{-11}$ presented in [47] and $(116 \pm 39) \times 10^{-11}$ estimated in [46]. Both included a wrong, too large, Landau-Yang theorem violating axial–vector contribution from [10], correcting for this we obtain our reduced value relative to [46].

The following tabular collects recent new/updated evaluations:

New contribution	Reference	$\Delta a_\mu \cdot 10^{11}$		
NNLO HVP	Kurz et al. 2014	12.4	±	0.1
NLO HLbL	Colangelo et al. 2014	3	±	2
New axial exchange HLbL	Pauk, Vanderhaeghen [66], FJ14 [1, 67]	7.55	±	2.71
Tensor exchange HLbL	Pauk, Vanderhaeghen 2014	1.1	±	0.1
New π^0 exchange HLbL	$\pi^0\gamma^*\gamma^*$ constraint from LQCD [68]	64.68	±	12.40
...				...
Old axial exchange HLbL	Melnikov, Vainshtein 2004	22	±	5
Old π^0 exchange HLbL	JN [46]	72	±	12
Total change		−5.6	±	12.85 [← 13]

The uncertainty from these contributions remains unchanged, while the central value is shifted downwards by almost 1 SD.

Possible interpretations of the 4 σ deviation: new physics?, a statistical fluctuation?, underestimated uncertainties (experimental, theoretical)? Do experiments measure what theoreticians calculate? The challenge for the future is to keep up with the future experiments, which will improve the experimental accuracy from $\delta a_\mu^{\rm exp} = 63 \times 10^{-11}$ [±0.54 ppm] at present to $\delta a_\mu^{\rm exp} = 16 \times 10^{-11}$ [±0.14 ppm] the next years. Next generation experiments require a factor 4 reduction of the uncertainty optimistically feasible should be a factor 2 we hope.

In view of the upcoming two **complementary** experiments, one at Fermilab working with ultra hot muons and the other at J-PARC operating with ultra cold muons (very different radiation effects), the big challenge is to keep up on the prediction side as much as possible. The deviation between theory and experiment can be scrutinized provided theory and the needed cross section data improves the same as the muon $g-2$ experiments. Primarily we need more/better data and/or progress in non-perturbative QCD, where the main obstacle (data, lattice QCD, RLA) is the hadronic light-by-light scattering contribution. Progress in evaluating HVP also depends on more data (BaBar, Belle, VEPP-2000, BESIII,...) and lattice QCD where recent progress is very promising (see figure 12). In both cases HVP as well as HLbL, lattice QCD will provide answers one day, but also low energy effective RL and DR approaches need be further developed. One has also to keep in mind that progress in calculations of radiative corrections [71–73] is mandatory in precision measurements of hadronic cross sections.

For future improvements of HLbL one urgently needs more information from $\gamma\gamma \to$ hadrons physics [74, 75] in order to have better constraints on modeling hadronic amplitudes (see [76–78] for theoretical studies). Some sample processes are collected in figure 15. Mostly experiments at e^+e^-–

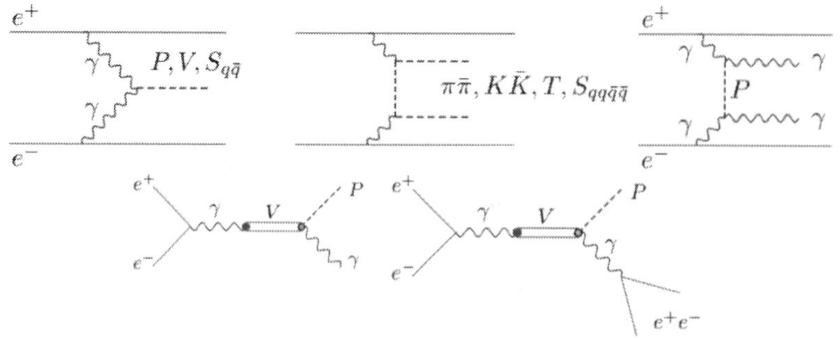

Figure 15. Sample processes relevant for the dispersive approach to HLbL.

facilities investigate single-tag events (higher rates, lower background). New data are expected from KLOE, KEDR exhibiting taggers and from BaBar, Belle, BES III which have high luminosity. More information is also expected from Dalitz–decay studies $\rho, \omega, \phi \to \pi^0(\eta)e^+e^-$ possible at Novosibirsk, CERN NA60, JLab, Mainz, Bonn, Jülich and BES. Unfortunately some of the interesting processes seem to be buried in the background. The background is a general problem in $\gamma\gamma \to$ hadrons physics.

The dispersive approach [8, 66] is able to allow for real progress since contributions which we have treated so far as separate contributions will be treated in an integral manner. An example is the $\gamma\gamma \to \pi\pi$ process which includes contributions attributed to the two–pion channel , the pion–loop, the scalar contribution as well as the tensor contribution. All-in-one can be gotten from the experimental data (see e.g. figure 3 of [79]). This also will settle such issues as the pion polarizability. A lot remains to be done while new $a_\mu^{\rm exp}$ is expected soon. For details see the Contribution by M. Procura.

Thanks to the organizers for the kind invitation and support to the "KLOE-2 Workshop on e^+e^- collision physics at 1 GeV" at Frascati, and for giving me the opportunity to present this talk.

References

[1] F. Jegerlehner, EPJ Web Conf. **118**, 01016 (2016)

[2] G. W. Bennett et al. [Muon g-2 Collaboration], Phys. Rev. D **73**, 072003 (2006)

[3] D. W. Hertzog, EPJ Web Conf. **118**, 01015 (2016)

[4] M. Hayakawa, T. Kinoshita, A. I. Sanda, Phys. Rev. Lett. **75**, 790 (1995); Phys. Rev. D **54**, 3137 (1996); M. Hayakawa, T. Kinoshita, Phys. Rev. D **57**, 465 (1998) [Erratum-ibid. D **66**, 019902 (2002)];

[5] J. Bijnens, E. Pallante, J. Prades, Phys. Rev. Lett. **75**, 1447 (1995) [Erratum-ibid. **75**, 3781 (1995)]; Nucl. Phys. B **474**, 379 (1996); [Erratum-ibid. **626**, 410 (2002)];
J. Bijnens, J. Prades, Mod. Phys. Lett. A **22**, 767 (2007)

[6] M. Knecht, A. Nyffeler, Phys. Rev. D **65**, 073034 (2002)

[7] M. Benayoun, P. David, L. DelBuono, F. Jegerlehner, Eur. Phys. J. C **72**, 1848 (2012); **73**, 2453 (2013)

[8] G. Colangelo et al., Phys. Lett. B **738**, 6 (2014)

[9] M. Knecht, S. Peris, M. Perrottet, E. de Rafael, JHEP **0211**, 003 (2002)

[10] K. Melnikov, A. Vainshtein, Phys. Rev. D **70**, 113006 (2004)

[11] A. Czarnecki, W. J. Marciano, A. Vainshtein, Phys. Rev. D **67**, 073006 (2003) [Erratum-ibid. D **73**, 119901 (2006)]

[12] R. R. Akhmetshin et al. [CMD-2 Collab.], Phys. Lett. B **578**, 285 (2004); V. M. Aulchenko et al. [CMD-2 Collab.], JETP Lett. **82**, 743 (2005) [Pisma Zh. Eksp. Teor. Fiz. **82**, 841 (2005)]; R. R. Akhmetshin et al., JETP Lett. **84**, 413 (2006) [Pisma Zh. Eksp. Teor. Fiz. **84**, 491 (2006)]; Phys. Lett. B **648**, 28 (2007)

[13] M. N. Achasov et al. [SND Collab.], J. Exp. Theor. Phys. **103**, 380 (2006) [Zh. Eksp. Teor. Fiz. **130**, 437 (2006)]

[14] A. Aloisio et al. [KLOE Collab.], Phys. Lett. B **606**, 12 (2005); F. Ambrosino et al. [KLOE Collab.], Phys. Lett. B **670**, 285 (2009)

[15] F. Ambrosino et al. [KLOE Collab.], Phys. Lett. B **700**, 102 (2011)

[16] D. Babusci et al. [KLOE Collab.], Phys. Lett. B **720**, 336 (2013)

[17] B. Aubert et al. [BABAR Collab.], Phys. Rev. Lett. **103**, 231801 (2009); J. P. Lees et al., Phys.Rev. **D86**, 032013 (2012)

[18] M. Ablikim et al. [BESIII Collab.], Phys. Lett. B **753**, 629 (2016)

[19] B. Aubert et al. [BABAR Collab.], Phys. Rev. D **70**, 072004 (2004); **71**, 052001 (2005); **73**, 012005 (2006); **73**, 052003 (2006); **76**, 012008, ibid. 092005, ibid. 092006, (2007); **77**, 092002 (2008)

[20] J. P. Lees et al. [BaBar Collab.], Phys. Rev. D **85**, 112009 (2012); **86**, 012008 (2012); **87**, 092005 (2013); **88**, 032013 (2013); **89**, 092002 (2014)

[21] M. Davier, Nucl. Part. Phys. Proc. **260**, 102 (2015)

[22] M. Davier, A. Höcker, B. Malaescu, Z. Zhang, Adv. Ser. Direct. High Energy Phys. **26**, 129 (2016)

[23] R. R. Akhmetshin et al. [CMD-3 Collaboration], Phys. Lett. B **723**, 82 (2013); **759**, 634 (2016)

[24] E. A. Kozyrev et al. [CMD-3 Collaboration], Phys. Lett. B **760**, 314 (2016)

[25] M. N. Achasov et al., Phys. Rev. D **90**, no.11, 112007 (2014)

[26] V. M. Aulchenko et al. [SND Collaboration], Phys. Rev. D **91**, no.5, 052013 (2015)

[27] M. N. Achasov et al. [SND Collaboration], Phys. Rev. D **93**, no.9, 092001 (2016); **94**, no.3, 032010 (2016); **94**, no.9, 092002 (2016); **94**, no.11, 112006 (2016); **94**, no.11, 112001 (2016)

[28] J. Z. Bai et al. [BES Collab.], Phys. Rev. Lett. **84**, 594 (2000); Phys. Rev. Lett. **88**, 101802 (2000); M. Ablikim et al., Phys. Lett. B **677**, 239 (2009)

[29] V. V. Anashin et al., Phys. Lett. B **753**, 533 (2016); arXiv:1610.02827 [hep-ex].

[30] B. Krause, Phys. Lett. B **390**, 392 (1997)

[31] A. Kurz, T. Liu, P. Marquard, M. Steinhauser, Phys.Lett. **B734**, 144 (2014)

[32] G. Colangelo et al., Phys.Lett. **B735**, 90 (2014)

[33] A. Anastasi et al. [KLOE-2 Collaboration], doi:10.1016/j.physletb.2016.12.016 arXiv:1609.06631 [hep-ex].

[34] R. Barate et al. [ALEPH Collab.], Z. Phys. C **76**, 15 (1997); Eur. Phys. J. C **4**, 409 (1998); S. Schael et al. [ALEPH Collab.], Phys. Rept. **421**, 191 (2005)

[35] M. Davier et al., Eur.Phys.J. **C74**, 2803 (2014)

[36] K. Ackerstaff et al. [OPAL Collab.], Eur. Phys. J. C **7**, 571 (1999)

[37] S. Anderson et al. [CLEO Collab.], Phys. Rev. D **61**, 112002 (2000)

[38] M. Fujikawa et al. [Belle Collab.], Phys. Rev. D **78**, 072006 (2008)

[39] M. Benayoun, P. David, L. DelBuono, F. Jegerlehner, Eur. Phys. J. C **75**, no.12, 613 (2015)

[40] F. Jegerlehner, R. Szafron, Eur. Phys. J. C **71**, 1632 (2011)

[41] M. Davier, A. Höcker, B. Malaescu, Z. Zhang, Eur. Phys. J. C **71**, 1515 (2011) [Erratum-ibid. C **72**, 1874 (2012)]

[42] M. Davier et al., Eur. Phys. J. C **66**, 127 (2010)

[43] K. Hagiwara, R. Liao, A. D. Martin, D. Nomura, T. Teubner, J. Phys. G G **38**, 085003 (2011)

[44] M. Davier, arXiv:1612.02743 [hep-ph].

[45] Z. Zhang, EPJ Web Conf. **118**, 01036 (2016) doi:10.1051/epjconf/201611801036

[46] F. Jegerlehner, A. Nyffeler, Phys. Rept. **477**, 1 (2009)

[47] J. Prades, E. de Rafael, A. Vainshtein, Adv. Ser. Direct. High Energy Phys. **20**, 303 (2009)

[48] X. Feng, K. Jansen, M. Petschlies, D. B. Renner, Phys. Rev. Lett. **107**, 081802 (2011)

[49] M. Golterman, K. Maltman, S. Peris, Phys. Rev. D **90**, no.7, 074508 (2014)

[50] S. Borsanyi et al., arXiv:1612.02364 [hep-lat].

[51] X. Feng et al., Phys. Rev. D **88**, 034505 (2013)

[52] P. Boyle, L. Del Debbio, E. Kerrane, J. Zanotti, Phys. Rev. D **85**, 074504 (2012)

[53] C. Aubin, T. Blum, M. Golterman, S. Peris, Phys. Rev. D **88**, 074505 (2013)

[54] A. Francis et al., arXiv:1411.3031 [hep-lat]
[55] R. Malak et al. [Budapest-Marseille-Wuppertal Collab.], PoS LATTICE **2014**, 161 (2015)
[56] B. Chakraborty, C. T. H. Davies, P. G. de Oliviera, J. Koponen, G. P. Lepage, arXiv:1601.03071 [hep-lat].
[57] F. Burger, X. Feng, K. Jansen, M. Petschlies, G. Pientka, D. B. Renner, EPJ Web Conf. **118**, 01029 (2016)
[58] F. Burger et al. [ETM Collab.], JHEP **1402**, 099 (2014)
[59] C. Aubin, T. Blum, Phys. Rev. D **75**, 114502 (2007)
[60] M. Della Morte, G. Herdoiza, H. Horch, B. Jäger, H. Meyer, H. Wittig, PoS LATTICE **2015**, 111 (2015) [arXiv:1602.03976 [hep-lat]]
[61] M. Della Morte, B. Jäger, A. Jüttner, H. Wittig, JHEP **1203**, 055 (2012)
[62] C. M. Carloni Calame, M. Passera, L. Trentadue, G. Venanzoni, Phys. Lett. B **746**, 325 (2015)
[63] G. Abbiendi et al., arXiv:1609.08987 [hep-ex]
[64] F. Jegerlehner, Nucl. Phys. Proc. Suppl. **181-182**, 135 (2008); *Hadronic effects in* $(g-2)_\mu$ *and* $\alpha_{\mathrm{QED}}(M_Z)$*: Status and perspectives*, In: *Radiative Corrections*, ed by J. Solà (World Scientific, Singapore 1999) pp 75–89 [hep-ph/9901386].
[65] A. Nyffeler, EPJ Web Conf. **118**, 01024 (2016)
[66] V. Pauk, M. Vanderhaeghen, Eur. Phys. J. C **74**, 3008 (2014)
[67] F. Jegerlehner, Talk at the MITP Workshop "Hadronic contributions to the muon anomalous magnetic moment", 1-5 April 2014, Waldthausen Castle near Mainz,
[68] A. Gérardin, H. B. Meyer, A. Nyffeler, arXiv:1607.08174 [hep-lat]
[69] J. Bijnens, J. Relefors, arXiv:1608.01454 [hep-ph].
[70] C. Gnendiger, D. Stöckinger, H. Stöckinger-Kim, Phys. Rev. D **88**, 053005 (2013)
[71] S. Actis et al., Eur. Phys. J. C **66**, 585 (2010)
[72] H. Czyż, Int. J. Mod. Phys. Conf. Ser. **35**, 1460402 (2014) doi:10.1142/S2010194514604025
[73] F. Jegerlehner, K. Kołodziej, arXiv:1701.01837 [hep-ph].
[74] D. Babusci et al., Eur. Phys. J. C **72**, 1917 (2012)
[75] A. Nyffeler, PoS CD **12**, 045 (2013) [arXiv:1306.5987 [hep-ph]]
[76] G. Mennessier, Z. Phys. C **16**, 241 (1983)
[77] B. Moussallam, Eur. Phys. J. C **73**, 2539 (2013)
[78] L. Y. Dai, M. R. Pennington, Phys. Rev. D **90**, no.3, 036004 (2014)
[79] F. Jegerlehner, Acta Phys. Polon. B **44**, no.11, 2257 (2013)

A proposal to measure the hadronic contribution to the g-2 in the space-like region *

Luca Trentadue[1,2,**]

[1] *Università degli Studi di Parma, Dipartimento di Fisica e Scienze della Terra " Macedonio Melloni "*
[2] *INFN, Sezione di Milano Bicocca, Milano, Italy*

Abstract. We propose a novel approach to determine the leading hadronic corrections to the muon g-2. It consists in a measurement of the effective electromagnetic coupling in the space-like region. This method may become feasible at flavor factories as well as with μ e scattering resulting in a determination potentially competitive with the dispersive approach via time-like data.

1 Introduction

The motivation of this work [1, 2] is due to a long-standing discrepancy between experiment and the Standard Model (SM) prediction for a_μ, the muon anomalous magnetic moment. For this reason the hadronic corrections have been kept under close scrutiny [3–6]. The hadronic contribution represents the largest uncertainty of the Standard Model (SM) value and is comparable with the experimental one. When the new results from the g-2 experiments at Fermilab and J-PARC will reach the unprecedented precision of 0.14 parts per million (or better) [7–9], the uncertainty of the hadronic corrections will become the main limitation of this formidable test of the SM.
Vacuum Polarization (VP) makes α_{em} running assuming a well defined effective value at any scale. Vacuum polarization and the effective charge are defined by:

$$e^2 \to e^2(q^2) = \frac{e^2}{1 + (\Pi(q^2) - \Pi(0))} \tag{1}$$

and

$$\alpha(q^2) = \frac{\alpha(0)}{1 - \Delta\alpha} \quad \Delta\alpha = -Re(\Pi(q^2) - \Pi(0)). \tag{2}$$

$\Delta\alpha$ takes contributions from leptonic and hadronic elementary states among these the non-perturbative $\Delta\alpha_{had}$

$$\Delta\alpha = \Delta\alpha_{lept} + \Delta\alpha_{had} + \Delta\alpha_{top} \tag{3}$$

*Work done in collaboration with: G. Abbiendi, C. Carloni Calame, U. Marconi, C. Matteuzzi, G. Montagna, O. Nicrosini, M. Passera, F. Piccinini, R. Tenchini and G. Venanzoni.

**e-mail: luca.trentadue@unipr.it

$$\Delta\alpha^{(5)}_{had}(M_z^2) = -\frac{\alpha\, M_z^2}{3\pi} Re \int_{4m_\pi^2}^{\infty} ds \frac{R(s)}{s(s - M_z^2 - i\epsilon)} \tag{4}$$

A direct measurement of the running of $\alpha_{em}(s/t)$ in space/time-like regions can show the running of α_{em}. It can provide a test of duality far away form resonances and ita has been done in the past by few experiments at e^+e^- colliders by comparing well known QED processes with some reference (either data or MonteCarlo) normalization:

$$(\frac{\alpha(q^2)}{\alpha(q_0^2)})^2 \simeq \frac{N_{signal}(q^2)}{N_{norm}(q_0^2)}. \tag{5}$$

N_{signal} can be any QED process as muon pairs production or others similar ones and N_{norm} can be the Bhabha process, a pure QED channel as $\gamma\gamma$ pair production as well as theory, or any other reference process. q_0 represents a scale chosen as starting point for the running.

2 a_{mu}^{HLO} calculation

The leading-order hadronic contribution to the muon g-2 is given by the well-known formula [6, 13]

$$a_\mu^{\text{HLO}} = \frac{\alpha}{\pi^2} \int_0^\infty \frac{ds}{s}\, K(s)\, \text{Im}\Pi_{\text{had}}(s + i\epsilon), \tag{6}$$

where $\Pi_{\text{had}}(s)$ is the hadronic part of the photon vacuum polarization, $\epsilon > 0$, and

$$K(s) = \int_0^1 dx \frac{x^2(1-x)}{x^2 + (1-x)(s/m_\mu^2)} \tag{7}$$

is a positive kernel function with m_μ the muon mass.
As the total cross section for hadron production in low-energy e^+e^- annihilations is related to the imaginary part of $\Pi_{\text{had}}(s)$ via the optical theorem, the dispersion integral in eq. (6) is computed integrating experimental time-like ($s > 0$) data up to a certain value of s [4, 17, 18]. The high-energy tail of the integral is calculated by using perturbative QCD [19]. In a recent work the KLOE collaboration has obtained [21, 22] the the first measurement of the running of the effective QED coupling constant $\alpha(s)$ in the time-like region $0.6 < s < 0.975$ GeV at DAΦNE by using the Initial State Radiation approach [20] in the process $e^+e^- \to \mu^+\mu^-\gamma$.

Alternatively, if we exchange the x and s integrations in eq. (6) we obtain [23]

$$a_\mu^{\text{HLO}} = \frac{\alpha}{\pi} \int_0^1 dx\,(x-1)\,\overline{\Pi}_{\text{had}}[t(x)]\,, \tag{8}$$

where $\overline{\Pi}_{\text{had}}(t) = \Pi_{\text{had}}(t) - \Pi_{\text{had}}(0)$ and

$$t(x) = \frac{x^2 m_\mu^2}{x-1} < 0 \tag{9}$$

is a space-like squared four-momentum. If we invert eq. (9), we get $x = (1-\beta)\,(t/2m_\mu^2)$, with $\beta = (1 - 4m_\mu^2/t)^{1/2}$, and from eq. (8) we obtain

$$a_\mu^{\text{HLO}} = \frac{\alpha}{\pi} \int_{-\infty}^0 \overline{\Pi}_{\text{had}}(t) \left(\frac{\beta-1}{\beta+1}\right)^2 \frac{dt}{t\beta}. \tag{10}$$

Equation (10) has been used for lattice QCD calculations of $a_\mu^{\rm HLO}$ [24]; while the results are not yet competitive with those obtained with the dispersive approach via time-like data, their errors are expected to decrease significantly in the next few years [25].

The effective fine-structure constant at squared momentum transfer q^2 is defined above and is $\Delta\alpha(q^2) = -{\rm Re}\overline{\Pi}(q^2)$. The purely leptonic part, $\Delta\alpha_{\rm lep}(q^2)$, can be calculated order-by-order in perturbation theory – it is known up to three loops in QED [26] (and up to four loops in specific q^2 limits [27]). As ${\rm Im}\overline{\Pi}(q^2) = 0$ for negative q^2, eq. (8) can be rewritten in the form [28]

$$a_\mu^{\rm HLO} = \frac{\alpha}{\pi}\int_0^1 dx\,(1-x)\,\Delta\alpha_{\rm had}[t(x)]\,. \tag{11}$$

Equation (11), involving the hadronic contribution to the running of the effective fine-structure constant at space-like momenta, can be further formulated in terms of the Adler function [29], defined as the logarithmic derivative of the vacuum polarization, which, in turn, can be calculated via a dispersion relation with time-like hadroproduction data and perturbative QCD [28, 30].

We are going to proceed differently, proposing to calculate eq. (11) by measurements of the effective electromagnetic coupling in the space-like region. The hadronic contribution to the running of α in the space-like region, $\Delta\alpha_{\rm had}(t)$ (see eq. (3)), can be extracted by comparing Bhabha scattering data with Monte Carlo predictions. The Leading Order (LO) Bhabha cross section receives contributions from t- and s-channel photon exchange amplitudes. At Next to Leading Order (NLO) in QED, it is customary to distinguish corrections with an additional virtual photon or the emission of a real photon (photonic NLO) from those originated by the insertion of the vacuum polarization corrections into the LO photon propagator.

Let us consider a few simple points. In fig. 1 (left) we plot the integrand $(1-x)\Delta\alpha_{\rm had}[t(x)]$ of eq. (11) using the output of the routine `hadr5n12` [33] (which uses time-like hadroproduction data and perturbative QCD). The range $x \in (0,1)$ corresponds to $t \in (-\infty, 0)$, with $x = 0$ for $t = 0$. The peak of the integrand occurs at $x_{\rm peak} \simeq 0.914$ where $t_{\rm peak} \simeq -0.108$ GeV^2 and $\Delta\alpha_{\rm had}(t_{\rm peak}) \simeq 7.86 \times 10^{-4}$ (see fig. 1 (right)). Such relatively low t values can be explored at e^+e^- colliders with center-of-mass energy $\sqrt{s}$ around or below 10 GeV where

$$t = -\frac{s}{2}(1 - cos\theta)\left(1 - \frac{4m_e^2}{s}\right), \tag{12}$$

θ is the electron scattering angle and m_e is the electron mass. Depending on s and θ, the integrand of eq. (11) can be measured in the range $x \in [x_{\rm min}, x_{\rm max}]$, as shown in fig. 2 (left). Note that to span low x intervals, larger θ ranges are needed as the collider energy decreases. In this respect, $\sqrt{s} \sim 3$ GeV appears to be very convenient, as an x interval $[0.30, 0.98]$ can be measured varying θ between $\sim 2°$ and $28°$. It is also worth remarking that data collected at flavor factories, such as DAΦNE (Frascati), VEPP-2000 (Novosibirsk), BEPC-II (Beijing), PEP-II (SLAC) and SuperKEKB (Tsukuba), and possibly at a future high-energy e^+e^- collider, like FCC-*ee* (TLEP) [34] or ILC [35], can help to cover different and complementary x regions.

Furthermore, given the smoothness of the integrand, values outside the measured x interval may be interpolated with some theoretical input. In particular, the region below $x_{\rm min}$ will provide a relatively small contribution to $a_\mu^{\rm HLO}$, while the region above $x_{\rm max}$ may be obtained by extrapolating the curve from $x_{\rm max}$ to $x = 1$, where the integrand is null, or using perturbative QCD.

The analytic dependence of the MC Bhabha predictions on $\alpha(t)$ (and, in turn, on $\Delta\alpha_{\rm had}(t)$) is not trivial, and a numerical procedure has to be devised to extract it from the data.This was not the case

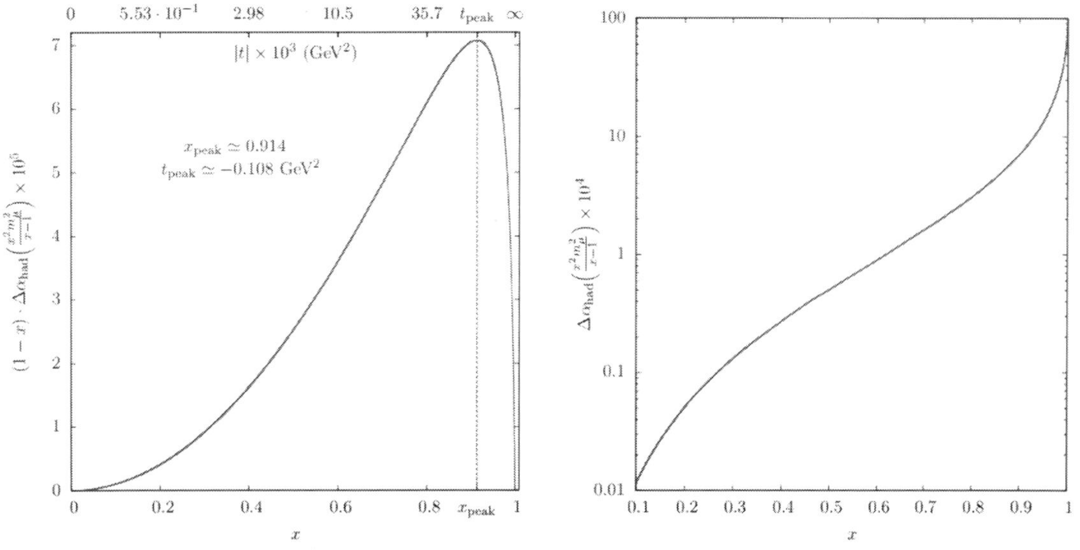

Figure 1. Left: The integrand $(1-x)\Delta\alpha_{\mathrm{had}}[t(x)] \times 10^5$ as a function of x and t. Right: $\Delta\alpha_{\mathrm{had}}[t(x)] \times 10^4$.

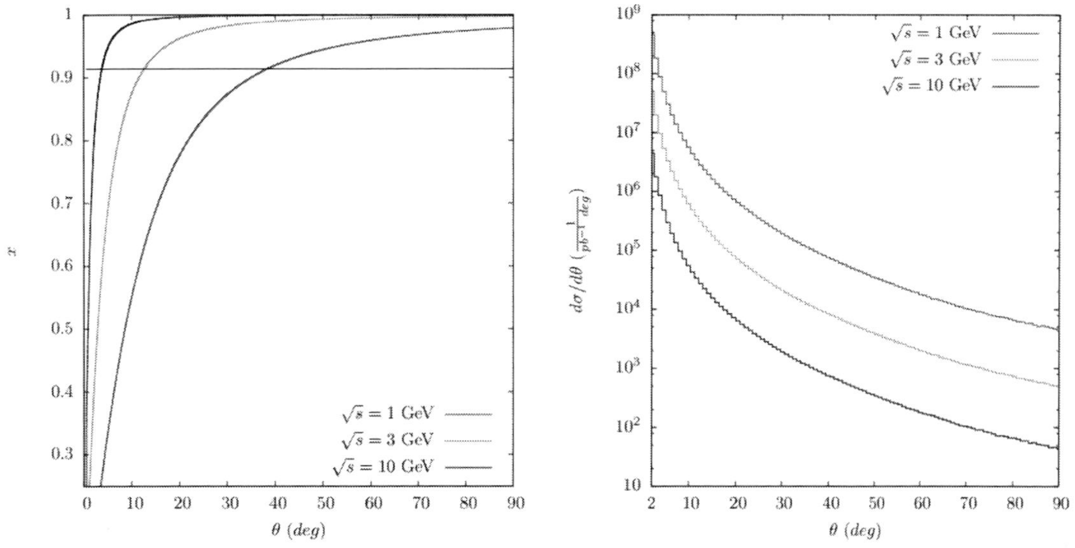

Figure 2. Left: Ranges of x values as a function of the electron scattering angle θ for three different center-of-mass energies. The horizontal line corresponds to $x = x_{\mathrm{peak}} \simeq 0.914$. Right: Bhabha differential cross section obtained with BabaYaga [31] as a function of θ for the same three values of $\sqrt{s}$ in the angular range $2^\circ < \theta < 90^\circ$.

for example in refs. [14, 15]: there $\alpha(t)$ was extracted from Bhabha data in the very forward region at LEP, where the t channel diagrams are by far dominant and $\alpha(t)$ factorizes (see for example [16]). In formulae, we have to find a function $\alpha(t)$ such that

$$\left.\frac{d\sigma}{dt}\right|_{\rm data} = \frac{d\sigma}{dt}\Big(\alpha(t), \alpha(s)\Big)\Big|_{\rm MC}, \tag{13}$$

where we explicitly kept apart the dependence on the time-like VP $\alpha(s)$ because we are only interested in $\alpha(t)$. We emphasise that, in our analysis, $\alpha(s)$ is an input parameter. Being the Bhabha cross section in the forward region dominated by the t-channel exchange diagrams , we checked that the present $\alpha(s)$ uncertainty induces in this region a relative error on the θ distribution of less than $\sim 10^{-4}$ (which is part of the systematic error).

We propose to perform the numerical extraction of $\Delta\alpha_{\rm had}(t)$ from the Bhabha distribution of the t Mandelstam variable. The idea is to let $\alpha(t)$ vary in the MC sample around a reference value and choose, bin by bin in the t distribution, the value that minimizes the difference with data. The procedure can be sketched as follows:

1. choose a reference function returning the value of $\Delta\alpha_{\rm had}(t)$ (and hence $\alpha(t)$) to be used in the MC sample, we call it $\bar{\alpha}(t)$;

2. for each generated event, calculate N MC weights by rescaling $\bar{\alpha}(t) \to \bar{\alpha}(t) + \frac{i}{N}\delta(t)$, where $i \in [-N, N]$ and $\delta(t)$ is for example the error induced on $\bar{\alpha}(t)$ by the error on $\Delta\alpha_{\rm had}(t)$. Being done on an event by event basis, the full dependence on $\alpha(t)$ of the MC differential cross section can be kept;

3. for each bin j of the t distribution, compare the experimental differential cross section with the MC predictions and choose the i_j which minimizes the difference;

4. $\bar{\alpha}(t_j) + \frac{i_j}{N}\delta(t_j)$ will be the extracted value of $\alpha(t_j)$ from data in the jth bin. $\Delta\alpha_{\rm had}(t_j)$ can then be obtained through the relation between $\alpha(t)$ and $\Delta\alpha_{\rm had}(t)$.

We finally find, for each bin j of the t distribution,

$$\left.\frac{d\sigma}{dt}\right|_{j,\rm data} = \frac{d\sigma}{dt}\Big(\bar{\alpha}(t) + \frac{i_j}{N}\delta(t), \alpha(s)\Big)\Big|_{j,\rm MC}. \tag{14}$$

We remark that the algorithm does not assume any simple dependence of the cross section on $\alpha(t)$, which can in fact be general, mixing s, t channels and higher order radiative corrections, relevant (or not) in different t domains.

In order to test our procedure, we perform a pseudo-experiment: we generate pseudo-data using the parameterization $\Delta\alpha^{I}_{\rm had}(t)$ of refs. [18, 36] and check if we can recover it by inserting in the MC the (independent) parameterization $\Delta\alpha^{II}_{\rm had}(t)$ (corresponding to $\bar{\alpha}(t)$ of eq. 14) of ref. [33] by means of the method described above. For this exercise, we use the generator BabaYaga in its most complete setup, generating events at $\sqrt{s} = 1.02$ GeV, requiring $10° < \theta_\pm < 170°$, $E_\pm > 0.4$ GeV and an acollinearity cut of $15°$. We choose $\delta(t)$ to be the error induced on $\alpha(t)$ by the 1-σ error on $\Delta\alpha_{\rm had}(t)$, which is returned by the routine of ref. [33], we set $N = 150$, and we produce distributions with 200 bins. We note that in the present exercise $\alpha(s)$ and all the radiative corrections both in the pseudo-data and in the MC samples are exactly the same, because we are interested in testing the algorithm rather than assessing the achievable accuracy, at least at this stage.

In fig. 3, $\Delta\alpha^{\rm extr}_{\rm had}$ is the result extracted with our algorithm, corresponding to the minimizing set of i_j: the figure shows that our method is capable of recovering the underlying function $\Delta\alpha_{\rm had}(t)$ inserted into the "data". As the difference between $\Delta\alpha^{I}_{\text{had}}$ and $\Delta\alpha^{\text{extr}}_{\text{had}}$ is hardly visible on an absolute scale, in fig. 3 all the functions have been divided by $\Delta\alpha^{II}_{\text{had}}$ to display better the comparison between $\Delta\alpha^{I}_{\text{had}}$ and $\Delta\alpha^{\text{extr}}_{\text{had}}$.

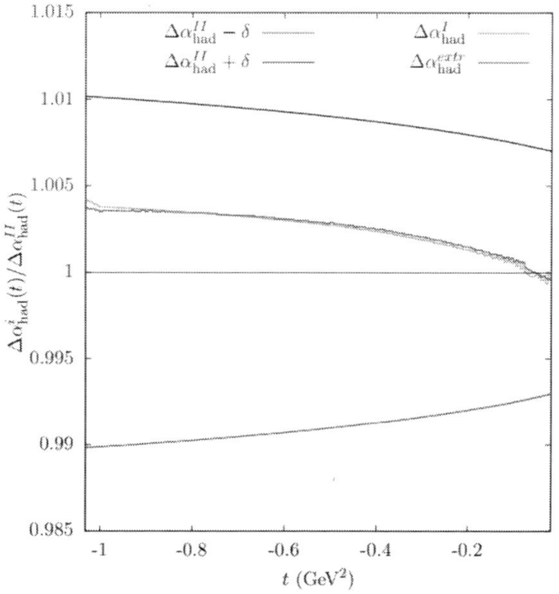

Figure 3. The extracted function $\Delta\alpha^{\text{extr}}_{\text{had}}(t)$ compared to the function $\Delta\alpha^{I}_{\text{had}}(t)$ used in the pseudo-data (see text). The functions $\Delta\alpha^{II}_{\text{had}}(t) \pm \delta(t)$ are shown to display the range spanned by the MC samples. All functions have been divided by $\Delta\alpha^{II}_{\text{had}}(t)$. The tiny difference between $\Delta\alpha^{I}_{\text{had}}$ and $\Delta\alpha^{\text{extr}}_{\text{had}}$ is due to the binning discretization.

In order to assess the achievable accuracy on $\Delta\alpha_{\text{had}}(t)$ with the proposed method, we remark that the LO contribution to the cross section is quadratic in $\alpha(t)$, thus we have

$$\frac{1}{2}\frac{\delta\sigma}{\sigma} \simeq \frac{\delta\alpha}{\alpha} \simeq \delta\Delta\alpha_{\text{had}} \tag{15}$$

Equation (15) relates the *absolute* error on $\Delta\alpha_{\text{had}}$ with the *relative* error on the Bhabha cross section. Let us stress here that:

1. From the theoretical point of view, the present accuracy of the MC predictions [32] is at the level of about 0.05%, which implies that the precision that our method can, at best, set on $\Delta\alpha_{\text{had}}(t)$ is $\delta\Delta\alpha_{\text{had}}(t) \simeq 2 \cdot 10^{-4}$. Any further improvement requires the inclusion of the NNLO QED corrections into the MC codes (see Ref. [32]).

2. Experimentally a measurement of a_{μ}^{HLO} from space-like data competitive with the current time-like evaluations would require an $O(1\%)$ accuracy. Statistical considerations show that a 3% fractional accuracy on the a_{μ}^{HLO} integral can be obtained by sampling the integrand $(1-x)\Delta\alpha_{\text{had}}[t(x)]$ in $\sim$ 10 points around the x peak with a fractional accuracy of 10%. Given the value of $O(10^{-3})$ for $\Delta\alpha_{\text{had}}$ at $x = x_{\text{peak}}$, this implies that the cross section must be known with relative accuracy of $\sim 2\times10^{-4}$. Such a statistical accuracy, although challenging, can be obtained at flavor factories, as shown in fig. 2 (right). With an integrated luminosity of $O(1)$,

$O(10)$, $O(100)$ fb^{-1} at $\sqrt{s} = 1, 3$ and 10 GeV, respectively, the angular region of interest can be covered with a 0.01% accuracy per degree. The experimental systematic error must match the same level of accuracy.

A source of experimental systematic errors comes from the machine luminosity, which can be normalized by calculating a theoretical cross section in principle not depending on $\Delta\alpha_{\rm had}$. We devise two possible options for the normalization process:

1. To use the $e^+e^- \to \gamma\gamma$ process, which has no dependence on $\Delta\alpha_{\rm had}$, at least up to NNLO order;
2. To use the Bhabha process at $t \sim 10^{-3}$ GeV^2 ($x \sim 0.3$), where the dependence on $\Delta\alpha_{\rm had}$ is of $O(10^{-5})$ and can be safely neglected.

It is worth quoting that a detailed analysis of the systematic errors involved in the measurement of the luminosity has been carried out at LEP by the OPAL collaboration reaching the final accuracy of $O(10^{-4})$ [15, 37].

3 An alternative approach

Always within the space-like approach, an alternative possibility has been investigated [2]. It consists in using a fixed target $\mu \to e$ scattering process and to analyse the elastic muon distribution in the forward region. A new experiment has to be devised to measure the running of the fine-structure constant in the space-like region by scattering high-energy muons on atomic electrons of a low-Z target through the process $\mu e \to \mu e$. The differential cross section of this process, measured as a function of the squared momentum transfer $t = q^2 < 0$, provides direct sensitivity to the leading-order hadronic contribution to the muon anomaly $a_\mu^{\rm HLO}$. By using a muon beam of 150 GeV, with an average rate of $\sim 1.3 \times 10^7$ muon/s, currently available at the CERN North Area, a statistical uncertainty of $\sim 0.3\%$ can be achieved on $a_\mu^{\rm HLO}$ after two years of data taking. As the Bhabha process also this direct measurement of $a_\mu^{\rm HLO}$ will provide an independent determination, potentially competitive with the time-like dispersive approach, to consolidate, with a firmer interpretation of the measurements of the future muon g-2 experiments, the theoretical prediction for the muon g-2 in the Standard Model.

A series of preliminary considerations on the detector have been discussed [2].
In order to perform the planned measurement to the required precision, a dedicated detector is necessary. We describe here a possible setup to measure the following observables:

- direction and momentum of the incident muon;
- directions of the outgoing electron and muon.

The CERN muon beam M2, used at 150 GeV, has the characteristics needed for such a measurement. The beam intensity appears to be adequate to provide the required event yield. The beam time structure allows to tag the incident muon while keeping low the background related to incoming particles (*e.g.* electrons). The electrons contamination is very small. The beam provides both positive and negative muons, which we plan to use.

The target consists of atomic electrons. To reach the required statistics, the target must consist of an adequate amount of material to give a sufficient number of electron scattering centres. The target has to be made of a low-Z material to minimize the impact of multiple scattering and the background due to bremsstrahlung and pair production processes. A promising idea for the detector is to use 20 layers of Be (or C) coupled to Si planes, spaced by intermediate air gaps, located at a relative distance of one meter from each other. Fig. 4 shows the basic layout, presently under study. The

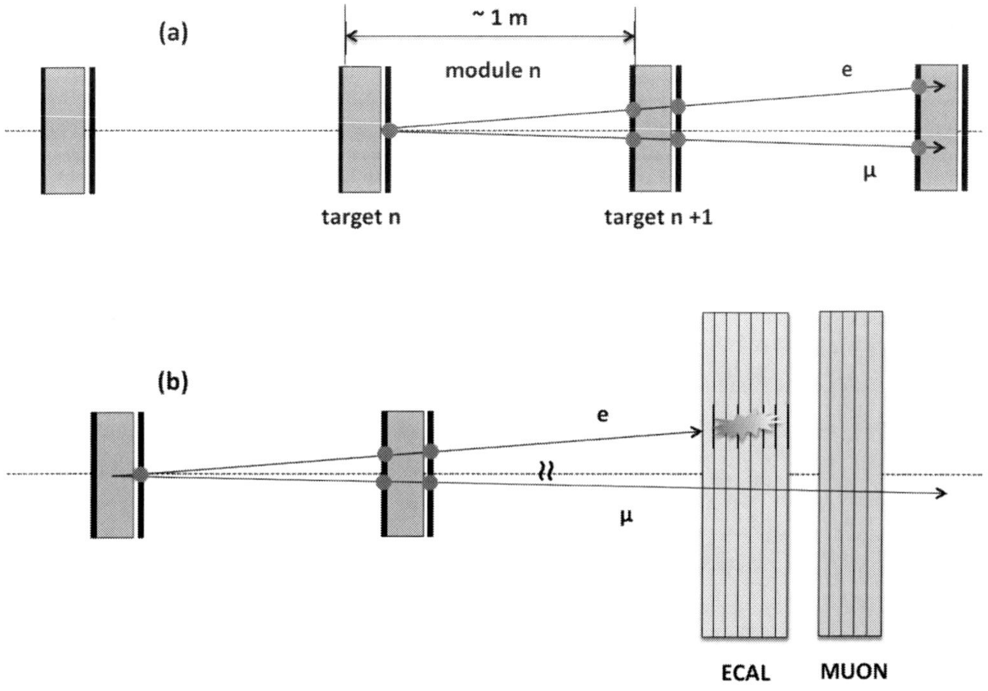

Figure 4. Scheme of a possible detector layout. (a) The detector is a modular system. Each module consists of a low-Z target (Be or C) and two silicon tracking stations located at a distance of one meter. (b) To perform the μ/e discrimination in the case of small scattering angles (both θ_μ and θ_e below 5 mrad) the detector is equipped with an electromagnetic calorimeter and a muon detector.

arrangement provides both a distributed target with low-Z and the tracking system. As downstream particle identifiers we plan to use a calorimeter for the electrons and a muon system for the muons (a filter plus active planes). This particle identifier system is required to solve the muon-electron ambiguity for electron scattering angles around (2–3) mrad (*cf.* Fig. 5). The preliminary studies of such an apparatus, performed by using GEANT4, indicate an angular resolution for the outgoing particles of ~ 0.02 mrad.

The detector acceptance must cover the region of the signal, with the electron emitted at extremely forward angles and high energies, as well as the normalization region, where the electron has much lower energy (around 1 GeV) and an emission angle of some tens of mrad. The boosted kinematics of the process allows the detector to cover almost 100% of the acceptance.

The incoming muons have to be tagged and their direction and momentum precisely measured. To this purpose, a detector similar to those used by COMPASS [39] or NA62 [40] can be employed.

- the angle of the scattered electron and muon are correlated as shown in Fig. 5 (for muons of 150 GeV). This constraint is extremely important to select elastic scattering events, rejecting background events from radiative or inelastic processes and to minimize systematic effects in the determination of t. Note that for scattering angles of (2–3) mrad there can be an ambiguity between the outgoing electron and muon, as their angles and momenta are similar. To associate them correctly it is necessary to identify the two particles by means of downstream dedicated detectors (calorimeter and muon detectors).

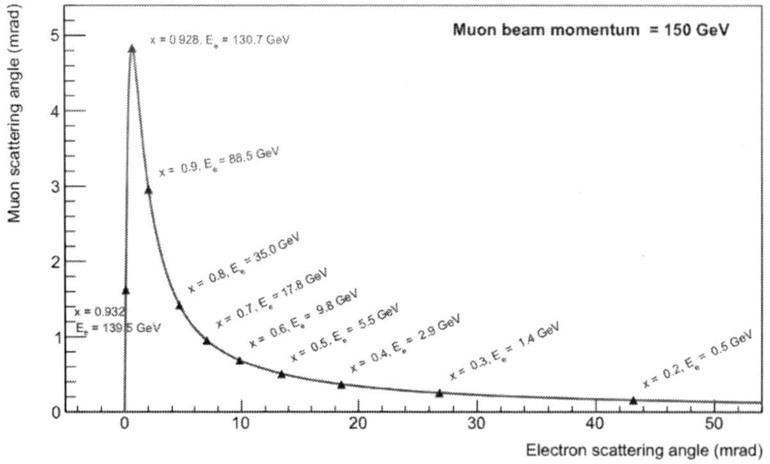

Figure 5. The relation between the muon and electron scattering angles for 150 GeV incident muon momentum.

We estimate the statistical sensitivity of this experiment (supplemented with large $|t|$ contributions derived from pQCD) to be $\sim$ 0.3% on the value of $a_\mu^{\rm HLO}$ after two years of data taking, using 30 experimental points in x, assuming a running time of 2×10^7 s/yr, and a muon beam of 150 GeV with an average intensity of $\sim 1.3 \times 10^7$ muon/s as the beam currently available at the CERN North Area [38].

A series of systematic uncertainties should be taken into account.

Significant contributions of the hadronic vacuum polarization to the $\mu e \to \mu e$ differential cross section are essentially restricted to electron scattering angles below 10 mrad, corresponding to electron energies above 10 GeV. The net effect of these contributions is to increase the cross section by a few per mille: a precise determination of $a_\mu^{\rm HLO}$ requires not only high statistics, but also a high systematic accuracy, as the final goal of the experiment is equivalent to a determination of the differential cross section with ~10 ppm systematic uncertainty at the peak of the integrand function (*cf*. Fig. 1).

Such an accuracy can be achieved if the efficiency is kept highly uniform over the entire q^2 range, including the normalization region, and over all the detector components. This motivates the choice of a purely angular measurement: an acceptance of tens of mrad can be covered with a single sensor of modern silicon detectors, positioned at a distance of about one meter from the target. It has to be stressed that particle identification (electromagnetic calorimeter and muon filter) is necessary to solve the electron-muon ambiguity in the region below 5 mrad. The wrong assignment probability can be measured with the data by using the rate of muon-muon and electron-electron events.

Another requirement for reaching very high accuracy is to measure all the relevant contributions to systematic uncertainties from the data themselves. An important effect, which distinguishes the normalization from the signal region, is multiple scattering, as the electron energy in this region is as low as 1 GeV. Multiple scattering breaks the muon-electron two-body angular correlation, moving events out of the kinematic line in the 2D plot of Fig. 5. In addition, multiple scattering in general causes acoplanarity, while two-body events are planar, within resolution. These facts allow multiple scattering effects to be modelled and measured by using data. An additional handle on multiple scattering could be the inclusion of a thin layer in the apparatus, made of the same material as the main target modules. This possibility will be studied in detail with dedicated simulations.

In experiments dedicated to high-precision measurements, several systematic effects can be explored within the experiment itself. In this respect the proposed modularity of the apparatus will help. A test with a single module could provide a proof-of-concept of the proposed methods.

From the theoretical point of view, the control of the systematic uncertainties requires the development of high-precision Monte Carlo tools, including the relevant radiative corrections to reach the needed theoretical precision. To this aim, QED radiative corrections at leading-logarithmic level resummed at all orders of perturbation theory and matched to the exact $O(\alpha)$ correction (as for instance implemented in the `BabaYaga` event generator [31] for Bhabha scattering) are mandatory, ensuring a theoretical precision at the level of $O(10^{-4})$ on the differential cross section. Moreover, by using the ratio of the cross sections in the signal and normalization regions, we expect that the theoretical uncertainties will be further reduced to the level of $O(10^{-5})$, due to partial cancellation of common radiative corrections. Work is in progress to extend `BabaYaga` to $\mu e \to \mu e$ scattering and to quantify the actual accuracy on the computation of the ratio of cross sections by means of dedicated Monte Carlo simulations. Any further improvement in the theoretical accuracy would require the matching of QED resummation with exact two-loop corrections, which are not available at present for the $\mu e \to \mu e$ process but are within reach.

4 Conclusions

We presented a novel approach to determine the leading hadronic correction to the muon g-2 based on measurements of the running of $\alpha(t)$ in the space-like region. An alternative formula for a_μ^{HLO} in the space-like region has been studied in detail. It emphasizes low values of $t \leq 1GeV^2$ which can be explored at low and at high energies machines.

We have also argued that this requires a measurement of the Bhabha cross section as of the $\mu \to e$ elastic distribution at relatively small angles, with an accuracy of the order 10^{-4} or better.

The $\mu \to e$ experiment is primarily based on a precise measurement of the scattering angles of the two outgoing particles as the q^2 of the muon-electron interaction can be directly determined by the electron (or muon) scattering angle.

The normalization of the cross section is provided by the very same $\mu e \to \mu e$ process in the low-q^2 region, where the effect of the hadronic corrections on the fine-structure constant is negligible. Such a simple and robust technique has the potential to keep systematic effects under control, aiming to reach a systematic uncertainty of the same order as the statistical one. For this purpose a preliminary detector layout has been described. By considering a beam of 150 GeV muons with an average intensity of $\sim 1.3 \times 10^7$ muons/s, a statistical uncertainty of $\sim 0.3\%$ can be achieved on a_μ^{HLO} in two years of data taking.

We have shown that an advantage of the muon beam is the possibility of employing a modular apparatus, with the target subdivided in subsequent layers. A low-Z solid target is preferred in order to provide the required event rate, limiting at the same time the effect of multiple scattering as well as of other types of muon interactions (pair production, bremsstrahlung and nuclear interactions).

We have shown that this approach, even if challenging, may become feasible both by using data collected at present flavor factories and at future high-energy e^+e^- colliders as well as with fixed target $\mu \to e$ elastic scattering.

The proposed determinations can become competitive with the accuracy of the present results obtained with the dispersive approach via time-like data. Reaching such an accuracy demands dedicated experiments and theoretical work for the next few years.

We propose a new approach to evaluate the leading-order hadronic contribution to the muon anomaly. The direct measurement of the hadronic contribution to anomalous muon magnetic moment will provide an independent determination, competitive with the time-like dispersive approach, and consolidate the theoretical prediction for the muon g-2 in the Standard Model. It will allow a firmer interpretation of the measurements of the future muon g-2 experiments at Fermilab and J-PARC.

This new approach may become a path within an unexplored region of the field theoretical dynamics. It may lead to a possibly long series of phenomenological results. The (crossed) t-channel dynamics, as complementary and independent with respect to the s-channel one, will permit an alternative new approach to the Standard Model precision physics.

5 Acknowledgements

I wish to thank Antonio Di Domenico for organizing and for involving me in this lively Workshop.

References

[1] C. Carloni Calame, M. Passera, G. Venanzoni, L. Trentadue, Phys. Lett. B **746** (2015) 325-329.
[2] G. Abbiendi, C.M. Carloni Calame, U. Marconi, C. Matteuzzi, G. Montagna, O. Nicrosini M. Passera, F. Piccinini, R. Tenchini, L. Trentadue, G. Venanzoni, Eur. Phys. J. C **77** (2017) no.3, 139 doi:10.1140/epjc/s10052-017-4633-z [arXiv:1609.08987 [hep-ex]].
[3] G.W. Bennett *et al.* [Muon g-2 Collaboration], Phys. Rev. D **73** (2006) 072003.
[4] F. Jegerlehner, A. Nyffeler, Phys. Rept. **477** (2009) 1.
[5] T. Blum *et al.*, arXiv:1311.2198 [hep-ph]; K. Melnikov, A. Vainshtein, Springer Tracts Mod. Phys. **216** (2006) 1; M. Davier, W.J. Marciano, Ann. Rev. Nucl. Part. Sci. **54** (2004) 115; M. Passera, J. Phys. G **31** (2005) R75; M. Knecht, Lect. Notes Phys. **629** (2004) 37.
[6] F. Jegerlehner, "The anomalous magnetic moment of the muon," Springer Tracts Mod. Phys. **226**, 2008.
[7] J. Grange *et al.* [Muon g-2 Collaboration], arXiv:1501.06858 [physics.ins-det].
[8] G. Venanzoni [Muon g-2 Collaboration], arXiv:1411.2555 [physics.ins-det].
[9] N. Saito [J-PARC g-2/EDM Collaboration], AIP Conf. Proc. **1467** (2012) 45.
[10] G. Venanzoni, Nuovo Cim. C **037** (2014) 02, 165; G. Venanzoni, Frascati Phys. Ser. **54** (2012) 52.
[11] G.V. Fedotovich [CMD-2 Collaboration], Nucl. Phys. Proc. Suppl. **181-182** (2008) 146.

[12] G. Colangelo, M. Hoferichter, A. Nyffeler, M. Passera, P. Stoffer, Phys. Lett. B **735** (2014) 90.

[13] C. Bouchiat, L. Michel, J. Phys. Radium 22 (1961) 121; L. Durand, Phys. Rev. **128** (1962) 441 [Erratum-ibid. **129** (1963) 2835]; M. Gourdin, E. De Rafael, Nucl. Phys. B **10** (1969) 667.

[14] A.B. Arbuzov, D. Haidt, C. Matteuzzi, M. Paganoni and L. Trentadue, Eur. Phys. J. C **34** (2004) 267.

[15] G. Abbiendi *et al.* [OPAL Collaboration], Eur. Phys. J. C **45** (2006) 1.

[16] A.B. Arbuzov, V.S. Fadin, E.A. Kuraev, L.N. Lipatov, N.P. Merenkov, L. Trentadue, Nucl. Phys. B**485** (1997) 457-502;
A. B. Arbuzov, K. I. Gach, V. Y. Gonchar, E. A. Kuraev, N. P. Merenkov and L. Trentadue, Phys. Lett. B **399** (1997) 312;
A. B. Arbuzov, V. S. Fadin, E. A. Kuraev, L. N. Lipatov, N. P. Merenkov and L. G. Trentadue, Phys. Lett. B **394** (1997) 218;
A. B. Arbuzov, E. A. Kuraev, N. P. Merenkov and L. Trentadue, Phys. Atom. Nucl. **60** (1997) 591 [Yad. Fiz. **60N4** (1997) 673];
A. B. Arbuzov, E. A. Kuraev, N. P. Merenkov and L. Trentadue, J. Exp. Theor. Phys. **81** (1995) 638 [Zh. Eksp. Teor. Fiz. **108** (1995) 1164];
V. S. Fadin, E. A. Kuraev, L. Trentadue, L. N. Lipatov and N. P. Merenkov, Phys. Atom. Nucl. **56** (1993) 1537 [Yad. Fiz. **56N11** (1993) 145].

[17] M. Davier, A. Hoecker, B. Malaescu, Z. Zhang, Eur. Phys. J. C **71** (2011) 1515 [Erratum-ibid. C **72** (2012) 1874].

[18] K. Hagiwara, R. Liao, A.D. Martin, D. Nomura, T. Teubner, J. Phys. G **38** (2011) 085003.

[19] R.V. Harlander, M. Steinhauser, Comput. Phys. Commun. **153** (2003) 244.

[20] A. B. Arbuzov, E. A. Kuraev, N. P. Merenkov and L. Trentadue, JHEP **9812** (1998) 009.

[21] A. Anastasi et al. arXiv:1609.06631v1 [hep-ex] 21 Sep 2016, Phys.Lett. B767 (2017) 485-492.

[22] See the talk of G. Venanzoni at this Workshop.

[23] B.E. Lautrup, A. Peterman, E. de Rafael, Phys. Rept. **3** (1972) 193.

[24] C. Aubin, T. Blum, Phys. Rev. D **75** (2007) 114502; P. Boyle, L. Del Debbio, E. Kerrane, J. Zanotti, Phys. Rev. D **85** (2012) 074504; X. Feng, K. Jansen, M. Petschlies, D.B. Renner, Phys. Rev. Lett. **107** (2011) 081802; M. Della Morte, B. Jager, A. Juttner, H. Wittig, JHEP **1203** (2012) 055.

[25] T. Blum, M. Hayakawa, T. Izubuchi, PoS LATTICE **2012** (2012) 022.

[26] M. Steinhauser, Phys. Lett. B **429** (1998) 158.

[27] P.A. Baikov, K.G. Chetyrkin, J.H. Kuhn and C. Sturm, Nucl. Phys. B **867** (2013) 182; C. Sturm, Nucl. Phys. B **874** (2013) 698; P.A. Baikov, A. Maier, P. Marquard, Nucl. Phys. B **877** (2013) 647.

[28] F. Jegerlehner, in Proceedings of "Fifty years of electroweak physics: a symposium in honour of Professor Alberto Sirlin's 70th birthday", New York University, 27-28 October 2000, J. Phys. G **29** (2003) 101.

[29] S.L. Adler, Phys. Rev. D **10** (1974) 3714.

[30] S. Eidelman, F. Jegerlehner, A. L. Kataev, O. Veretin, Phys. Lett. B **454** (1999) 369.

[31] G. Balossini, C.M. Carloni Calame, G. Montagna, O. Nicrosini, F. Piccinini, Nucl. Phys. B **758** (2006) 227.

[32] S. Actis *et al.*, Eur. Phys. J. C **66** (2010) 585.

[33] S. Eidelman, F. Jegerlehner, Z. Phys. C **67** (1995) 585; F. Jegerlehner, Nucl. Phys. Proc. Suppl. **181-182** (2008) 135.

[34] M. Bicer *et al.* [TLEP Design Study Working Group Collaboration], JHEP **1401** (2014) 164 [arXiv:1308.6176 [hep-ex]].

[35] G. Aarons *et al.* [ILC Collaboration], Physics at the ILC, arXiv:0709.1893 [hep-ph].
[36] K. Hagiwara, A.D. Martin, D. Nomura, T. Teubner, Phys. Lett. B **649** (2007) 173; Phys. Rev. D **69** (2004) 093003.
[37] G. Abbiendi et al. [OPAL collaboration], Eur. Phys. J. C**14** (2000) 373-425.
[38] L. Gatignon, private communication.
[39] P. Abbon *et al.* [COMPASS Collaboration], Nucl. Instrum. Meth. A **577** (2007) 455
[40] Mathieu Perrin-Terrin for the NA62 Collaboration, Proceed. 24th IWVD, Santa Fe, June 2015, pp.016.

AUTHOR INDEX